The Age of Clean Label Foods

Charis M. Galanakis
Editor

The Age of Clean Label Foods

Springer

Editor
Charis M. Galanakis
Galanakis Laboratories, R&I Department
Chania, P.C. 73131, Greece

ISBN 978-3-030-96697-3 ISBN 978-3-030-96698-0 (eBook)
https://doi.org/10.1007/978-3-030-96698-0

This Springer imprint is published by the registered company Springer Nature Switzerland AG
The registered company address is: Gewerbestrasse 11, 6330 Cham, Switzerland

Preface

Nowadays, consumers are more and more interested in processing, origin, health concerns, and ingredients of the food products. These consumer demands are leading food and beverage manufacturers toward developing new products by removing synthetic additives (e.g., emulsifiers) and at the same time incorporating natural ingredients. In response to consumer demands, the food industry has recently been directed to "clean label" products, having fewer additives. This trend is promoted with the general term "clean label" without any formal definition. However, due to its high market appearance and industrial relevance, "clean label" has become a trendy concept for the food sector leading to innovations in food product development. To this line, many big food companies have committed to removing artificial food additives, and subsequently formulating products with simple labels using recognizable ingredients is a growing commercial necessity. Nevertheless, the substitution of chemical preservatives with natural ingredients is a challenge for food technologists as synthetic additives extend the shelf life at a low cost. Besides, the safety, convenience, and sensory characteristics of the food products should not be compromised.

Food Waste Recovery Group (www.foodwasterecovery.group) has developed different initiatives, including consultation reports, workshops, reference modules, webinars, e-courses, publications, and multiple books in the broad fields of agricultural sciences, bioresources, food, and environment. Following the above considerations, our group has developed a new book that fills the gap existing in the current literature by providing information with regard to the incorporation of "clean label" ingredients in foods as well as the development of respective products. The final goal is to support the current industrial efforts and market applications.

The book consists of seven chapters. Chapter 1 comprises an introduction to the book by giving an overview of food labeling regulations and discussing relevant definitions of "Clean Label" and related terms, industrial activities, and relevant challenges. "Clean Labeling" is an initiative driven by consumer demands regarding easily understandable food labels and foods containing natural and familiar ingredients and free from additives. However, although food businesses acknowledge such demands and are increasingly developing "Clean Label" products, it is essential to note that there are no generally accepted standards for "Clean label." Other related terms include, for instance, certified organic foods, free-form, and no-additives, among others.

Emulsifiers are essential food additives with the ability to stabilize the foods comprising immiscible phases. In addition to this, they perform several other functions like lubrication, crystal modifier, texture enhancer, shelf-life enhancer, moisture barrier, etc. Chapter 2 provides details about various natural emulsifiers, their origins, and possible applications in multiple foods, pharmaceutical and cosmetic products. The behavior of different emulsifiers (e.g., proteins, polysaccharides, phospholipids, and bio-emulsifiers) toward interface and factors affecting (pH, ionic strength, temperature) their functionality as well as emulsion stability are also discussed in this chapter.

Synthetic food additives have long been used in the food industry to ensure oxidative stability, increase shelf life, or impart a specific flavor to foods. However, their potential links to human health risks are discussed among the consumers. Chapter 3 addresses the latest findings on the natural antioxidants and flavorings for clean label foods. It will also discuss the food safety and food structure (sensory) challenges that need to be taken into consideration when "chemicals" are replaced with "natural" ingredients in food and beverage production.

Lipids are an essential component of our daily food intake; however, their overconsumption has been associated with obesity and increased risk of diabetes, coronary heart disease, and certain forms of cancer. Fats are also responsible for many high-quality attributes in foods, mainly related to texture, stability, taste, and aroma. In Chap. 4, clean label ingredients to replace fat in foods, including fat substitutes and fat mimetics, are discussed. A description of different clean label reduced-fat foods as sauces and bakery, meat, or dairy products is provided as well.

Over the last years, there has been extensive research on aqueous foams. However, with time non-aqueous foam started to gain popularity in research work due to its favorable attributes since health-conscious consumers tend to consume products with reduced saturated fats. Chapter 5 provides a concise overview of the new promising areas in food science with a focus on the stability of oil foams, mainly formed by surface-active colloidal particles, as well as the factors (time, temperature, cooling rate, shear rate, and contact angle) affecting the ole foam stability. The critical role of ole foam in manufacturing unique texturized food products with less saturated fatty acids and improved mouthfeel is also discussed.

Traditional plasticizers are used for packaging purposes in order to keep the food quality over extended periods, but at the same time, they generate potential health and environmental issues. The development of clean label or bio-based food packaging materials (which are nontoxic and free from any chemical) has been suggested as an alternative strategy. Chapter 6 discusses clean label interventions in active and intelligent food packaging, giving emphasis on cost-effective and reliable bio-based and antioxidant-rich solutions.

The need for transparency in health and food safety always confronts customers with questions such as product content, the origin of production, production processes, additives used, and the status of ethical factors in the production process. To this line, investigating the relationship between consumers' behavior and the clean label of food products is crucial for both consumers and food producers. Chapter 7 deals with consumer behavior upon clean label products, revising the most undesirable ingredients for consumers in the composition of products. Thereafter, it presents modern methods to detect food fraud and to mislabel, prior to exploring different uses of nontraditional raw materials in food products and discussing the main principles of creating attractive consumer products.

Conclusively, the current book assists food technologists, scientists, engineers, and chemists activated in the food science field, but also professionals working in the food industry. It also concerns food industry employees, researchers working with food applications and food processing, as well as professionals and those who are interested in the development of innovative food products. It could also be used by academic institutes all around the world as a textbook and ancillary reading in undergraduates and postgraduate level multidiscipline courses dealing with food science and technology.

At this point, I want to take this opportunity to thank all the authors of this book for their high-quality work in bringing together important aspects of clean label foods. Their acceptance of editorial guidelines, collaboration, and dedication to the book's concept is highly appreciated. I would also like to acknowledge the book manager Sofia Valsendur and the acquisition editor Daniel Falatko. Last but not least, those collaborative efforts contain hundreds of thousands of words and could contain some errors. This is why constructive comments and even criticism are always welcome, so please do not hesitate to contact me to propose any changes.

Chania, Greece
Taif, Saudi Arabia
Vienna, Austria

Charis M. Galanakis

Contents

Chapter 1
Introduction

Marina Cano-Lamadrid, Juan Miguel Valverde, Leontina Lipan, Ángel Antonio Carbonell-Barrachina, and Esther Sendra

1.1 Concepts Related to "Clean Label" and State of the Art in Scientific Production

1.1.1 Definition

Nowadays, the concept of "Clean Label" is a priority for food and beverage companies, but it is an unregulated and undefined descriptor. Due to the lack of legal definitions and specific regulations (Busken 2015; Varela and Fiszman 2013), the interpretation of "Clean Label", in particular some terms such as "artificial" and "natural" ingredient, is subjective both to consumers and food companies. What is more, definitions depend on the party involved (food and beverage manufacturers, ingredient suppliers, retailers, and consumers), each having their definitions of the meaning of "Clean Label". It seems that consumers' interpretation is the most important one if they repeatedly purchase the product (Katz 2011)

Several definitions and interpretations of "Clean Label" have been proposed by market trend reports and scientific literature as follow:

Marina Cano-Lamadrid and Juan Miguel Valverde contributed equally with all other contributors.

M. Cano-Lamadrid · L. Lipan · Á. A. Carbonell-Barrachina · E. Sendra (✉)
Department of Agro-Food Technology, Research Group "Food Quality and Safety, CSA", Escuela Politécnica Superior de Orihuela (EPSO), Universidad Miguel Hernández de Elche (UMH), Alicante, Spain
e-mail: marina.canol@umh.es; leontina.lipan@goumh.umh.es; angel.carbonell@umh.es; esther.sendra@umh.es

J. M. Valverde
Department of Agro-Food Technology, Research Group "Postharvest Group of Fruit and Vegetables", Escuela Politécnica Superior de Orihuela (EPSO), Universidad Miguel Hernández de Elche (UMH), Alicante, Spain
e-mail: jm.valverde@umh.es

C. M. Galanakis (ed.), *The Age of Clean Label Foods*,
https://doi.org/10.1007/978-3-030-96698-0_1

- Borra (2010) suggests that "Clean Label" involves not only ingredient statement, but also the company's sustainability (source of raw material, production conditions and processing, including how it gets to stores and homes) (Borra 2010).
- Roller (2010) defines it as those present the following claims: "all natural", "organic", "no antibiotics", "no-GMO ingredients", "free range", "locally grown". These concepts influence the consumer opinion about if the product presents a "Clean Label" or not (Roller 2010).
- Rozin et al. (2012) says that consumers tend to define natural foods as those with fewer additives and alterations, which might be a "Clean Label" (Rozin et al. 2012).
- Busken (2015) and Varela and Fiszman (2013) mention that a "Clean Label" is characterized due to the lack of ingredients that induce contradiction in the ingredient list such as allergens, and additives others. It includes ingredients perceived as natural, harmless, and simple which consumers know and use them (Busken 2015; Varela and Fiszman 2013).
- Varela and Fiszman (2013) defines "as being produced free of 'chemicals' and additives. Also, as formulating with familiar ingredients, and being produced by using traditional technologies" (Brockman et al. 2013).
- Hutt and Sloan (2015) define a "Clean Label" as that in which natural ingredients are listed, or no high fructose corn syrup and artificial ingredients, preservatives, and colorants are used in food processing (Hutt and Sloan 2015).
- Lefferts (2017) mentions that "Clean Label" is usually interpreted as food products with short-ingredient list, without synthetic chemicals, and with understandable and familiar ingredients (Lefferts 2017).
- Cargill (2017) explains that a "Clean Label" is based on a shorter and simple ingredient list containing as less processed ingredients as possible. It should not include chemicals and E-numbers elements (Cargill 2017).
- Asioli et al. (2017) indicated that "Clean Label" should be understood as a product in which ingredient list is distinguished by several specifications such as short- and simple-ingredient list, with no synthetic and familiar ingredients. Also, they are, among other things, defined by 'free from', and a multifactor item, including environment and sustainability (Asioli et al. 2017).
- Aschemann-Witzel et al. (2019) indicate that it is referred to as a food product with minimally processed ingredients defined by 'natural' and 'free from'. They are based on the perception of consumers which classify ingredients depending on appearing natural (favorable) and un-natural (avoided) (Aschemann-Witzel et al. 2019).
- Ingredion (2020) specify that the meaning depends on the point of view. As to consumers, a "Clean Label" means a food product as 'natural', 'organic' and/or 'free from additives'. Regarding producers, it spells food products in which consumers widely accept ingredients. Therefore, the ingredient list may include short, simple and little processed ingredients. Also, no-including chemicals and E-numbers (codes for additives in the European Union) in the ingredient list is recommended (Ingredion 2020).

Taking all the definitions mentioned above into account, it can be said that the interpretation of a "Clean Label" is not objective. It depends on the familiarity of the consumer with the food ingredients and/or production method. Additionally, it does not exist a specific place in the supermarket dedicated to clean-label foods, and it is also not seen the words "Clean Label" on the product label. However, what is clear is that most of the definitions agree that a "Clean Label" must contain a short and simple ingredient list, and words such as natural, organic, or free from additives. On the other hand, words that sound like chemicals or E-numbers must be avoided.

Table 1.1 shows the terminology related to "Clean Label", some of the most common terms are defined in Fig. 1.1.

Based on scientific literature, the origin of "Clean Label" is shown in Fig. 1.2. First of all, "free from" claim seems to be the basis of "Clean Label" trend, specifically "GMO free".

Additionally, Fig. 1.3 shows the evolution of "Clean Label" trend, including E-numbers, non-GMO, organic, natural, and simple ones.

Consumers demand foods which are more natural and organic, and minimally processed. What is more, 'free from' or 'no-' allergens and/or additives are demanded (Ingredion 2020). Conceptually, terms such as health and sustainability motive the current trends on the food market. It is essential to indicate that the consumer expectation mainly is to consume more plant-based food products. Therefore, manufactures are focused on offering "Clean Label" foods (Aschemann-Witzel et al. 2019). Several food companies are including claims such as 'Free from' synthetic colorants, flavorings, and additives. 'Natural ingredients' are other example food companies include promoting "Clean Label" products. The desire to find food products more natural is due to the increasing demand for these types of products.

Generally, consumers avoid ingredients that are not ease to pronounce (Song and Schwarz 2009) or unfamiliar (Moskowitz et al. 2012) since those ingredients are unhealthy. Similarly, Dickson-Spillmann et al. (2011) studied the risk perception of chemicals on the consumers' attitudes by chemical and synthetic ingredients. Those ingredients' presence increased the risk perception, while the risk perception was positively correlated with the natural foods tendency (Dickson-Spillmann et al. 2011).

Clean Label is a global megatrend, expanding into new segments and geographies. The COVID-19 epidemic has increased consumer's consciousness on health and well-being. The consumer is more connected, informed and more concerned about their purchase decisions on themselves, their family and the environment (Ingredion 2020).

Moreover, it is essential to highlight that consumer expectation on Clean Label presents three cores. These translate that a product with a "Clean Label" must contain: (a) a natural, familiar, and simple ingredients list; (b) nutrient-dense ingredients; and (c) produced sustainably, considering animal welfare, without compromising the needs of future generations.

Table 1.1 Concepts included in "Clean Label" definitions from several sources. Numbers in brackets are the sources of definitions listed in the reference list

	Terminology	2010 (4)	2010 (5)	2012 (6)	2013 (1)	2013 (2)	2013 (7)	2015 (8)	2017 (9)	2017 (10)	2017 (11)	2019 (12)	2020 (13)
To include	Sustainability/ environment	X									X		
	Raw material	X											
	All natural/natural		X	X	X	X		X				X	X
	Organic		X										X
	No antibiotics		X										
	No-GMO		X										
	Free/free from/no		X				X	X			X	X	X
	Simple				X	X				X	X	X	X
	Traditional						X						
	High-fructose syrup							X					
	Shorter ingredient list								X	X	X		X
	Accepted by consumers												X
	Locally grown		X										
To avoid	Artificial							X	X		X		
	Allergenic				X	X							
	Additives/E-numbers			X	X	X	X			X			X
	Industrially				X	X				X		X	X
	Unfamiliar				X	X			X		X	X	
	Chemical sounding				X	X	X		X	X	X		X
	Synthetic								X				
	Preservatives												X

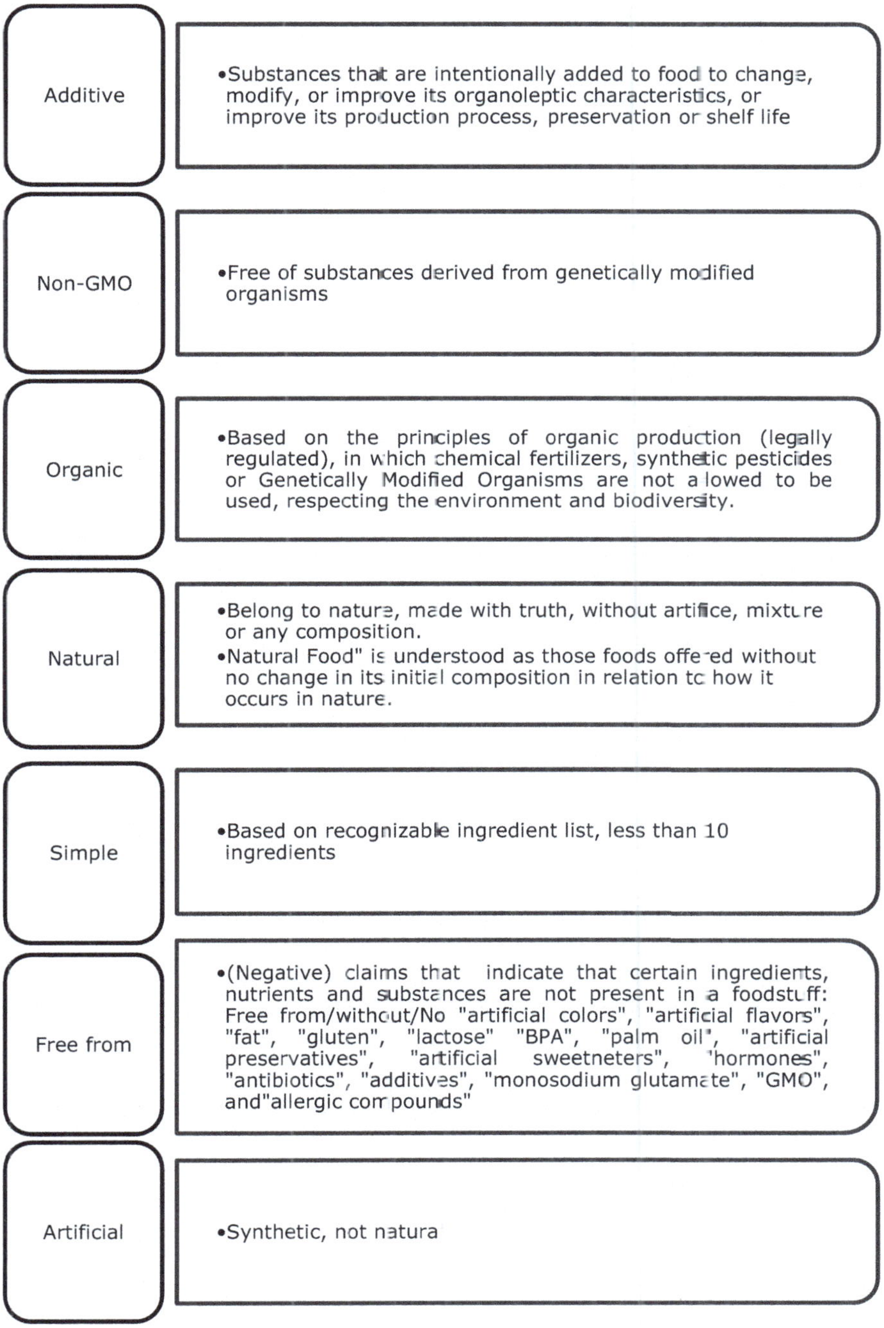

Fig. 1.1 General definitions of concepts related to "Clean Label"

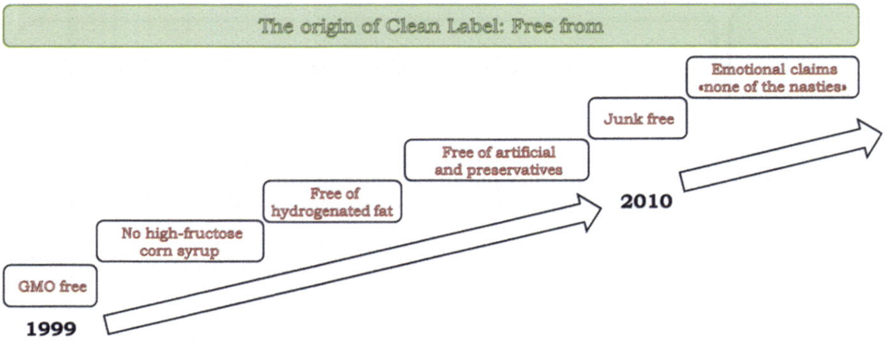

Fig. 1.2 Origin of "Clean Label"

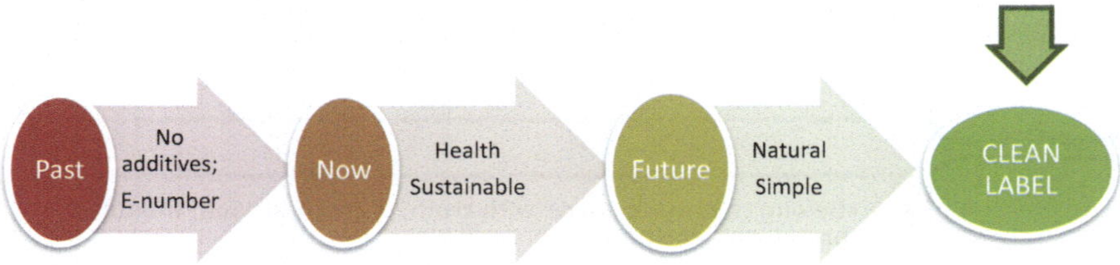

Fig. 1.3 Evolution of "Clean Label" Trend (Ingredion 2020)

1.1.1.1 Scientific Production

In Fig. 1.4, it can be observed the increase of scientific documents including Clean, Food and Label words (Scopus 2020). This means that in the last 10 years, the production per year increased more than 4-times.

Among countries, United States is by far the highest producer of documents including food "Clean Label" (Scopus 2020), producing at least three times more than the rest of the countries. Among European countries, Belgium and Ireland are the highest producers of documents related to "Clean Label" Foods.

What is more, "Clean Label" movement is a food manufacture's strategy, which also involves Food Agencies. Many of these Food Agencies work as funding sponsors for different scientific research, being Food Institutional Research Measurement the sponsor agency with the most scientific research done regarding "Clean Label" up to now (Scopus 2020).

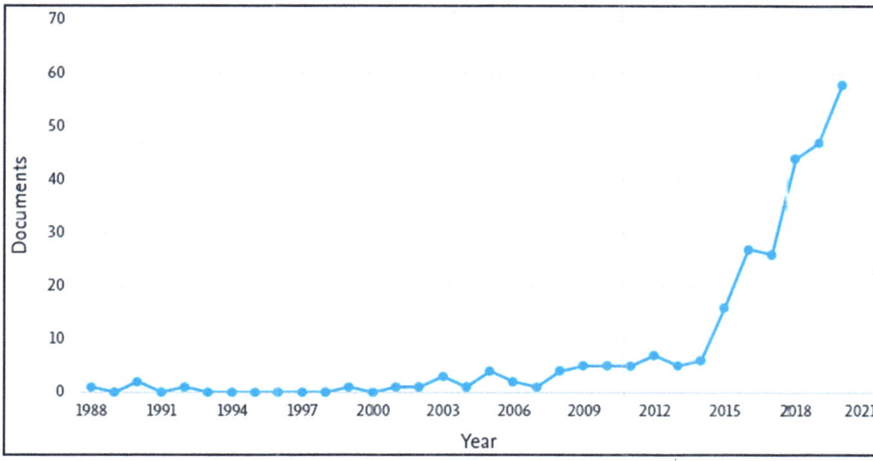

Fig. 1.4 Documents based on "Clean Label" Foods by year (Scopus 2020)

1.2 Food Labelling Regulations Worldwide

Nowadays, a wide variety of regulations, standards and guidelines on nutritional labelling exists around the world. Not only governments develop them, food industry is also involved. In some countries/regions, governments and food industry have not developed any policy or regulation; while in others, specific and well-written compulsory guides have been developed (FAO 2014).

1.2.1 Codex Alimentarius

The *Codex Alimentarius* (Food Code) is a set of standards, guidelines and codes of practice adopted by the *Codex Alimentarius* Commission (CAC). The CAC is the central part of the Joint FAO/WHO Food Standards Program and was established by FAO and WHO to protect consumer health and promote fair food trade practices. The kick-off meeting was held in 1963. The *Codex Alimentarius* contributes to the safety, quality and fairness of this international food trade. It is essential to indicate that its standards and texts are not mandatory. However, they are a truly international reference that although not compulsory is usually taken as reference in different countries (Food and Agriculture Organization off the United Nations, Organization WH 2017a). Due to the food products are following their specifications, consumers should trust the safety and quality. They need to be translated into national legislation or regulations to be enforceable.

1.2.1.1 Nutrition and Labelling

Thanks to the current technology applied in the food industry, it is possible to formulate novel food products according to consumers' expectations and preferences. Nowadays, consumer must face the defiance of labelling comprehension of what exactly they are buying and consuming, due to the diversity of options. Labelling data, about ingredients, nutrients, additives, etc. are essential for consumers by aiding them to decide when purchasing food products. In this regard, *CodexAlimentarius* offers a collection of standards and guidelines about the compositional requirements of food products to ensure that food products are nutritionally safe and can be traded. Moreover, *Codex* also provides support on general labelling about health or nutrient claims, using terms such as "fat free", "low fat", "high fat", "sugar free", etc.

The *Codex* Committee on Nutrition and Foods for Special Dietary uses (CCNFSDU) focuses on technical and regulatory topics for foods products to prevent nutritional deficiencies and diseases related to diet. On the other hand, the *Codex* Committee on Food Labelling (CCFL) provides standards and procedures regarding nutrition information on food packages to allow consumers to make informed food selections. The Nutrition and Food Labelling committees as well as industry and consumer associations leaders in member countries are key tools guiding the food system and achieving overall unanimity on nutrition topics (Food and Agriculture Organization off the United Nations, Organization WH 2017b).

1.2.1.2 Codex Documents on Food Labelling

As shown in Table 1.2, the Codex Alimentarius has made a great effort to clarify and provide recommendations on food labelling (WHO 2007) to aid companies, public authorities, and consumers in a better communication and understanding of the information provided to consumers. Standards, but also guidelines have been published.

1.2.2 EU Regulations on Food Labelling

Food regulations and policies assure safe food to consumers and accurate and honest information is also established. In the case of European Union, the specific law about labelling provides comprehensive information about food products' content and composition. Therefore, it can be stated that labelling helps consumers to make an informed decision about the foods they want to purchase (European Commission n.d.).

Table 1.2 Codex Documents on Food Labelling (Food and Agriculture Organization off the United Nations, Organization WH 2017c, d, e)

Reference		Title	Committee	Last modification
STANDARDS				
CXS 106-1983	*General Standard*	*"for Irradiated Foods"*	CCFH	2003
CXS 107-1981		*"for the Labelling of Food Additives"*	CCFA	2016
CXS 1-1985		*"for the Labelling of Prepackaged Foods"*	CCFL	2018
CXS 146-1985		*"for the Labelling of and Claims for Prepackaged Foods (Special Dietary Uses)"*	CCFL	2009
CXS 180-1991		*"for Labelling of and Claims for Foods (Special Medical Purposes)"*	CCNFSDU	1991
CXS 192-1995		*"for Food Additives"*	CCFA	2019
GUIDELINES				
CXG 1-1979		*"on Claims"*	CCFFP	2009
CXG 2-1985		*"on Nutrition Labelling"*	CCFL	2017
CXG 3-1989		*"for Simple Evaluation of Dietary Exposure to Food Additives"*	CCFA	2014
CXG 8-1991		*"on Formulated Complementary Foods for Older Infants and Young Children"*	CCNFSDU	2017
CXG 9-1987		*"General Principles for the Addition of Essential Nutrients to Foods"*	CCNFSDU	2015
GUIDELINES				
CXG 10-1979		*"Advisory Lists of Nutrient Compounds for Use in Foods for Special Dietary Uses intended for Infants and Young Children"*	CCNFSDU	2015
CXG 13-1991		*"for the Preservation of Raw Milk by Use of the Lactoperoxidase System"*	CCMMP	1991
CXG 23-1997		*"for Use of Nutrition and Health Claims"*	CCFL	2013
CXG 24-1997		*"for Use of the Term" "Halal"*	CCFL	1997
CXG 32-1999		*"for the Production, Processing, Labelling and Marketing of Organically Produced Foods"*	CCFL	2013
CXG 55-2005		*"for Vitamin and Mineral Food Supplements"*	CCNFSDU	2005
CXG 76-2011		*Compilation of Codex texts relevant to the labelling of foods derived from modern biotechnology*	CCFL	2011

(continued)

Table 1.2 (continued)

Reference		Title	Committee	Last modification
MISCELLANEOUS				
CXA 4-1989		*Classification of Foods and Animal Feeds*	CCPR	1993
CXA 6-2019		*List of Codex Specifications for Food Additives*	CCFA	2019

1.2.2.1 General Regulation

Some general labelling specifications should be considered (European Commission n.d.):

- The presentation of allergens such as gluten, lactose and nuts will be clear and harmonized in the ingredient list. It is recommended to focus on font and style of pre-packed foods.
- Allergen information is compulsory for non-prepacked food (including in restaurants and cafes)
- Legibility of the mandatory information must be present a minimum font size.
- The majority of processed food requires determined nutrition information.
- It is compulsory to indicate origin information for fresh meat (including ones from goats, poultry, pigs, and sheep.
- Both formed meat and fish one should be declared.
- Vegetal refined oil and fats must specify the origin.
- Labeling legislation must be complied as well as the form of sale (online, distance-selling or in a shop).
- List of engineered nanomaterials is included in the ingredient list.
- It is recommended to strengthen the current rules to prevent misleading practices.
- Indication of substitute ingredient for 'Imitation' foods.
- If the product is defrosted, it is necessary to declare this information.

1.2.2.1.1 Unfair Business-to-Consumer Commercial Practices

The information provided in food labelling should be clear and avoid misleading consumers. However, and due to marketing strategies, several terms are commonly included in food labels such as 'natural' 'simple' and others that are not specific, and not clearly defined. Food labelling information is included in official control schemes run by public authorities aiming to avoid consumers' misinformation and food fraud. Misleading actions with respect to food information are defined/clarified in Article 6 of D 2005/29/EC (EUR-Lex 2005):

> *A commercial practice shall be regarded as misleading if it contains false information and is therefore untruthful or in any way, including overall presentation, deceives or is likely to deceive the average consumer, even if the information is factually correct, in relation to one*

or more of the following elements, and in either case causes or is likely to cause to take a transactional decision that he would not have taken otherwise:

(a) the existence or nature of the product.

(b) the main characteristics of the product, such as its availability, benefits, risks, execution, composition, accessories, after sale customer assistance and complaint handling, method and date of manufacture or provision, delivery, fitness for purpose, usage, quantity, specification, geographical or commercial origin or the results to be expected from its use, or the results and material features of tests or checks carried out on the product.

The Commission staff working document guidance on the implementation/application of directive 2005/29/EC on unfair commercial practices detailed its application to specific sectors as following:

- Environmental claims
- The expressions environmental claims' and 'green claims' raise the fact that a good or a service has a positive or no negative impact on the environment or is less harmful to the environment than other goods or services. This can be achieved in all steps of food chain: manufacturing, product composition, acquisition, by reduction in energy or pollution expected from its use. On the contrary, if this claim is not effective or cannot be verified, this practice is called 'greenwashing'.
- 'Greenwashing' claim
- In this sense, the expression 'greenwashing' is refers to all forms of business-to-consumer commercial practices regarding the environmental characteristics of goods or services. This aspect can include all types of statements, information, symbols, logos, graphics, brand names, interaction with colors packaging, labelling, and advertising (in mass and social media). Moreover, any organization can use it if it qualifies as a "trader" and joins commercial practices with respect to consumers.

1.2.2.1.2 Food Information to Consumers

The information included in this section is based on Regulation (EU) No 1169/2011 (EUR-Lex 2018a). The Regulation clarifies in the Article 1, 2, 9, 17, 18 and 20 the most relevant information about the topic involved, and it is the following:

- Article 1 stresses the fact that the Regulation provides the basis to protect consumers though food information, taking into consideration differences in consumer perceptions and information needs, while ensuring the functioning of the market.
- General principles, requirements and responsibilities regarding food information (including food labelling) are established in this Regulation. The rights of consumers to be informed and the procedures to provide food information are also stated. Procedures take into consideration future developments and needs regarding food information.

- Article 2
- This article defines the terms 'label' and 'labelling' *"'label' means any tag, brand, mark, pictorial or other descriptive matter, written, printed, stencilled, marked, embossed or impressed on, or attached to the packaging or container of food".*

'Labelling' means any words, particulars, trademarks, brand name, pictorial matter or symbol relating to a food and placed on any packaging, document, notice, label, ring or collar accompanying or referring to such food.

- Article 7 states that the information provided shall be accurate, not be misleading and should be clear and easy to understand by consumers.
- Article 9 states that food product name and the list of ingredients, should be included in the label.
- Article 17 clarifies which should be the name of the food: the legal name, and in case of no legal name the customary name has to be used. If there is no customary name, a descriptive name of the food has to be provided
- Article 18 provides instructions regarding the list of ingredients. A clear heading, such as the word 'ingredients' should precede the list, and ingredients should be listed in descending order of weight used in the formulation at the time of use.
- Article 20 details those constituents that do not need to be included in the ingredients' list:

(a) those constituents of a food ingredient, that had been temporarily separated during processing and that are later on reintroduced without exceeding their original proportion; dairy fat separated and further reintroduced for milk standardization would be an example of this exception.

(b) those additives and enzymes: *"(i) whose presence in a given food is solely due to the fact that they were contained in one or more ingredients of that food, in accordance with the carry-over principle referred to in points (a) and (b) of Article 18(1) of Regulation (EC) No 1333/2008, provided that they serve no technological function in the finished product; or (ii) which are used as processing aids;"*

(c) "carriers and substances which are not food additives but are used in the same way and with the same purpose as carriers, and which are used in the quantities strictly necessary";

(d) "substances which are not food additives but are used in the same way and with the same purpose as processing aids and are still present in the finished product, even if in an altered form";

(e) water used only for reconstituting concentrated or dehydrated ingredients; or used in liquid mediums not normally consumed.

- Article 21 states that it is compulsory that a clear and highlighted reference to certain substances or products causing allergies or intolerances has to be included in food labels. Such compounds should be included in the list of ingredients. Making a clear reference to the name of the product (nuts, celery…) or substance (lactose, gluten…). Such products and compounds are included in Annex II and have to be clearly highlighted to be distinguished from other ingredients by means of a higher font size, style or color.

1.2.2.1.3 Nutrition and Health Claims Made on Foods

Food claims help consumers to associate a product with nutritional, physiological, or other health advantages over similar or different products to which such nutrients and other substances are not added. In this way, consumers are aware of their total intake of individual nutrients or other substances. Consequently, it is compulsory to require restrictions for those products holding claims to avoid potential undesirable effect. In this context, aspects such as the presence of specific substances, or the nutrient profile of a product, are suitable standards for determining whether the product can hold a claim or not.

The Regulation (EC) No 1924/2006 includes information about nutrition and health claims made on foods (EUR-Lex 2014a). The most relevant definitions included in the Article 2 of the above-mentioned Regulation are shown:

- 'Claim'

  Any message or representation, which is not mandatory under Community or national legislation, including pictorial, graphic or symbolic representation, in any form, which states, suggests or implies that a food has particular characteristics.

- 'Nutrition claim'

  Any claim which states, suggests or implies that a food has particular beneficial nutritional properties due to: (a) the energy (calorific value) it (i) provides; (ii) provides at a reduced or increased rate; or (iii) does not provide; and/or (b) the nutrients or other substances it (i) contains; (ii) contains in reduced or increased proportions; or (iii) does not contain.

- 'Health claim'

  Any claim that states, suggests or implies that a relationship exists between a food category, a food or one of its constituents and health.

- 'Reduction of disease risk claim'

  Any health claim that states, suggests or implies that the consumption of a food category, a food or one of its constituents significantly reduces a risk factor in the development of a human disease.

Additionally, some key principles for all claims are included in the Article 3. *"The use of nutrition and health claims shall not:*

i. *be false, ambiguous or misleading;*
ii. *give rise to doubt about the safety and/or the nutritional adequacy of other foods;*
iii. *encourage or condone excess consumption of a food;*
iv. *state, suggest or imply that a balanced and varied diet cannot provide appropriate quantities of nutrients in general;*
v. *refer to changes in bodily functions which could give rise to or exploit fear in the consumer, either textually or through pictorial, graphic or symbolic representations".*

Table 1.3 shows a community list of permitted claims included in the Article 13 (European Commission 2016).

Table 1.3 Authorized nutrition claims (European Commission 2016)

AUTHORIZED CLAIMS ABOUT THE AMOUNT OF ENERGY
Low energy value, reduced energy value, no energy intake, light, reduced energy content.
AUTHORIZED CLAIMS ON THE PRESENCE OF SUGARS
Low sugar content, sugars free, no added sugars, reduced sugar content.
AUTHORIZED CLAIMS ON THE PRESENCE OF FATS
Low fat content, fat free, low in saturated fat, no saturated fat, reduced content of fat or saturated fat, source of omega-3 fatty acids, high content of omega-3 fatty acids, high content in monounsaturated fats, high content of polyunsaturated fats, high unsaturated fat content.
AUTHORIZED CLAIMS ON THE PRESENCE OF SALT
Low sodium/salt content, very low sodium/salt content, no sodium or no salt, reduced salt content.
AUTHORIZED CLAIMS ON THE PRESENCE OF FIBER
Fiber source, high fiber content.
AUTHORIZED CLAIMS ON THE PRESENCE OF PROTEINS
Protein source, high protein content.
AUTHORIZED CLAIMS ABOUT THE PRESENCE OF VITAMINS AND MINERALS
Source of [name of vitamins] or [name of minerals], high in [name of vitamins] or [name of minerals], reduced content of [micronutrient name], higher content of [nutrient name].

What is more, given the high number of applications for the authorization of food claims, the European Commission, aiming to clarify and aid in this process, published a list of authorized health claims, other than those referring to the reduction of disease risk and to children's development and health in Regulation (EU) No 432/2012 (EUR-Lex 2012a). Article 13 of the Regulation includes a long list of approved health claims, which may be made on foods. The Annex included in this Regulation contains a LIST OF PERMITTED HEALTH CLAIMS. The list includes "nutrient, substance, food or food category. Claim. Conditions of use of the claim. Conditions and/or restrictions of use of the food and/or additional statement or warning. EFSA Journal number. Relevant entry number in the Consolidated List submitted to EFSA for its assessment".

1.2.2.2 Regulation Related to Terms Included in "Clean Label"

In recent years, the trends, and patterns of consumption in society have undergone an important evolution, incorporating new variables in purchase decision making, such as long-term health, environmental sustainability, animal welfare or ethical values in food production. This phenomenon has been magnified and accelerated by the explosion of social media platforms as a source of information for the consumer. Moreover, several influencers (own and outside the food sector) have started and promoted trends with an important impact on consumer perception about certain attributes or product characteristics (sometimes without supporting evidence) (AECOC n.d.).

The regulatory framework on the use of these attributes, either directly or generally, it is broad, complex and, in some cases, diffuse, encompassing all types of legislation (European, Spanish or Autonomic Region), private regulations, decisions or official positions, among others.

1.2.2.2.1 Trends in Agri-Food Production

Ecological, BIO, Organic

The concept of 'natural' is commonly linked to 'organic', being the first one a non-defined term and the second regulated by law. Organic production in growing worldwide as well as the demand for organic foods, which are among the best candidate foods for "Clean Label" use. Ecological, Bio and Organic concepts are included in the Regulation (EU) 2018/848 and Regulation (EU) 834/2007 (EUR-Lex 2018b). European Regulations on organic production lay down principles and rules of organic production, certification, and use of indications, labelling and advertising. Rules regarding official controls affecting organic production and labelling are included in Regulation (EU) 2017/625 (Article 1i) (EUR-Lex 2017).

The Regulation clarifies in the Article 5 the general obligations concerning the competent authorities and the organic control authorities, focusing mainly on organic production as a sustainable management system based on:

> *(a) respect for nature's systems and cycles and the sustainment and enhancement of the state of the soil, the water and the air, of the health of plants and animals, and of the balance between them;*
>
> *(b) the preservation of natural landscape elements, such as natural heritage sites;*
>
> *(c) the responsible use of energy and natural resources, such as water, soil, organic matter and air;*
>
> *(d) the production of a wide variety of high-quality food and other agricultural and aquaculture products that respond to consumers' demand for goods that are produced by the use of processes that do not harm the environment, human health, plant health or animal health and welfare;*
>
> *(e) ensuring the integrity of organic production at all stages of the production, preparation and distribution of food and feed;*
>
> *(f) the appropriate design and management of biological processes, based on ecological systems and using natural resources which are internal to the management system, using methods that: (i) use living organisms and mechanical production methods; (ii) practice soil-related crop cultivation and land-related livestock production, or practice aquaculture which complies with the principle of the sustainable exploitation of aquatic resources; (iii) exclude the use of GMOs products produced from GMOs, and products produced by GMOs, other than veterinary medicinal products; (iv) are based on risk assessment and the use of precautionary measures and preventive measures, where appropriate;*
>
> *(g) the restriction of the use of external inputs; where external inputs are required or the appropriate management practices and methods referred to in point (f) do not exist, the external inputs shall be limited to: (i) inputs from organic production; in the case of plant reproductive material, priority shall be given to varieties selected for their ability to meet the specific needs and objectives of organic agriculture; (ii) natural or naturally-derived substances; (iii) low solubility mineral fertilizers;*

(h) the adaptation of the production process, where necessary and within the framework of this Regulation, to take account of the sanitary status, regional differences in the ecological balance, climatic and local conditions, stages of development and specific husbandry practices; (i) the exclusion from the whole organic food chain of animal cloning, of rearing artificially induced polyploid animals and of ionising radiation;

(j) the observance of a high level of animal welfare respecting species-specific needs".

It is important to highlight that the term Ecological can be substituted in the labelling for Biological, Eco, Bio or Organic. These products are generally perceived as healthier, safer or of higher quality than conventional products, although, at present there is no scientific evidence to support such claims (positioning of the European Food Safety Authority) (Kneafsey et al. 2013). Main traits of organic production are environmental, biodiversity and animal welfare protection as seen in the listed articles.

Sustainable

There is no specific regulation of this attribute, so there is no legal definition of this word. It is understood that a sustainable product is one that has been generated in accordance with the environment and is different in each country, even within the same food category (Sánchez-Bravo et al. 2020, 2021). The concept of sustainability has many approaches and nuances that are interconnected. In this sense, its definition should be regulated and its intrinsic characteristics clarified. This implies for sustainable products: Acceptable ecological footprint, fair social conditions, reduced packaging, absence or minimal presence of additives and basic ingredient lists. In the present chapter, the focus is given only to those technical aspects of sustainability that may be measured and certified. There are different certificates related to sustainability, both generic and by sector or activity. For example: ISO 14000 (environmental management), ISO 26000 (social responsibility), UNE 15500 (management system for the production of fruits and vegetables respectful with the environment), MSC (sustainable fishing), etc.

It needs to be pointed out that 'sustainable', as well as 'natural' are widely used terms, which use needs an in deep critical analysis in search of clear definitions to avoid misunderstandings and contradictions. They are not always compatible, as an example: sustainability and natural concepts are sometimes contradictory. Some practices, such as the extraction of small quantities of natural flavors, sweeteners, or phytonutrient involving the use of considerable amounts of resources to grow plants (land, water, energy) may be natural but not sustainable. Could it be by far "more sustainable" to produce the key substances via fermentation processes (using yeasts or other microorganisms engineered) thousand times more efficient?. A careful and more specific use of those terms would be capital to clarify the message to consumers.

Biodynamic

There is no specific regulation of this attribute, so there is no legal definition of the attribute. Biodynamic agriculture is a method of agriculture considered ecological, based on Rudolf Steiner's theories, founder of anthroposophy. Certification of products and obtaining the registered trademark "Biodynamic Product" depend on the private group Demeter.

In Europe, to obtain biodynamic certification, the farm must be previously certified in organic farming under European Regulations. Biodynamic production is not based on commonly accepted scientific evidence.

Seasonal

Eating seasonal and local food is a message sent by non-governmental organizations, mainly promoted in social media to reduce the environmental impact by shifting current dietary intakes unto more sustainable diets (Macdiarmid 2014).

There is no specific regulation of this attribute, so there is no legal definition of the attribute. The scientific literature proposed 2 definitions for seasonal food, one related with the place where the product is produced and the other based on the place the product is produced and consumed (DEFRA 2012):

- "*Produced in season (Global): Food that is outdoor grown or produced during the natural growing/production period for the country or region where it is produced. It need not necessarily be consumed locally to where it is grown.*
- *Food that is produced and consumed in the same climatic zone without high-energy use for climate modification or storage*".

Consequently, they are fresh products, usually agricultural and fisheries that, naturally and due to their biological cycle, are at the optimum moment of production at some time in the country or region of consumption (respecting the seasonality of the productions).

1.2.2.2.2 Trends in Environmental/Social Responsibilities

Local, Proximity, km 0

There is no specific regulation of this attribute, no legal definition of the attribute. It refers to products that are produced and processed within a relatively short distance between producer and consumer, near or nearby. Most groups and administrations assume that proximity products or km 0 do not exceed the 100 km radius between production and sale.

The intrinsic characteristics of this type of products are:

- Cooler and with a longer shelf life not elapsed at the time of purchase.
- Greater consumer confidence due to proximity to its origin.
- Consumption models that reduce polluting emissions (by transport).

- They promote the local economy, connecting producers and consumers from the same territory.

The European Commission published statements supporting the qualities of local food and short chain. "They contribute to the true reduction of CO_2 emissions and support the idea of creating quality, seasonal, fresh and local food that meets expectations environmental, climatic and social" (Kneafsey et al. 2013).

Slow Food

There is no specific regulation of this attribute, no legal definition of the attribute, but there is a code of use for Slow Food trademarks, text approved by the Sixth International Congress in 2012. Slow Food is an organization that works for the preservation of local food cultures and traditions, aiming to avoid their lost, to counteract the rise of accelerated rhythms of life and to fight against the general lack of interest in the food that is consumed, its origin and the way in which our food decisions affect the world around us. Its three principles are Good, Clean and Fair:

- "GOOD: a diet with seasonal, fresh and tasty foods that satisfy the senses and that belong to our local culture;
- CLEAN: food production and consumption that does not harm the environment, animal welfare or our health;
- FAIR: affordable prices for consumers, fair pay and conditions for small-scale producers".

Fair Trade

There is no general legislation that directly regulates this attribute. The standards for this type of trade are established by FAIRTRADE Labeling Organizations International (FLO). It is defined as a type of alternative trade, promoted by various NGOs, the UN social and political movements, which promotes a voluntary and fair commercial relationship between producers and consumers.

Fairtrade includes several concepts, mainly better prices and decent working conditions for producers, assurance of local sustainability, and fair trading conditions for farmers and workers in the developing world. Companies involved in Fair Trade schemes are asked to pay sustainable prices (which must never be below the market price), Fairtrade aims to address main issues of conventional trade, which traditionally goes against the interest of weakest producers and countries. Fair trade aims to improve their life quality, position and to empower them.

Fair trade label indicates that the product carrying the label, not the entire company practices, has been certified to offer a better deal to the farmers and workers involved.

Recyclable, Biodegradable

Recycling is any recovery operation through which waste materials are transformed from new, by man's action, in products, materials or substances, whether it is with the original purpose or for any other purpose. In the European Union there are several directives for this term:

- Directive (EU) 2018/852 of the European Parliament and of the Council amending Directive 94/62/EC on packaging and packaging waste (EUR-Lex 2018c).
- Directive (EU) 2019/904 of the European Parliament and the Council on reducing the impact of certain plastic products on the environment (EUR-Lex 2019).

On the other hand, each European country publish their own standards, for example Spain: UNE-EN 13432:2001/AC:2005 Packaging—Requirements for packaging recoverable through composting and biodegradation.

Some recycling symbols internationally accepted are listed below:

- Mobius Loop

- This symbol highlights that the product can be recycled, but not recycled or accepted in all recycling collection systems. Occasionally, this symbol includes a percentage in the middle of the figure indicating that the packaging contains a certain percentage of recycled material. On the other hand, some logos are intended to identify the plastic material used in the manufacture of the product or container (Fig. 1.5a).
- Green Dot

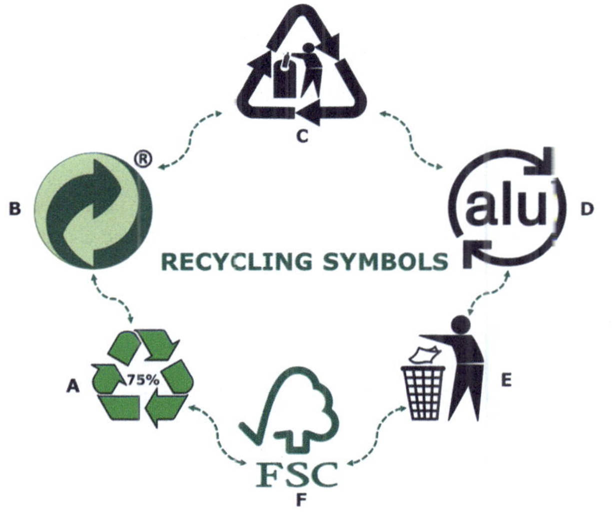

Fig. 1.5 Understanding recycling labels (Jeans 2017; Europe PRO 2020; FS Council 2020; recyclenow 2016). (**a**) Mobius Loop, (**b**) Green Dot, (**c**) Glass, (**d**) Recyclable aluminum, (**e**) Tidyman, and (**f**) Paper, card, and wood

- This symbol (Fig. 1.5b) reflects that the producer has made a financial contribution regarding the recovery and recycling of packaging. It is used on packaging in some European countries and does not necessarily mean that the packaging is recyclable, recycled, or recycled.
- Glass
- This symbol demands consumers to recycle the glass container. It is a widely recognized symbol meaning that glass bottles and jars should be disposed into a bottle bank, remembering to separate colors, or even use glass household recycling collection when available (Fig. 1.5c).
- Recyclable aluminum
- This symbol indicates that the item is made from recyclable aluminum. Recyclable steel: This symbol means that the product is made of steel, so the local authorities collect steel cans for recycling. While other steel or metal items might be manipulated by the household recycling center (Fig. 1.5d).
- Tidyman
- This symbol from Keep Britain Tidy, which is the UK independent environmentally charity aware the consumers not to waste. It does not relate to recycling but suggest or remind to be a good citizen, placing the items in the most appropriate place (Fig. 1.5e).
- Paper, card, and wood
- The Forest Stewardship Council (FSC) logo categorizes products made of wood belonging to well managed forests independently certified in accordance to the rules of the FSC. For instance, products certified to be industrially compostable according to the European standard EN 13432/14955 may hold the 'seedling' logo (Fig. 1.5f).

1.2.2.2.3 Commercial/Consumer Driven Trends

Geographical Indications

The EU Geographical indications system defends (Fig. 1.6) the products that derive from specific regions and have particular traits or experience a reputation linked to the production territory (EUR-Lex 2012b). Product names can be granted with a 'geographical indication' (GI) if they have a specific link to the place where they are made. The GI credit allows consumers to trust and distinguish quality products belonging to a specific area and aid farmers or producers in selling their products with an added value. Moreover, GI establish intellectual property rights for specific products, whose qualities are manly linked to the area of production.

There are different quality schemes, and the differences between, for example, Protected Designation of Origin (PDO) and Protected Geographical Indication (PGI) are mainly related to the amount of the product's raw materials needed from that specific area or about how much of the production process has to belong to the specific region:

Fig. 1.6 Recognizing the EU quality schemes on labels (European Commission 2018). (**a**) Protected Designation of Origin (PDO), (**b**) Protected Geographical Indication (PGI), and (**c**) Traditional specialty guaranteed (TSG)

- The products registered as Protected Designation of Origin (PDO) are those most connected to the place in which they are made (Fig. 1.6a).
- Protected Geographical Indication (PGI) highlights the connection between the specific geographic region and the name of the product, where a particular quality, reputation or other characteristic is essentially attributable to its geographical origin (Fig. 1.6b).
- Traditional specialty guaranteed (TSG) emphasizes the traditional aspects related to the manufacturing and composition, without being related to a specific geographical area. A product registered as a TSG is protected against falsification and misuse (Fig. 1.6c).

Vegetarian/Vegan

There is no general legislation that directly regulates this attribute. The Switzerland-based V-Label GmbH coordinates international cooperation on the V-Label. Vegan is any food that is not animal origin and whose production and transformation have not been used food or substances of animal origin, neither in their natural form, nor in their processed form. This includes additives, catalysts, flavorings, enzymes, and adjuvants.

Any food that meets the requirements of the definition of vegan food is vegetarian, but this last term can include milk, eggs, honey, beeswax, propolis, lanolin (and its derivatives) and additives.

The consumers of vegan and vegetarian products is a vast segment in continuous growth, due to different reasons such as food intolerance, religious, environmental protection, health benefits or ethical reasons.

Kosher

There is no general legislation that directly regulates this attribute. The most widely recognized kosher certification agency is the OU (Orthodox Union) Kosher. Kosher foods are those that meet the religious standards based on the Talmud and the Torah and, therefore, are suitable for consumption by the believing Jewish community.

Halal

There is no general legislation that directly regulates this attribute. The regulation of the Halal concept comes from the fulfilment of the precepts of Sharia or Islamic law. These requirements have been incorporated into the various Halal certification schemes. Some several Halal guarantees marks or seals cover specific requirements depending on the country of destination of the product. Each country recognizes specific entities and guarantees based on its standards and interpretations they make of Islamic law. There are various seals managed by entities that certify Halal products.

1.2.2.2.4 Trends in Free-Form or No-

Gluten Free

In the EU this attribute is directly regulated by Regulation No 828/2014, regarding the requirements for the transmission of information to consumers about the absence or reduced presence of gluten in food (EUR-Lex 2014b).

The "gluten-free" declaration may only be used in foods that do not contain more than 20 mg/kg of gluten. The 'very low gluten' claim can only be used when the food has been specifically processed to reduce their gluten content, and do not contain more than 100 mg/kg of gluten in the food as sold to the final consumer. The certification systems are private and voluntary: for example, in Europe the barred herringbone license system "ELS" is applied.

Lactose Free

There is no harmonized standard at European level. The specific criteria in force for this type of products have been established on the basis of the scientific opinion of the European Food Safety Authority (EFSA), on September 10, 2010, on the maximum levels in lactose intolerance and galactosemia. "Lactose-free" products are those that, as sold to the final consumer, contain less than 0.01%, or 0.01 g/100 g or 100 ppm of lactose in their composition. Products that declare "low lactose content" can contain up to 1% lactose in their composition (100 times more than Lactose Free products).

The seal of Adilac-No Lactose exists at a private level and for voluntary use. This seal identifies the products lactose free, and guarantees that ingredients comply with all the requirements demanded by Adilac in its Regulations for use.

No Additives

There is no specific regulation at European level. They may not contain directly, or through any of its ingredients, the following substances: Sweeteners, Colorants, Preservatives, Antioxidants, bulking agents, Acidulants, Acidity correctors, Anti-caking agents, Defoamers, Freight agents, Emulsifiers, Molten salts, Hardeners, Flavor enhancers, Foaming, Gelling agents, Coating agents, Moisturizers, Modified starches, Packaging gases, Propellant gases, Raising agents, Sequestrants, Stabilizers, Thickeners, Treatment agents for flours, Contrast enhancers.

In many cases, a generic mention is not made (No additives) but rather the "claim" is restricted to a specific category (No preservatives, No colorants, etc.). The concept "Free from" is equivalent to "Without". This attribute closely linked to the "Clean Label" concept.

Within the strong media pressure on ultra-processed foods, the consumer considers that the product is healthier when it does not have, or has very few, additives. The authorized additives, used in the authorized amounts, are safe for consumers. This list is constantly being revised to incorporate new scientific evidence into the risk assessment and, if necessary, measures are taken on specific substances.

No Palm Oil

There is no specific regulation at European level. Products labeled as "No Palm Fat" or "No Palm Oil" may not contain Palm Fat or its derivatives as part of the product, its ingredients or the additives used. The claim "No Palm Fat" does not imply, by definition, that the product has better nutritional quality, is more sustainable or safer. It is important not to be confused with the "Green Palm RSPO" certificate (Certified Sustainable Palm Fat), which guarantees the Palm Fat's sustainable origin in the product.

Bisphenol A (BPA)

Bisphenol A (BPA) is an industrial chemical compound that is used in many food products. The monomer makes up the polycarbonate plastic present in many containers, such as reusable water bottles. In the case of authorizing its use, current regulations establish a maximum specific migration limit (SML) of 0.6 mg of BPA per kg of food. The BPA Free attribute implies the absence of a compound in the composition of the products or in the packaging materials of food products (EUR-Lex 2018d).

1.2.3 USA Regulation on Food Labelling

The Food and Drug Administration (FDA) is the agency in charge of assuring that the United States' food products are safe, healthy and well labeled. The agency standards are developed to protect both the foods produced in USA and from foreign countries. The Federal Food, Drug, and Cosmetic Act (FD&C Act) and the Fair Packaging and Labelling Act control food products under the FDA's authority (U.S. Food & Drug Administration 2013).

Manufacturers, distributors, and importers address to the FDA their questions about the appropriate labelling of their food products. The FDA guides the companies regarding the required statements allowed to appear on food labels under these laws and their regulations. To streamline the process of legal action and to avoid delays, the manufacturers and importers may learn about the applicable laws and regulations for distribution in the United States before offering their foods.

The Nutrition Labeling and Education Act (NLEA), which modified the FD&C Act requires that most foods must bear nutrition labeling and food labels must assume nutrient content claims and some health messages to fulfill with specific requirements. The final regulations have been already launched and presented in this guidance, however; companies and people with interest on this matter must be always updated because regulations are often changed. Thus, the food industry's obligation is to stay up to date with the legal requirements for food labeling. For this, all new regulations are available in the Federal Register (FR) prior to their active date and compiled annually in Title 21 of the Code of Federal Regulations (CFR).

Currently, there is no universal standard/regulation/law to facilitate international food labelling. An example of the variability of legislation is about the GMO compulsory labelling. If modified genetic material is more than 1%, it is compulsory to be labeled in Brasil, Australia and New Zealand. Otherwise, only if the modified genetic material is more than 5% is compulsory in United States, Canada, Japan, Thailand and Philippines (Borges et al. 2018; Hakim et al. 2020; Premanandh et al. 2012; Smyth 2018).

The high increase in health expenses due to non-communicable diseases and an ageing population has raised concerns on public authorities, the academy, and the food industry, searching for healthy foods for consumers. In this scenario FOSHU Foods (Food with specified health uses) have been developed. Japanese FOSHU category is only given to a food after demonstrating a positive effect of these foods in preventing some specific diseases through clinical and epidemiological trials. There are established rigorous procedures for the evaluation and approval of FOSHU products, such procedures must be constantly evaluated. There is a characteristic logo used in the package of Foods having FOSHU category, such logo is highly recognized and valued by consumers (Duran and Valenzuela 2010).

1.3 Most Common Applications of Clean Label

1.3.1 Additive-Free Current Application

The most common trend in 'Clean Label' is the absence of additives or at least the absence of its coded number in the ingredients list. Many additive names are quite common names perceived as ordinary or safe (many organic acids, gelling agents…). According to European Regulations, the additives named in the labelling must indicate: the functional group followed by the E-code or full name of the compound (*i.e.* acidity regulator: E-330, or acidity regulator: citric acid). In the present section, the functional groups of additives are presented.

The Online Edition: "Combined Compendium of Food Additive Specifications" provides the most recent specifications for food additives evaluated by Joint FAO/WHO Expert Committee on Food Additives (JECFA) (Joint FAO/WHO Expert Committee on Food Additives 2019). The food additive functional classes are based on the *Codex* Class Names and the International Numbering System (INS) for Food Additives (CAC/GL 36-1989). The additive classes (groups) are listed below (FAO 2019):

- **Acidity regulator:** It is a kind of food additive, through which it is possible to control the acidity or alkalinity of any food.
- **Anticaking agent:** It allows reducing the adherence of particles to one another.
- **Antifoaming agent:** It can be added in order to avoid or reduce foaming.
- **Antioxidant:** This additive's role is prolonging the shelf-life of foods opposite the deterioration caused by oxidation.
- **Bleaching agent:** This additive can be used to decolorize any food without pigments.
- **Bulking agent:** This additive contributes to increase the bulk of a food without contributing to energy value of food.
- **Carbonating agent:** It could be defined as additive which is used to supply carbonation in any food.
- **Carrier:** This additive is utilized for dissolving, diluting, dispersing, and/or also modifying physically some food additive or nutrient without altering its function.
- **Color:** This additive can be added for re-establishing or changing color of food.
- **Color retention agent:** This additive could be used to stabilize, enhance, or retain any food color.
- **Emulsifier:** It allows keeping several phases in a food and in a uniform emulsion.
- **Emulsifying salt:** This additive can rearrange proteins and avoid fat separation at the same time in the manufacture of processed food.
- **Firming agent:** This additive is used to keep, firm and crisp tissues of vegetables or use gelling agents to produce an accurate gel.
- **Flavor enhancer:** This additive can be used to enhance the flavor and/or odor of a food.

- **Flour treatment agent:** This additive can be used to improve baking quality or color in flour or dough.
- **Foaming agent:** It can be added to form or maintain a uniform dispersion.
- **Gelling agent:** This additive, through formation of a gel, gives a food texture
- **Glazing agent:** This additive can be added to the surface or a food and consequently adds shine and provides a protective layer.
- **Humectant:** An additive that can be added to prevent drying out by counteracting a dry atmosphere's effect.
- **Packaging gas:** This additive gas is introduced into a pack before, during or after filling with food in order to protect the food, for instance: oxidation or spoilage.
- **Preservative:** A kind of additive, through which the shelf-life of a food can be prolonged opposite deterioration caused by microorganisms.
- **Propellant:** It is an additive gas which a food can be expels from a container.
- **Raising agent:** This food additive is used to liberate gas and consequently, to increase volume in a dough or batter.
- **Sequestrant:** This additive allows controlling the readiness of a cation.
- **Stabilizer:** This additive makes possible to keep a uniform dispersion of several ingredients.
- **Sweetener:** This additive (different than a mono- or disaccharide sugar) adds a sweet taste to a food.
- **Thickener:** This additive can be used to increase the viscosity of a food.

1.3.2 Initiatives on Clean Label of Foods

1.3.2.1 The Clean Label Project

The foundation of Clean Label Project is rooted in bringing truth and transparency to consumer product labeling through data and science (Clean Label Project 2020). Beyond category specific benchmarking, Clean Label Project establishes thresholds and rewards brands with products with overall results in the top 33% of their category. However, Clean Label Project has a limited number of categories currently benchmarked. The Clean Label Project Code of Practice Certification aims to establish pass/fail criteria for brands with products in categories that Clean Label Project has not yet benchmarked but still want to communicate their commitment to product purity. Table 1.4 shows the requirements of the current Analytical Equipment Administrator and Fig. 1.7 includes the testing requirements which are compulsory by category.

Table 1.4 The Clean Label Project: analytical equipment for the determination of contents of specific compounds which absence needs to be guaranteed (Clean Label Project n.d.)

Assay	Analytes	Equipment	Detection limit
Heavy metals	As, Cd, Pb, Hg	ICP-MS	<4 ppb
Pesticides	Glyphosate, among others	LC-MSMS	<10 ppb
BPA	BPA	LC-MSMS	<40 ppb
Antibiotics	Tetracycline, ampicillin		<10 ppb
Residual solvents	1,2-dichoroethane, 2-propanol, acetone, acetonitrile, benzene, chloroform, cyclohexane, diethyl ether, ethanol, ethyl acetate, heptane, hexane, m/p-xylene, methanol, methylene chloride, o-Xylene, pentane, toluene, trichloroethene, isobutane, n-butane, n-propane	GC-FID	USP limits
Phthalates	Benzyl butyl, dibutyl, dihexyl, bis(2-ethylhexyl), bis(7-methyloctyl)	LC-MSMS	<40 ppb
Micotoxins	Aflatoxin G2, aflatoxin G1, aflatoxin B2, aflatoxin B1, Ochratoxin A	LC-MSMS	<10 ppb
Acrylamide	Acrylamide	LC-MSMS	<40 ppb
Parabens	Methyl, ethyl, isopropyl, propyl, total butyl, benzyl	LC-MSMS	<500 ppb

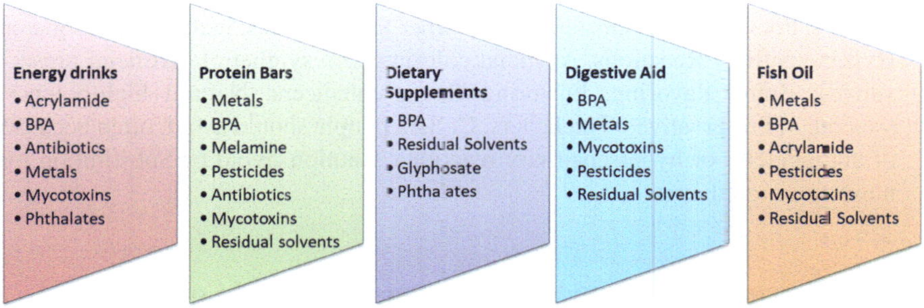

Fig. 1.7 Testing requirements compulsory by cathegory to meet a "Clean Label" according to Clean Label Project (n.d.)

1.3.2.2 Recommended Priorities for Addressing Food Ingredients in Clean-Label Programs of the Center for Science in the Public Interest (CSPI)

Founded in 1971, the Center for Science in the Public Interest (CSPI) is an independent, science-based consumer advocacy organization aiming to improve the food system to support healthy eating and promote the importance of the link between diet and health to public authorities, industries, and consumers.

As it has been previously mentioned, certified organic foods are usually also clean-label. However, the organic food products must meet strict standards are no standards are developed for clean-label foods yet. Clean-label programs differ widely in terms of forbidden additives and ingredients covered, the food and beverage products covered, and the program's transparency. Nowadays, there are more and more companies open to remove additives such as artificial colorings and flavorings from part of their products. For instance, many food companies published the lists of chemicals that they will not use in their food products. In many cases these lists cover the names of the substances, however some times are non-specific and non-clear (e.g. "preservatives" or "artificial ingredients"). In this scenario, CSPI has published recommendations to address clean-label programs.

CSPI Recommended Priorities for Addressing Food Ingredients in Clean-Label Programs, at different levels are detailed as following:

- Level 1: Major Health Concerns Due to the Excessive Amounts Consumed and Number of People Affected. Added sugars and salt. Cut back.
- Level 2: Small Risk of Serious Adverse Effects (e.g., Cancer) Considering Usage. Artificial Colorings (Carcinogens), flavorings, sweeteners or preservatives, Fat/Oil, Meat substitute. Avoid or Substitution by nutritious ingredients.
- Level 3: Slight or Possible Risk of Serious Adverse Effects (e.g., Cancer) Considering Usage. Artificial and natural flavorings, Artificial colorings, Artificial preservatives, Bread/Flour additives, Emulsifier, Fat Substitute, Fillers or Thickeners, Substances with multiple uses, Plant ingredients. Caution Avoid or Substitution by nutritious ingredients.
- Level 4: Poorly Tested Substances and Other Safety Issues, including Allergic or Hypersensitivity Reactions. Artificial colorings (no synthetic), Artificial preservatives, Butter flavoring, Flavorings/flavor enhancers, Natural high-potency sweetener, Preservatives, Thickeners. Certain people should avoid, often because of allergenicity or hypersensitivity reactions. Caution Avoid or Substitution by nutritious ingredients.

1.3.2.3 Recommended Considerations According to Food Companies

Many food ingredient companies offer food manufacturers solutions to reduce additives, use natural ingredients, and other formulation challenges arising from "Clean label" issues. There are also private companies involved in clean label research. Recently, a study of more than 2800 consumers in 9 European countries about the insights on shopper profiles, the importance of front-of-pack claims and ingredient labelling to better understand the "Clean Label" (Ingredion 2020) was carried out. This study served to classify the ingredient acceptability, tolerance, and restriction to guide the food industry in product formulation (Table 1.5).

Table 1.5 Ingredient acceptability, tolerance, and restriction by European consumers (Ingredion 2020)

Accepted	Tolerated	Restricted
• Natural flavors	• Yeast extract	• Guar gum
• Natural colors	• Ascorbic acid	• Maltodextrin
• Sugar, glucose	• Corn syrup	• Monosodium glutamate
• Vegetal oils	• Pectin	• Xanthana gum
• Potato starch	• Whey protein	• Hydrogenated fats
• Starch	• Lecithin	• Carrageenan
• Corn starch	• Potato fibre	• Mono/di-glyceride fatty acids
• Gelatine	• Artificial flavors	• Caseinate
• Soya protein	• Artificial colors	• MCC (microcrystalline cellulose gel)
• Rice starch	• Modified starch	• CMC (sodium carboxymethyl cellulose)
• Preservatives	• Tapioca starch	• HPMC (hydroxypropyl methylcellulose)

1.3.2.4 Summary and Identified Alternatives

As a summary on the described initiatives of "Clean Label" foods, Table 1.6 presents main identified 'non-clean label' ingredients and their suggested alternatives to comply with "Clean Label" products as suggested by such initiatives and authors of this chapter.

1.4 Future Trends in "Clean Label"

After going to all initiatives and regulations included in the present chapter, some general agreement has been identified. Table 1.7 shows the proposed concepts that must be considered part of a "Clean Label".

Also, some terms commonly present in food labels have been identified as not included in "Clean Label", Table 1.8 shows the proposed concepts that are not included in "Clean Label" requirements.

1.5 Conclusion

An overview of food labelling regulation is presented in this chapter. A regulation meant to fulfill the definition of "Clean Label" is missing at present; there is no widely accepted definition of "Clean Label". However, consumers demand clear and 'clean' information in food labels, and food companies are already attending such demand within and unregulated framework. Thus, an interaction among policy makers, food manufacturers and consumers is required to reach understanding and application of the term "Clean Label". Main concepts to be included in "Clean

Table 1.6 Summary of non-clean-label additives/ingredients and "Clean-label" alternatives

Non-clean label ingredient	Functionality	Priorities for addressing	Alternative
Flour: all-purpose, bleached, enriched	Main ingredient for texture	Level 3	Whole grain flours/tubers, ancient grain, seed flours
Chemically modified starches		Level 3, tolerated	Functional native starches
	Gluten-free and wheat-free mixes	Level 4	Flax, oats, and legumes
Sugar	Sweetener, texture	Level 1, accepted	Low GI sugars from palm, coconut, chicory, agave, dates, D-Psicose (allulose)
High fructose corn syrup (HFCS)	Sweetener	Level 1, tolerated	Honey or agave
Synthetic antioxidants TBHQ y BHA	Shelf-life by preserving color, flavor, and general age-related spoilage	Level 2,3,4	Herbal plant and by-product extracts rosemary extract, liquorice extract, green tea, and acerola extract, vitamin E extract (mixed tocopherols)
Azoic colorants	Color	Level 2, 3, 4, tolerated	Fruit, vegetable powders and extracts, turmeric, spirulina, beets, carrots
Caramel color	Color	Level 4, accepted	Caramelized sugar, turmeric root powder, apple juice powder
Carrageenan	Texture, stabilizer	Level 3, restricted	Gellan gum, oat fiber, citrus fiber, sunflower lecithin, konjac (Asian yam)
Emulsifiers *mono-* and *di-glycerides*	Emulsification, stabilizer	Level 3, restricted	Flaxseed, tiger nut, chia, native starch
Glycerin	Humectant, carrier	Level 3,	Fruit juices, date, fig, prune paste, rice dextrin
Dairy fats	Texture, taste, shelf life	Level 2	Avocado, coconut, palm
Partially hydrogenated fats	Texture, coating	Level 2, 3, restricted	Fractionated palm, interesterification
Potassium sorbate, sodium benzoate	Bacteria, yeast, mold inhibitor	Level 2, 3, 4, accepted	Cultures of whey, rice and seeds
Propylene glycol	Flavor carrier	Level 3	Ethyl alcohol/water
Plant proteins	Label claims	Level 3	Potato, pea, rice and brown-rice, sprouted rice, lentil, coconut
Silicon dioxide	Anticaking flow agent	Level 3	Rice hull

(continued)

Table 1.6 (continued)

Non-clean label ingredient	Functionality	Priorities for addressing	Alternative
Soluble polydextrose	Fiber	Level 3	Chicory root fiber
Organic acids	Antioxidant, acidulant, flavoring	Level 3	Fruit juices and concentrates
Texturing and thickening agents	Texture, emulsification, stabilizer	Level 3, restricted	Native starches and grains
Guar gum	Texture, emulsification, stabilizer	Level 3, restricted	Psyllium seed husk
Maltodextrins	Texture, emulsification, stabilizer	Level 2, 3, restricted	Pectin, dates.
Monosodium glutamate	Flavoring	Level 2, 3, restricted	Mushrooms, soy
Xantan gum	Texture, emulsification, stabilizer	Level 3, restricted	Psyllium seed husk, cornstarch, agar agar
Hydrogenated fats	Stabilizer	Level 2, 3, restricted	Wax-based oleogels, polymer gels, and structured emulsions
Caseinate	Edible coating	Level 3, restricted	Whey and soy protein isolates
Cellulose (MCC, CMC, HPMC)	Texture, emulsification, stabilizer	Level 3, restricted	Citrus fiber
Inorganic nitrates	Bacteria inhibitor, color, flavor, preservation	Level 2, accepted	Celery powder
Additives	Extending shelf-life, adding depth of flavor, texturing, intensifying aroma, optimizing organoleptic properties, assisting fermentation	Level 2, 3, 4, restricted	Enzymes
Minimally processed products	No chemical preservatives	Accepted	Cold-pressing and high-pressure processing (HPP) techniques

Table 1.7 Agreement on requirements for a "Clean Label"

Non additive/E-numbers	Essential requirement
Non GMO	Essential requirement
Fewer chemical	Alternatives (e.g Table 1.6)
Free allergen	Alternatives (gluten, lactose, egg) (e.g. Table 1.6)
Non toxics	Clean label Project; or, <% LOD (limit of detection below those established by regulations)
Sustainable	Easy to assume by Food Industry: ISO 14000, ISO 26000, Fair Trade FLO

Table 1.8 Non-proposed elements in the concept of "Clean Label"

Nutrition claim	Regulated
Organic	Regulated
Origin, locally	Partially regulated
Vegan, kosher, halal, demeter	Regulated by private institutions.

Label" definition may be: free from additives/E-numbers, free from GMO, reduced in chemicals, free from allergens, free from toxics, and sustainable. In addition to such "free from" claims, regulations might also define and support terms such as "sustainable", "natural", or "locally grown" among others, as it is already done with 'organic' food. Last, but not least, "Clean Label" products should be made available to all consumers within all products categories. Most relevant strategies for the re-formulation of foods to meet the requirements of "Clean Label" are addressed in the book's following chapters.

References

AECOC (n.d.) Libro blanco de atributos diferenciados de producto [29/12/2020]. https://www.aecoc.es/libro-blanco-de-atributos-diferenciados-de-producto/2019

Aschemann-Witzel J, Varela P, Peschel AO (2019) Consumers' categorization of food ingredients: do consumers perceive them as 'clean label' producers expect? An exploration with projective mapping. Food Qual Prefer 71:117–128

Asioli D, Aschemann-Witzel J, Caputo V, Vecchio R, Annunziata A, Næs T, Varela P (2017) Making sense of the "clean label" trends: a review of consumer food choice behavior and discussion of industry implications. Food Res Int 99:58–71

Borges BJP, Arantes OMN, Fernandes AAR, Broach JR, Fernandes PMB (2018) Genetically modified labeling policies: moving forward or backward? Front Bioeng Biotechnol 6(181)

Borra S (2010) Clean food labeling & sustainability benefit claims: what do consumers want? What are the legal requirements? What if compliance is not enough? In: Annual meeting, Institute of Food Technologists; July 17-20; Chicago, III

Brockman C, Edwards A, Adams-Hutt C, Mulry M (2013) Natural and clean label trends 2013: who's driving the agenda? From simple truth to open. Nature [29/12/2020]. https://www.foodnavigator-usa.com/Article/2013/06/21/Natural-Clean-Label-Trends-20132013

Busken DF (2015) Cleaning it up what is a clean label ingredient? Cereal Foods World 60:112–113

Cargill (2017) Staying ahead of the curve in the drive for "clean label" products [29/12/2020]. https://www.cargill.com/doc/1432106811290/clean-label-white-paper.pdf

Clean Label Project (2020) The consumer standard for food and consumer product quality and safety [08/01/2021]. https://cleanlabelproject.org2020

Clean Label Project (n.d.) Code of practice: certification [08/01/2021]. https://cleanlabelproject.org/wp-content/uploads/CLP_Code-of-Practice_Certification-v1-10-04-19.pdf2019

DEFRA (2012) Understanding the environmental impacts of consuming foods that are produced locally in season

Dickson-Spillmann M, Siegrist M, Keller C (2011) Attitudes toward chemicals are associated with preference for natural food. Food Qual Prefer 22(1):149–156

Duran CR, Valenzuela BA (2010) The Japanese experience with FOSHU foods, the true functional foods? Rev Chil Nutr 37(2):224–233

EUR-Lex (2005) Directive 2005/29/EC of the European parliament and of the council of 11 May 2005 concerning unfair business-to-consumer commercial practices in the internal market and amending Council Directive 84/450/EEC, Directives 97/7/EC, 98/27/EC and 2002/65/EC of the European Parliament and of the Council and Regulation (EC) No 2006/2004 of the European Parliament and of the Council [29/12/2020]. http://data.europa.eu/eli/dir/2005/29/oj

EUR-Lex (2012a) COMMISSION REGULATION (EU) No 432/2012 of 16 May 2012 establishing a list of permitted health claims made on foods, other than those referring to the reduction of disease risk and to children's development and health [29/12/2020]. http://data.europa.eu/eli/reg/2012/432/oj

EUR-Lex (2012b) Regulation (EU) No 1151/2012 of the European parliament and of the council of 21 November 2012 on quality schemes for agricultural products and foodstuffs [29/12/2020]. http://data.europa.eu/eli/reg/2012/1151/oj

EUR-Lex (2014a) Regulation (EC) No 1924/2006 of the European parliament and of the council of 20 December 2006 on nutrition and health claims made on foods [29/12/2020]. http://data.europa.eu/eli/reg/2006/1924/2014-12-13

EUR-Lex (2014b) COMMISSION IMPLEMENTING REGULATION (EU) No 828/2014 of 30 July 2014 on the requirements for the provision of information to consumers on the absence or reduced presence of gluten in food [29/12/2020]. http://data.europa.eu/eli/reg_impl/2014/828/oj

EUR-Lex (2017) Regulation (EU) 2017/625 of the European Parliament and of the Council of 15 March 2017 on official controls and other official activities performed to ensure the application of food and feed law, rules on animal health and welfare, plant health and plant protection products [29/12/2020]. http://data.europa.eu/eli/reg/2006/1924/2014-12-13

EUR-Lex (2018a) Regulation (EU) No 1169/2011 of the European parliament and of the council of 25 October 2011 on the provision of food information to consumers, amending Regulations (EC) No 1924/2006 and (EC) No 1925/2006 of the European Parliament and of the Council, and repealing Commission Directive 87/250/EEC, Council Directive 90/496/EEC, Commission Directive 1999/10/EC, Directive 2000/13/EC of the European Parliament and of the Council, Commission Directives 2002/67/EC and 2008/5/EC and Commission Regulation (EC) No 608/2004 [29/12/2020]. http://data.europa.eu/eli/reg/2011/1169/2018-01-01

EUR-Lex (2018b) REGULATION (EU) 2018/848 OF THE EUROPEAN PARLIAMENT AND OF THE COUNCIL of 30 May 2018 on organic production and labelling of organic products and repealing Council Regulation (EC) No 834/2007 [29/12/2020]. http://data.europa.eu/eli/reg/2018/848/2018-06-14

EUR-Lex (2018c) Directive (EU) 2018/852 of the European Parliament and of the council of 30 May 2018 amending Directive 94/62/EC on packaging and packaging waste [29/12/2020]. http://data.europa.eu/eli/dir/2018/852/oj

EUR-Lex (2018d) COMMISSION REGULATION (EU) 2018/213 of 12 February 2018 on the use of bisphenol A in varnishes and coatings intended to come into contact with food and amending regulation (EU) No 10/2011 as regards the use of that substance in plastic food contact materials [29/12/2020]. http://data.europa.eu/eli/reg/2018/213/oj

EUR-Lex (2019) Directive (EU) 2019/904 of the European Parliament and of the council of 5 June 2019 on the reduction of the impact of certain plastic products on the environment [29/12/2020]. http://data.europa.eu/eli/dir/2019/904/oj

Europe PRO (2020) The Green Dot [29/12/2020]. https://www.pro-e.org/the-green-dot-trademark

European Commission (2016) Nutrition and health claims [29/12/2020]. https://ec.europa.eu/food/safety/labelling_nutrition/claims/register/public/?event=search

European Commission (2018) Quality schemes explained [29/12/2020]. https://ec.europa.eu/info/food-farming-fisheries/food-safety-and-quality/certification/quality-labels/quality-schemes-explained_en

European Commission (n.d.) Food information to consumers—legislation [08/01/2021]. https://ec.europa.eu/food/safety/labelling_nutrition/labelling_legislation_en2020

FAO (2014) Government and voluntary policies on nutrition labelling: a global overview [08/01/2021]. http://www.fao.org/3/i0576e/i0576e04.pdf

FAO (2019) Norma General del Codex para los Aditivos Alimentarios (GSFA, Codex STAN 192-1995). In: AhlaRdlCdC (ed), Alimentarius [08/01/2021]. http://www.fao.org/gsfaonline/reference/techfuncs.html

Food and Agriculture Organization off the United Nations, Organization WH (2017a) Codex Alimentarus—International Food Standards. Codex Texts [08/01/2021]. http://www.fao.org/fao-who-codexalimentarius/codex-texts/en/

Food and Agriculture Organization off the United Nations, Organization WH (2017b) Codex Alimentarus-International Food Standards. Nutrition and Labelling [29/12/2020]. http://www.fao.org/fao-who-codexalimentarius/thematic-areas/nutrition-labelling/en/

Food and Agriculture Organization off the United Nations, Organization WH (2017c) Codex Alimentarus-International Food Standards. Standards [29/12/2020]. http://www.fao.org/fao-who-codexalimentarius/codex-texts/list-standards/en/

Food and Agriculture Organization off the United Nations, Organization WH (2017d) Codex Alimentarus-International Food Standards. Guidelines 2017 [29/12/2020]. http://www.fao.org/fao-who-codexalimentarius/codex-texts/guidelines/en/

Food and Agriculture Organization off the United Nations, Organization WH (2017e) Codex Alimentarus-International Food Standards. Miscellaneous [29/12/2020]. http://www.fao.org/fao-who-codexalimentarius/codex-texts/miscellaneous/es/

FS Council (2020) Forest for all forever [08/01/2021]. https://www.fsc-uk.org/en-uk/about-fsc

Hakim MP, Zanetta LDA, de Oliveira JM, da Cunha DT (2020) The mandatory labeling of genetically modified foods in Brazil: consumer's knowledge, trust, and risk perception. Food Res Int 132:109053

Hutt C, Sloan A (2015) Coming clean: what clean label means for consumers and industry. In: Global food forum clean label conference

Ingredion (2020) The clean label guide [29/12/2020]. https://emea.ingredion.com/Campaign/Clean-Label-Guide.html

Jeans KB (2017) Recyclability symbols [29/12/2020]. https://buddhajeans.com/encyclopedia/recyclability-symbols/

Joint FAO/WHO Expert Committee on Food Additives (2019) Combined compendium of food additive specifications [29/12/2020]. http://www.fao.org/3/a0691e/A0691E03.htm

Katz B (2011) Cleaning up processed foods. Institute of Food Technologists, Chicago. [29/12/2020]. https://www.ift.org/news-and-publications/food-technology-magazine/issues/2011/december/features/cleaning-up-processed-foods

Kneafsey M, Venn L, Schmutz U, Balázs B, Trenchard L, Eyden-Wood T et al (2013) Short food supply chains and local food systems in the EU. A state of play of their socio-economic characteristics. European Union, Brussels

Lefferts L (2017) Clean labels—public relations or public health?: Center for Science in the Public Interest CSPI; [29/12/2020]. https://cspinet.org/sites/default/files/attachment/Clean%20Label%20report%20%281%29.pdf

Macdiarmid JI (2014) Seasonality and dietary requirements: will eating seasonal food contribute to health and environmental sustainability? Proc Nutr Soc 73(3):368–375

Moskowitz HR, Beckley JH, Resurreccion AVA (2012) So what are the practical considerations in actually running a test? What do I need to know? What does the rest of the company need to know? In: Sensory and consumer research in food product design and development. Wiley, New York, pp 321–363

Premanandh J, Maruthamuthu M, Sabbagh A, Al Muhairi S (2012) Prevalence of genetically modified foods (GM foods) in the United Arab Emirates. Food Control 25(1):10–12

recyclenow (2016) Recycling symbols explained [29/12/2020]. https://www.recyclenow.com/recycling-knowledge/packaging-symbols-explained

Roller S (2010) Clean food labeling & sustainability benefit claims: what do consumers want? What are the legal requirements? What if compliance is not enough? In: Annual meeting, Institute of Food Technologists; July 17-20; Chicago, III

Rozin P, Fischler C, Shields-Argelès C (2012) European and American perspectives on the meaning of natural. Appetite 59(2):448–455

Sánchez-Bravo P, Edgar Chambers V, Noguera-Artiaga L, López-Lluch D, Chambers E IV, Carbonell-Barrachina ÁA et al (2020) Consumers' attitude towards the sustainability of different food categories. Foods 9:1608

Sánchez-Bravo P, Chambers VE, Noguera-Artiaga L, Sendra E, Chambers E IV, Carbonell-Barrachina ÁA (2021) Consumer understanding of sustainability concept in agricultural products. Food Qual Prefer 89:104136

Scopus (2020) [08/01/2021]. https://www.scopus.com/term/analyzer.uri?sid=5feee792e15d59e77ad6e840855a28da&origin=resultslist&src=s&s=TITLE-ABS-KEY%28food%2c+clean+and+label%29&sort=plf-f&sdt=b&sot=b&sl=36&count=281&analyzeResults=Analyze+results&txGid=76817e5faa28f64b62b569c8fa2d0952

Smyth SJ (2018) Genetically modified food regulations in Canada. Reference module in food science. Elsevier, Amsterdam

Song H, Schwarz N (2009) If it's difficult to pronounce, it must be risky. Psychol Sci 20(2):135–138

U.S. Food & Drug Administration (2013) Guidance for industry: food labeling guide [29/12/2020]. http://wayback.archive-it.org/7993/20180907182523/https://www.fda.gov/Food/GuidanceRegulation/GuidanceDocumentsRegulatoryInformation/LabelingNutrition/ucm247920.htm

Varela P, Fiszman SM (2013) Exploring consumers' knowledge and perceptions of hydrocolloids used as food additives and ingredients. Food Hydrocoll 30(1):477–484

WHO (2007) World Health Organization, nations FaAOoTU. Codex Alimentarus—Food labelling, 5th edn

Chapter 2
Natural Emulsifiers as Clean Label Ingredients

Sana Riaz, Usman Amin, and Abid Aslam Maan

2.1 Introduction

Processed foods have occupied a significant part of our everyday life and are made with a combination of mutually immiscible phases like a mixture of lipids, carbohydrates, and proteins. These insoluble mixtures are thermo-dynamically unstable and ultimately disturb the microbial stasis, appearance, flavor, and texture. In order to stabilize this immiscible combination of different phases, emulsifiers are commonly used. (McClements 2015; Hasenhuettl and Hartel 2008). Emulsifiers facilitate and stabilize the immiscible phases by reducing the interfacial tension between them. They perform several other functions by interacting with other food components such as lubricant, crystal modifier, texture enhancer, a shelf life enhancer, and moisture barrier, etc. (Orthoefer and Kim 2019).

Emulsifiers are amphiphilic compounds having a non-polar and polar head on the same molecule. Emulsifiers work via a hydrophilic head, which is attached to the aqueous phase, and a sizeable lipophilic tail, which prefers to associate with the oil phase. Therefore, emulsifiers relatively align themselves at oil/water or air/water interface at which they tend to lower interfacial/surface tension, respectively. Lipophilic tails are comprised of 16 carbon (palmitic) or longer fatty acids. While Short chains (like C12; lauric) can be suitable emulsifiers but produce soapy or

S. Riaz · U. Amin
National Institute of Food Science and Technology, University of Agriculture Faisalabad-Pakistan, Faisalabad, Pakistan
e-mail: Usman.amin@uaf.edu.pk

A. A. Maan (✉)
National Institute of Food Science and Technology, University of Agriculture Faisalabad-Pakistan, Faisalabad, Pakistan

Department of Food Engineering, University of Agriculture Faisalabad-Pakistan, Faisalabad, Pakistan
e-mail: abid.maan@uaf.edu.pk

C. M. Galanakis (ed.), *The Age of Clean Label Foods*,
https://doi.org/10.1007/978-3-030-96698-0_2

other unpleasant flavors upon hydrolysis. While, in the case of protein, the lipophilic tail is composed of lipophilic amino acids like leucine, phenylalanine, and isoleucine. The polar group may be composed of amino, carboxy, hydroxy, or phosphoric acid. The polar head may exist in a different functional group, which may either contain a positive or negative charge. Depending upon their charge on the active group, they can behave as cationic, anionic, nonionic, or amphoteric surfactants. Anionic surfactants carry a negatively charged molecule, attached with a small positive counterion.

Conversely, cationic surfactant contains a positively charged molecule with a slight negative counterion. However, amphoteric surfactants have both negative and positive charges on the same molecule. Nonionic surfactants do not possess any charge, but a polar heteroatom causes a dipole with an electron-dense and electron-depleted region.

The most commonly cited classification system for emulsifiers is hydrophile-lipophile balance (HLB). Emulsifiers having low HLB values are considered lipophilic, while hydrophilic emulsifiers have high HBL. Oil in water (O/W) emulsions are mostly stabilized with hydrophilic emulsifiers, while water in oil (W/O) emulsions are stabilized via lipophilic emulsifiers (Yamashita et al. 2017; Hasenhuettl and Hartel 2008).

2.2 Natural Emulsifiers

Consumer demand for natural ingredients has forced food manufacturers to replace synthetic chemical emulsifiers with natural emulsifiers. Currently, many types of natural emulsifiers are being used in food industries; their major types are given below:

2.2.1 Proteins

Proteins are usually small molecules (≈10–50 kDa) that rapidly adsorb to interface and form thin electrically charged interfacial layers (Bouyer et al. 2012). Molecular properties of proteins (a charge, molar mass, polarity) depend on a protein source, extraction, purification, and storage method (Lam and Nickerson 2013). Proteins (extracted from plants and animals) are usually amphiphilic (Lam and Nickerson 2015). Animal proteins include bovine milk proteins (whey, casein), egg proteins (ovalbumin), fish and meat proteins (gelatin), while plant-based proteins are generally extracted from wheat, maize, legumes, oilseeds, etc. (Chen 2015; Karaca et al. 2015). Proteins find applications in different food (sauces, sausages, desserts, beverages, dressings, etc.) and non-food (pharmaceutics and cosmetics) products. In such applications, their functionality sometimes needs to be modified through chemical, physical and enzymatic means (Clare and Daubert 2011; Mirmoghtadaie et al. 2016; Lam and Nickerson 2013).

Emulsifying properties of different food proteins can be measured by their physicochemical properties such as number, sequence, and type of amino acids and their peptide chains (Papalamprou et al. 2010). Surface activity and water solubility of proteins depend on a number of polar and non-polar ends on the surface. Very few non-polar molecules tend to reduce surface activity, while an excess of them cause low solubility at the oil-water interface. Therefore, an optimum number of polar and non-polar head groups are required to attain maximum surface activity and water solubility (Damodaran et al. 2007). Moreover, environmental conditions like pH, temperature, and ionic strength tend to change the protein structure that ultimately affects the emulsifier's stability (Dickinson 2010). Adsorption of proteins on the interface can produce a net negative or positive charge depending upon pH. A similar charge on the interface of dispersed droplets (in emulsions) results in better stability owing to high electrostatic repulsion (McClements 2015). They can also stabilize the emulsions by making a thick interfacial layer through steric hindrance (Ozturk and McClements 2016).

2.2.1.1 Animal Proteins

2.2.1.1.1 Casein

Casein is a major milk protein that exhibits a random and highly flexible structure (Fig. 2.1). Its open form makes it more hydrophobic and can be isolated through isoelectric precipitation (pH 4.6) or proteolytic coagulation (Huppertz et al. 2018a). Bovine casein micelles dry matter contains 94% protein and 6% small compounds known as colloidal calcium phosphate (CaP) that are attached with casein side-chain phosphoserine residues (De Kruif et al. 2012; Ranadheera et al. 2016). Casein

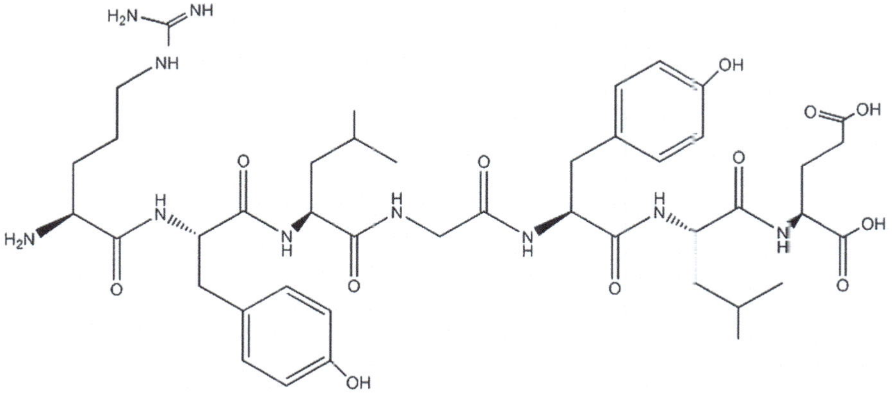

Fig. 2.1 Structure of casein

clusters comprise of four fractions: $\alpha S1$, $\alpha S2$, β and κ-casein, that account for 38%, 10%, 36%, and 12% of total casein. Naturally, αS1- and αS2-caseins have unfolded structures with prolonged coil-like configuration, whereas β- and κ-caseins are molten globular proteins that contain dense structures having highly flexible side chain and a high degree of hydration. All casein fractions contain signal peptides on N-terminals and Pro-Gln rich peptides near C-terminus that provide a binding site for CaP (McMahon and Oommen 2013; Holt 2016). α- and β-casein comprises of 8–10 and 5 phosphoserine group (PO4 residues), respectively, but β-casein is more hydrophobic as compared to other fractions. κ-casein is only glycosylated casein having one phosphoserine residue, mostly containing galactose. κ-casein helps in casein micelles stabilization and protects different casein subunits from precipitation, as it is highly stable in calcium ions presence (Huppertz et al. 2018a; Elzoghby et al. 2011).

Casein has been widely used as an ingredient in numerous food products to improve their functional and physical properties, like thickening, textural, emulsification, foaming, and most prominently, to enhance their nutritional quality. Moreover, casein represents good carrier properties for delivering sensitive ingredients in pharmaceutical and food industries (Ranadheera et al. 2016).

Li et al. (2020a) investigated the effect of casein amount (0.5, 1.5, and 2.5%) on whippability and stability of milk cream. The results indicated that an increased amount of casein tended to decrease serum loss and increased viscosity, stability, and whippability. Another study also reported that the addition of casein in low-fat cream leads to reduce in droplet size, an increase in viscosity, and stability of cream (Ziaeifar et al. 2018). Villanueva et al. (2018) evaluated the effect of caseinate and soy protein on rheological attributes of gluten-free dough. Protein addition decreased the dough stickiness while decreased the pasting value and viscoelasticity of the dough. Wusigale et al. (2020) demonstrated that casein, together with pectin, could stabilize the acidified dairy products. Casein has been reported to increase (oxidative and physical) stability of fish O/W nano-emulsions and resulted in emulsions with small particles and thin interfacial layer (Yang et al. 2019a). Several studies have reported the applications of casein in the microencapsulation of bioactive compounds (Vaucher et al. 2019; Yan et al. 2020; Ranadheera et al. 2016; Troncoso Recio 2019).

2.2.1.1.2 Whey Protein

Whey proteins represent approximately 20% of whole milk protein (~80% caseins) and exhibit a high nutritional profile due to the high level of essential amino acids. These are globular proteins having spherical confirmation persuaded by the tertiary structure in which hydrophilic groups are positioned outside while hydrophobic groups are suppressed inwards (Adjonu et al. 2014). Whey protein comprises of different fractions; β-lactoglobulin (60%), α-lactalbumin (22%), bovine serum albumin (BSA) (5.5%) and immunoglobulins (9%) (Kilara and Vaghela 2018).

β-lactoglobulin (β-Lg) is a major protein present in whey possessing a molecular weight (MW) of about 18.3 kDa. Its secondary and tertiary structures comprise of β-sheet (43–50%), β-turn (15–20%), and α helix (10–15%). Moreover, it is sensitive to pH and has an IEP at 4.0–5.2 (Boland 2011). However, α-lactalbumin (α-La) contains MW of 14.2 kDa, and IEP is at pH 4.8–5.1. The presence of a high level of calcium and hydrophilic groups makes it highly soluble in water even at its IEP (Jambrak et al. 2014).

BSA is present in meager amounts in almost all types of animal milk than other whey proteins, i.e., β-Lg and α-La. Cow milk contains about 0.4 g/L of BSA, covering 1.5% of whole milk proteins while 8% of total whey proteins. BSA is a nonglycosylated, flexible, and highly structured protein. BSA has been reported as heart-shaped and consists of three helical fractions (I, II, and III) (Deeth and Bansal 2019). Immunoglobulins (Ig) are another minor protein found in whey that are glycoproteins containing antibody-like properties. IgA, IgM, and IgG are major immunoglobulins present in milk and whey. IgG is present in monomeric form having MW of B60 kDa, IgA is found as dimers (MW of B370 kDa) and tetramers, and IgM occurs as a pentamer with MW of 9,001,000 kDa (Deeth and Bansal 2019). The heat stability of whey proteins at a constant time is in the order of α-lactalbumin > β-lactoglobulin > serum albumin/lactoferrin > immunoglobulins (Dissanayake and Vasiljevic 2009).

Whey proteins have been used in various food applications such as liquid meat replacers, sports beverages, ice cream, bakery products, salad dressing, infant foods, and several other dairy products (Wagoner and Foegeding 2017). Danesh et al. (2017) studied the effect of whey proteins and enzymatically modified whey proteins on ice cream properties. Results concluded that whey proteins (with/without enzyme modification) decreased the viscosity, fat content, overrun, and flow index. Whereas firmness, melting resistance, viscosity, and sensory properties increased with an increase in whey protein concentration. Sołowiej et al. (2020) evaluated the effect of substitution of emulsifying salt with whey protein isolate (WPI) in cheese preparation. Results represented that WPI not only stabilized the emulsion but also enhanced the textural attributes, viscosity, meltability and spreadability, whereas decreased stickiness and firmness of cheese. WPI has been reported to stabilize the O/W emulsion and improve astaxanthin stability during *in vitro* digestion (Boonlao et al. 2020). WPH (whey protein hydrolysate), when used as emulsifier and wall material for encapsulation of fish oil incorporated in fortified mayonnaise, improved oxidative stability of mayonnaise during 2 months storage (Rahmani-Manglano et al. 2020). Scalone et al. (2020) reported that WPH act as a precursor in the production of 2,5-Dimethylpyrazine and other aroma compounds, which leads to improving the aroma of cookies and bread. Szafrańska et al. (2020) demonstrated that increasing the amount of whey protein concentrates (2–8 g/100 g) in cheese sauce tends to increase viscosity, hardness, stickiness, and stability. Whey protein has been widely studied for encapsulation of vitamin D3 (Liu et al. 2020), β-carotene (Rodrigues et al. 2020), curcumin (Lv et al. 2020), iron (Durán et al. 2020), and eugenol (Talón et al. 2019).

2.2.1.1.3 Gelatin

Gelatin (Fig. 2.2) is referred to as a denatured protein that is obtained through acid or alkaline hydrolysis of collagen extracted from the skin, bones, and connective tissues of animals. In gelatin manufacturing, raw material containing collagen is pretreated with acid or base to break crosslinks that produce two forms of gelatin partly. Type A gelatin resulting from acid hydrolysis of fig and fish skin and is lightly crosslinked (Schrieber and Gareis 2007; Hussain et al. 2020). Type B gelatin is produced via an alkaline process and is heavily cross-linked. Type A gelatin has an IEP at pH 7–9, whereas type B gelatin has an IEP at pH 4.8–5.2 (Haug and Draget 2011). Gelatin is a polyampholyte compound exhibiting hydrophobic, anionic, and cationic groups with a ratio of 1:1:1 (Liu et al. 2015). It has been extensively utilized in cosmetic, food, photographic, and pharmaceutical applications due to its unique technical and functional properties. Gelatin is being widely used in the food industry as an ingredient to enhance product stability, elasticity, and consistency (Abdullah et al. 2018). Gelatin is cold-setting and thermo-reversible, which allows it to occur as a gel or solution depending on temperature. Its conversion from gel to solution occurs below human body temperature, which gives it a significant advantage over other gelling agents (Otoni et al. 2012).

Gelatin has various applications as an emulsifier in food and pharmaceuticals. Huang et al. (2020) evaluated the application of fish gelatin as an emulsifier in coffee whitener. Results reported that lubrication, whiteness, and physical stability increased with an increase in gelatin. A study showed that the addition of gelatin in frozen dough prevents the distortion of ice crystals during 60 days of storage. The bread prepared from that dough showed enhanced dough volume, more uniform, and softer crumb (Yu et al. 2020). Yu et al. (2019) also reported that gelatin in bread prevents bread staling and improves bread quality. Zhao et al. (2020) assessed the role of gelatin in the encapsulation of probiotics (*L. reuteri*). Gelatin and caseinate composite improved the survival rate of probiotics at room temperature. Duan et al. (2018) represented the application of gelatin in ice cream and beer. They concluded that gelatin improved viscosity, emulsion stability, foam stability, and mouthfeel in beer and ice cream. Gelatin has also been reported to stabilize simple (O/W) (Anvari and Joyner 2017) and multiple (W/O/W) emulsions (Zhu et al. 2018). It has also

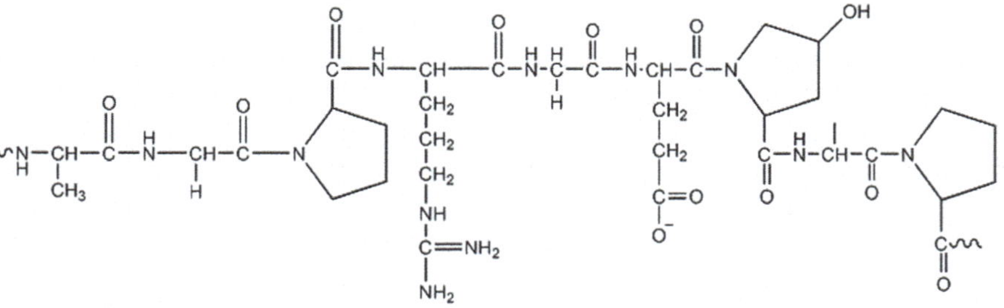

Fig. 2.2 Chemical structure of gelatin

been reported to enhance the stability and digestion of encapsulated β-carotene (Guedes Silva et al. 2021). The applications of gelatin in food packaging have been extensively studied (Nazmi and Sarbon 2019; Stejskal et al. 2020; Tian et al. 2020; Xiong et al. 2020).

2.2.1.1.4 Ovalbumin

Ovalbumin (OVA) is a significant protein of egg white (about 54% of total protein content) and well known for its excellent properties such as foaming and gelling properties. OVA is a glycosylated protein comprising 385 amino acids (>50% hydrophobic) covalently linked with a single carbohydrate group (Xu et al. 2020). OVA can be used as a foaming and emulsifying agent, which also plays a role in texture and structure development (Xiong et al. 2016). The effect of ovalbumin on steamed bread properties was investigated by Sang et al. (2018). The authors reported that OVA interacts with gluten, resulting in dough softer, larger dough volume, and less elasticity. OVA also served as a lubricant and inert filler in bread. Feng et al. (2016) stated that curcumin encapsulation in OVA-dextran nanogel improved the bioavailability in gastrointestinal trac *in vitro*. Encapsulation increased broadcastability (%) from 18.9 to 52.54. In a study, OVA was used as a stabilizer in copper nanoclusters biosensor for the detection of folic acid. It improved the stability of the biosensor by preventing oxidation and the growth of larger nanoparticles (Li et al. 2019). OVA has also been reported for encapsulation of vitamin D3 (Xiang et al. 2020), emulsion stabilizer (Li et al. 2020b; Niu et al. 2016; Zhang et al. 2020), and as texture improver in bread (Nozawa et al. 2016).

2.2.1.2 Plant Proteins

2.2.1.2.1 Cereal Proteins

Proteins extracted from wheat and corn are the most commonly used among cereal proteins. They play a role in the structural development of various cereal products. Wheat protein primarily consists of gluten (~85% including gliacins and glutenins) and a small quantity of non-gluten proteins (~15%) (Wouters and Delcour 2019). Gliadins consist of a single polypeptide chain, having MW of 30–75 kDa that can link with each other and glutenins via hydrophobic interaction. Glutenins are composed of high MW fractions that make an internal framework and low MW fractions that exhibit structures similar to gliadin (Day 2011). Non-gluten proteins include albumins and globulins. Liu et al. (2018) reported gluten prepare and stabilizes the O/W emulsion as mayonnaise replacer. The resulted emulsions possessed sensory and textural attributes (slimness, creaminess, viscoelasticity, and consistency) similar to mayonnaise. It also presented high thermal stability as compared to mayonnaise. Deepika et al. (2020) demonstrated the effect of gluten on properties of PVA films and showed that gluten improved the mechanical and biodegradable properties of films.

Zein, the major protein of corn, occupies 35–60% of the total protein of corn. Zein is generally rich in glutamic acid (21–26%), leucine (20%), proline (10%), and alanine (10%), but deficient in essential amino acids (tryptophan and lysin). It has the potential to physically bind hydrophobic compounds that make it an attractive compound for food applications. Zein is ethanol-soluble (75–90%) and water-insoluble. Zein has been reported to stabilize an oil-in-glycerol emulsion-gel fortified with β-Carotene as a healthy substitute for margarine in cake preparation. The resulting cake exhibited similar texture and sensory attributes with standard mayonnaise cake (Chen et al. 2016). Zein used as a fat replacer in mayonnaise preparation showed that fat substitution with zein up to 40 % provided good rheological and sensory properties (Gu et al. 2016). Zein has also been reported for stabilization of foams and emulsions (Nephomnyshy et al. 2020) as a polymer material for films and coatings (Sun et al. 2020) as well as in the production of gluten-free products (Sosa et al. 2019; Berta et al. 2019). Zein has been extensively studied for pharmacological applications, especially in drug delivery (Cheng et al. 2019; Gagliardi et al. 2019).

Rice contains about 10–16% protein that is further comprised of glutelin 32.5%, globulin 30.9%, albumin 24.9%and prolamine 11.6% (Phongthai et al. 2017). Perrechil et al. (2021) reported rice protein application concentrates as a coating material for flaxseed oil microencapsulation. Rice protein showed poor emulsifying (1.15 m^2/g) and low encapsulating power (<1%). The emulsifying properties of rice protein isolate from various waxy and non-waxy rice cultivars were evaluated. According to results, waxy rice cultivar stabilized the O/W emulsion even at pH 7 and could be used as emulsifier in food products (Mun et al. 2016). Rice bran protein was observed to prevent bread staling along with the increase in nutrition, crumb porosity, and sensory properties of gluten-free bread (Phongthai et al. 2016).

2.2.1.2.2 Legume Proteins

The podded plants like peas, lentils, peanuts, beans, etc., are known as legumes. Legumes possess a very high amount of proteins that are 2–3 times higher than cereals. Legumes mainly contain two storage proteins: globulins and albumins (Alting and Van De Velde 2012). The central portion of the total storage proteins of legumes is occupied by globulins (about 70%). Globulins are multi-fractional proteins comprising 2 significant subunits known as hexameric (300–370 kDa) and trimeric 83 (175–180 kDa), which contain sedimentation coefficients of 11 and 7 (S 84 values), respectively. Albumins comprise 10–30% of whole legume proteins and include lower sedimentation coefficient (2*S*) and MW (Nozawa et al. 2016). Chickpea, pea, and soy proteins have been extensively used as natural emulsifiers in many products. The incorporation of soy protein hydrolysates and isolates in ice cream improved the rheology, melting rate, and fat coalescence (Chen et al. 2019b). Patel et al. (2020) reported soy protein isolate and a starch conjugate as an excellent material for anthocyanins' encapsulation. According to Chen et al. (2019a), the emulsifying capacity and stability of emulsions are significantly improved with increasing pea protein concentration. The application of pea, lentil, and chickpea protein isolates as an emulsifier in salad dressing was studied by Ma et al. (2016). The results revealed that the protein isolates are an efficient replacer of egg protein as they provide better functional properties than egg proteins. Modified chickpea proteins have also been reported to stabilize O/W emulsions (Glusac et al. 2020; Felix et al. 2020) (Table 2.1).

2.2.2 *Polysaccharides*

Polysaccharides are highly hydrophilic compounds and mainly consist of polar monosaccharide units that make them poor emulsifiers. They can work as emulsion stabilizers by enhancing the viscosity of the aqueous phase and restrict the movement of droplets of the dispersed phase (McClements 2015). Enzymatic or chemical modification can be carried out to make these polysaccharides surface-active agents, e.g., modified starches and Millard carbohydrates and proteins complex (Evans et al. 2013). However, natural polysaccharides exhibit good emulsifying properties as they bound covalently with lipid or protein, which make them slightly hydrophobic, i.e., glycoproteins and glycolipids (Bai et al. 2016). Galactomannans, gum arabic, modified starches, pectin, and modified cellulose are typical examples of such emulsifiers. Nevertheless, modified cellulose and starches are not considered natural because of their chemical modification process.

Emulsifying polysaccharides' properties depend on their polarity, molar mass, hydrophobicity, electric charge, and a number of branches. Polysaccharides are relatively large molecules and form a thick interfacial layer, which generates strong steric repulsion. Hence, such emulsion is less affected by ionic strength and pH of the environment as compared to proteins (McClements 2015).

2.2.2.1 Gum Arabic (GA)

Gum arabic is a natural polysaccharide obtained from *acacia seyal*. It comprises of anionic hydrophilic polysaccharide chains covalently attached with a backbone of the hydrophobic peptide. It contains a highly branched structure having β1,3-linked galactose residue as the central core. GA is one of the most widely used polysaccharide-based emulsifiers in the food industry, mainly in beverages. Instead of adequate emulsifying capacity, its variable supply and high price have restricted its applications. The requirement of a high emulsifier to lipid ratio (1:1) is the main disadvantage of gum arabic (Xue and Zhong 2014; Piorkowski and McClements 2014). China et al. (2020) evaluated the effect of GA on the properties of biscuits. The results indicated that increasing the amount of GA (0, 0.3, 0.5%) tended to increase the hardness, crispiness, and overall acceptability of biscuits. In another

Table 2.1 Characteristics of common protein emulsifiers from plant and animal origin

Type of protein	Source	Fractions	Molecular structure	Molecular weight (kDa)	Isoelectric point	Denaturation temperature (°C)	References
Casein	Milk	$\alpha S1$-casein	Flexible, random coiled, unstructured	23.6	4.4–4.8	84	Huppertz et al. (2018b), Horne (2020)
		$\alpha S2$-casein		25.2	4.9		
		β-casein		24	4.8–5		
		κ-casein		19.1	3.5		
Whey	Milk	α-lactalbumin	Monomeric globular protein	14.2	4.8–5.1	67	Deeth and Bansal (2019), Kilara and Vaghela (2018)
		β-lactoglobulin		18.4–36.9	4–5.2	75	
		Serum albumin		69	4.8	62	
		Immunoglobulin		150	5	65	
Gelatin	Animal collagen	Type A	Flexible protein	10–400 depending upon processing conditions	7–9	10–30	Meng and Cloutier (2014)
		Type B			4.8–5.2		
Ovalbumin	Egg white		Globular protein	45	4.5	70	Xu et al. (2020)
Zein	Corn	α-zein	Globular	19–22	6.2	58.5	Chang et al. (2017), Kasaai (2018)
		β-zein		10–27			
Gluten	Wheat	Gliadin	Complex globular protein	28–55	7.5	<80	Wang et al. (2015)
		Glutenin		32–35			
Soy protein	Soybean	β-conglycinin (7S)	Complex globular protein	150–200	4.6	78.6–90.8	Nishinari et al. (2018)
		Glycinin (11S)		300–380		74–85	
Pea protein	Peas	Legumin (11S)	Complex globular protein	150–190	4.8	69–76	Amagliani et al. (2017)

study, it was also reported to improve the texture, volume, and springiness of gluten-free bread (Chakraborty et al. 2020). Cai et al. (2019) demonstrated that encapsulation of thyme essential oil (TEO) in GA decreased the release time while increased the stability of TEO. GA did not alter the properties of TEO as it does not interact with EO, just acts as an emulsifier. GA also has a wide application in the development of biodegradable packaging films and coatings (Xu et al. 2019a; Yang et al. 2019c).

2.2.2.2 Pectin

Pectin is a natural polysaccharide obtained from cell walls of different plants such as sugar beet, citrus, apple, and okra. It consists of a D-galacturonic acid backbone linked via α-(1-4) glycosidic bond and side chains made up of different monosaccharides such as fructose, glucose, xylose, mannose, rhamnose, and arabinose (Fig. 2.3) (Nesic and Seslija 2017). Its emulsifying ability depends on the composition of side chains, acetyl and feruloyl groups, and protein moiety (Ngouémazong et al. 2015). Commercial and natural pectin exhibits a high degree of polymerization and MW ranging from 100 to 400 kDa (Khotimchenko 2020). Pectin comprises of hydrophilic polysaccharide that inhibits steric repulsion and a hydrophobic phenolic or protein site which facilitate adsorption on the surface of oil droplet (BeMiller 2019; Schmidt et al. 2015). Pectin (Fig. 2.3) is also a widely used natural emulsifier in the food industry for the stabilization of O/W emulsions. Okra pectin has been used as an emulsifier in chocolate and as a substitute for lecithin. It was observed that pectin improved sensory and textural properties of chocolate (increased viscosity, slowed flow rate, decreased adhesiveness, etc.) (Datsomor et al. 2019). Dubrovskaya et al. (2017) reported that pectin increased the crumb compressibility

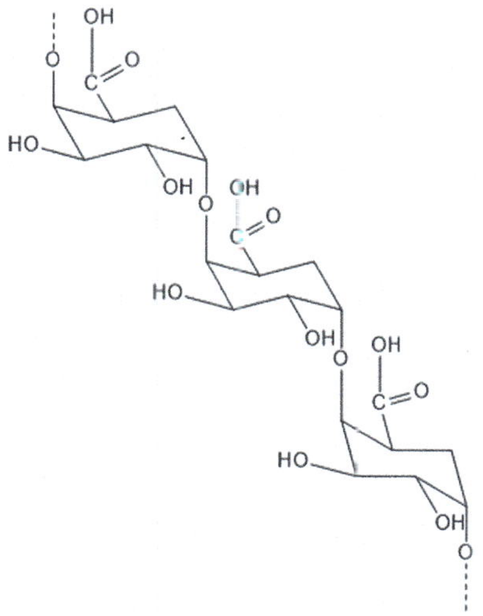

Fig. 2.3 Chemical structure of pectin

and volume of gluten-free bread. Pectin is an excellent biopolymer for the target delivery of antitumor drugs due to its high stability (Khotimchenko 2020).

2.2.2.3 Alginate

Alginates are naturally occurring linear polysaccharides obtained from brown seaweeds such as *Laminaria digitata, Macrocystis pyrifera, Laminaria japonica, Ascophyllum nodosum, Laminaria hyperborea*. Some bacteria of Pseudomonas and Azotobacter genera also extracellularly produce alginate. It is chemically composed of different ratios of β-D-mannuronic acid (M) and its C-5 epimer α-L-guluronic acid (G) residues attached via 1,4-glycosidic linkages (Takeshita and Oda 2016). Alginates have been widely observed for pharmaceutical and biomedical applications owing to their biocompatibility, low toxicity, easy availability, and biodegradability (Hariyadi and Islam 2020). Alginate has been widely used for emulsion stabilization, such as in pickering emulsions (Dai et al. 2018; Zhao et al. 2018a; Yang et al. 2019b) and O/W emulsions (Su et al. 2018; Zhao et al. 2018b). Gama et al. (2019) observed the positive response in the functional properties of peanut-based beverage (flow behavior, physical stability, and viscosity) due to propylene glycol alginate addition. Yang et al. (2017) reported the film-forming potential of alginate together with bacterial cellulose. The results described that alginate improves the mechanical strength, swelling capacity, and antimicrobial properties of films. Supartini and Mushollaeni (2017) demonstrated the effect of type (synthesized from *Turbinaria* sp. and *Sargassum* sp.) and concentration of alginate on bread nutrition and quality. According to results, *Turbinaria* sp. alginate with 1.5% concentration produced bread with optimum quality and nutrition.

2.2.2.4 Xanthan Gum (XG)

XG is an ionic exopolysaccharide synthesized by *Xanthomonas campestris* bacteria and constituted of a main chain (β1,4-D-glucose) and side chains, each originating from 2 glucose residues. Side chains are composed of β-D-glucuronic acid, β-D-mannose and α-D-mannose (1:2:2) as terminal residues. Acetate group and Pyruvic acid groups may also be present as non-terminal and terminal residues, correspondingly (Kumar et al. 2018). The pyruvate and acetate contents and molecular weight can vary with fermentation conditions. The rigid polymer chain in water can present as single, double or triple helix which intermingles with other chains to make a complex and flexible network. This structure provides the gum its unique thickening properties and thixotropic behavior. The unusual rheological attributes of XG make it an excellent emulsion stabilizer (Abu Elella et al. 2020; Bouyer et al. 2012).

XG has been widely used in food industries as thickener, emulsifier and stabilizer owing to its properties including producing high viscosity (even at low quantity), high stability and good solubility in cold and hot water (Habibi and Khosravi-Darani 2017). Lara et al. (2020) reported the role of XG based coatings in extending the shelf life of lotus root (fresh cut). XG coating effectively reduced the enzymatic browning and microbial spoilage specially the growth of *B. subtilis*. Muhammad et al. (2020a) investigated the effect of various stabilizers (carrageenan, xanthan gum or alginate) on cinnamon-cocoa beverage stability. The outcomes revealed that XG is most effective stabilizer even at 0.1% as compared to other stabilizers. It increased the viscosity from 2.47 to 70.44 mPa s at a low shear rate (10/s). It also prevented sedimentation and did not alter the antioxidant properties of cinnamon. Muhammad et al. (2020b) reported that XG stabilized the shellac nanoparticles even at stomach pH as well as improved the thermal stability and release rate of bioactive compounds.

2.2.3 *Phospholipids*

Phospholipids are the most commonly used food emulsifiers due to their amphiphilic nature. Phospholipids are naturally present in plants, animals, and microbial cell membranes (Erickson 2008). Chemically phospholipids are comprised of non-polar tails containing a phosphoric acid and two fatty acids attached with a polar head backbone made up of glycerol. They can bind with interfacial layers and serve as emulsifiers due to the presence of polar and non-polar groups on the same molecules (Pichot et al. 2013).

2.2.3.1 Lecithin

Lecithin is the most widely used phospholipid-based emulsifier in food industries and can be obtained from eggs, sunflower, rapeseeds, soybean, cotton seeds, milk, and canola seeds (Klang and Valenta 2011). Lecithin is generally a combination of different phospholipids (phosphatidylinositol, phosphatidylcholine, and phosphatidylethanolamine) along with other lipids like sterol, triglycerides, and glycolipids (Fig. 2.4) (Guiotto et al. 2013; Erickson 2008).

Naturally, lecithin exhibits a good HLB value of about 7, reflecting their solubility in both water and lipid phase. The stability of lecithin-based emulsions depends on the type of phospholipids used, their pH, and ionic strength. Most of the lecithin derivatives are not good emulsifiers due to their less or intermediate HBL and lipophilic nature. However, their emulsifying properties can be improved by combining them with other natural emulsifiers such as whey proteins (Frede et al. 2014), mono glycerol (Zou and Akoh 2013), and casein (Xue and Zhong 2014), etc. Moreover, chemical or enzymatic removal of one fatty acid tail from lecithin to form lysolecithin can also improve lecithin performance. Lysolecithin generally dissolves in the water phase before homogenization due to high HLB value than native lecithin (Casado et al. 2012).

Lecithin has been used in the pharmaceutical, nutritional, food, and cosmetic industries. In the food industry, lecithin is used to perform multiple functions like an

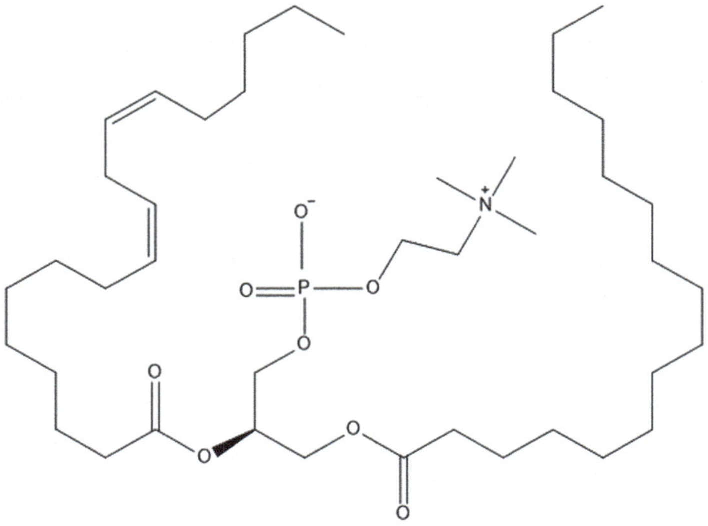

Fig. 2.4 Chemical structure of lecithin

emulsifier, wetting agent, release agent, anti-spattering agent, browning agent, crystallization controller (in manufacturing of spreads and margarine) as lubricant and preservative coatings (List 2015). Lecithin has been extensively used to stabilize emulsion (Koo et al. 2019; Xie and Dunford 2017). Soy lecithin (1–5%) and saponins (0.5–2.5%) were investigated for their emulsifying potential in coffee whitener. The increased concentration of emulsifiers decreased the whiteness and droplet size (0.5–0.15 μm by saponin and 0.8–0.14 μm by soy lecithin) and increased the heat stability (Chung et al. 2017). Gavahian et al. (2018) reported the emulsifying capacity of lecithin in salad dressings. Ahmed et al. (2020) evaluated the impact of soy lecithin and chia seeds on the properties of dough. It was observed that they significantly enhanced the nutritional profile and bread volume while decreased water loss, crumb firmness, and bread staling during storage. Various studies have reported the application of lecithin in the encapsulation of bioactive compounds, including vit E (Ozturk et al. 2014), whey peptides (Mohan et al. 2016), lycopene (Cheng et al. 2017), and curcumin (Prasad et al. 2020).

2.2.4 *Bio-emulsifiers and Bio-surfactants*

The compounds which contain both hydrophilic and hydrophobic groups on the same molecule and can be obtained from plant and microorganisms are known as bioemulsifiers/biosurfactants (Paniagua-Michel et al. 2014). They adsorb at water-oil interfaces, minimize interfacial tension, and develop a protective coating due to the presence of both polar and non-polar groups on the same compound. They are biodegradable, natural, sustainable, and relatively less toxic emulsifiers. Biosurfactants and bio-emulsifiers have been categorized based on their molecular weight. Biosurfactants have low molecular weight, such as lipoproteins (iturin, fengycin, surfactin) and glycolipids (trehalose lipids, sophorolipids, rhamnolipids) extracted from plants or microbial fermentation. Bioemulsifiers have high molecular weight and can be isolated from microbes, including proteins, lipoproteins, polysaccharides, and lipopolysaccharides (Moldes et al. 2020; Kralova and Sjöblom 2009). Bio-emulsifiers can adsorb at an oil-water interface but cannot efficiently reduce surface tension due to their high molecular weight (Uzoigwe et al. 2015).

2.2.4.1 Mannoproteins (MP)

Mannoproteins have been classified under bioemulsifiers due to the presence of high molecular weight glycoproteins of about 15 kDa. They comprise the central part of the yeast cell wall and consist of protein (5–20%) backbone attached via covalent bond mannose (70–90%). Mannoproteins can be isolated from different yeast strains through simple techniques (Li and Karboune 2019; De Iseppi et al. 2019; Sugiah et al. 2019; Mulyadewi et al. 2019). Mannoproteins are highly stable to a wide pH range (3–11) and NaCl concentration of up to 5%. Hydrophobic protein and hydrophilic mannose chain increase their amphiphilic character and capacity to reduce surface tension. Their mannose to protein ratio governs the HLB and amphiphilic nature of mannoproteins (Li and Karboune 2019). According to Li and Karboune (2019), mannoproteins have less emulsifying properties than commercial lecithin. Sugiah et al. (2019) evaluated the emulsifying capacity of mannoprotein extracted from *Candida apicola*. The emulsion with protein level 1.68 g/mL showed maximum emulsification activity (42 %) with the potential to use ad bioemulsifier in food products. Silva Araújo et al. (2014) extracted mannoproteins from yeast (*Saccharomyces uvarum*) and reported their application as emulsifiers in mayonnaise. Stabilizing and emulsifying activity (%) increased with increasing MP contents. de Melo et al. (2015) described that MP showed more nutritional, emulsifying, and sensory properties in salad dressing than soy lecithin. The effect of MP concentration (0.2, 0.25, and 0.3 g/L) on blueberry wine properties were evaluated (Wu et al. 2015). The results revealed that MP incorporation inhibited anthocyanin depletion, maintained the color, enhanced alcohol level, and improved the taste of wine.

2.2.4.2 Saponins

Saponins are amphiphilic glycosidic biosurfactants and can be extracted from numerous plants as secondary metabolites. Hydrophobic aglycone group attached with polar sugar chain provides saponins surface-active character (Liu et al. 2017). Quillaja saponin is the most commercially used saponin in the food industry and is isolated from the bark of *Quillaja Saponaria* (Fig. 2.5) (Yang et al. 2013). It is an

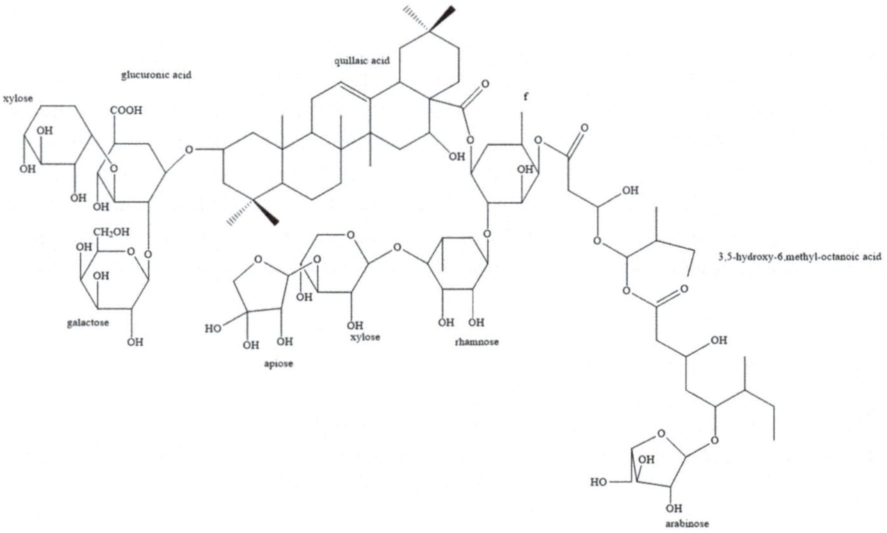

Fig. 2.5 Chemical structure of quillaja saponins

excellent emulsifier to stabilize O/W emulsions with small droplets (Ozturk et al. 2014). Quillaja saponin based emulsion is highly stable to environmental stresses such as wide pH range, temperature, and ionic strength. Quillaja saponins can also protect oil droplets from aggregation when the lipid phase crystallizes, which is essential for preventing partial coalescence and producing solid lipid nanoparticles (Salminen et al. 2014).

Saponins (0.5–2.5%) investigated for emulsifying potential in coffee whiteners decreased the whiteness and droplet size and increased heat stability (Chung et al. 2017). Jurado Gonzalez and Sörensen (2020) studied the effect of temperature, pH, ethanol, and salt concentration on the foaming properties of saponins. Results revealed that foaming properties have a direct relation with temperature while the other three factors showed no effect. Bai et al. (2016) demonstrated that Quillaja saponin is an effective emulsifier in nanoemulsion compared to soy lecithin and gum Arabic. Saponins have been used for stabilization of emulsion-based delivery systems of ω-3 (Salminen et al. 2014), vitamin D (Ozturk et al. 2015), and vitamin E (Lv et al. 2019).

2.2.4.3 Rhamnolipids

Rhamnolipids are also glycosidic bioemulsifiers obtained from microorganisms (*Pseudomonas* sp., *Pseudomonas aeruginosa*, *Burkholderia* sp.). They comprise of the lipophilic hydrocarbon chain and a hydrophilic rhamnose chain that facilitate to reduce surface tension (Fig. 2.6) (Clements et al. 2019). Chemical and structural properties of rhamnolipids depend on microbial strain, substrate type, and

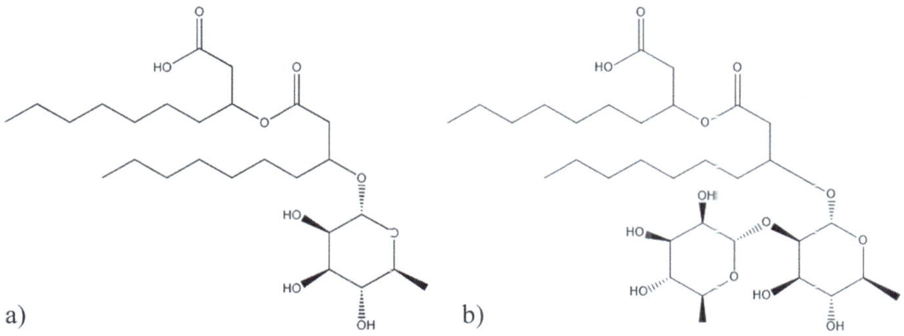

Fig. 2.6 Structure of mono-rhamnolipid (**a**) and di-rhamnolipid (**b**)

fermentation method. They are stable to wide range of pH (5–9), temperature (upto 90 °C) and ionic strength (100 mM NaCl) (Dammak et al. 2020). Rhamnolipids possess low to medium HLB (4–8), therefore, they are individually suitable for W/O emulsion stabilization. Instead, for stabilization of O/W emulsion, it requires combination with other hydrophilic compounds or chemical alteration (Li et al. 2018). Rhamnolipids have been used in the food industry to improve the properties of bread, buttercream and frozen confectionary, as an antimicrobial agent (Nitschke and Costa 2007) and as the encapsulation of β-carotene (Wei et al. 2020). Abbasi et al. (2020) reported that rhamnolipids tend to improve stability and viscosity of sesame based O/W beverage.

2.2.4.4 Sophorolipids (SLs)

Sophorolipids are glycolipids which can be synthesized from yeast via fermentation (e.g. *Torulopsis bombicola*, *T. apicola* or *Candida bombicola*). Chemically, SL are composed of a sophorose group (2 glucose linked via β-1,2 bond) attached with hydroxylated fatty acid by glycosidic linkages (Jiménez-Peñalver et al. 2020; Dammak et al. 2020). Its high yield, amphillic nature and safe production strains has triggered its utilization in various fields i.e. cosmetics, pharmaceuticals and foods. Hipólito et al. (2020) investigated the antifungal and sanitizing effect of SL. The observations showed that it significantly reduce the food spoilage microorganisms including *Aspergillus niger, A. flavus, A. parasiticus, Fusarium oxysporum etc.* Several other studies also reported its antimicrobial effects (Chen et al. 2020; Fontoura et al. 2020). SL improved the bio-acceptability of encapsulated curcumin as well as improved the stability of nanoparticles (Yuan et al. 2021) (Table 2.2).

Table 2.2 Food applications of various natural emulsifiers

Emulsifier	Source	Application	References
Animal proteins			
Casein	Milk	Improved stability, viscosity and whippability of milk cream	Li et al. (2020a), Ziaeifar et al. (2018)
		Stabilization of O/W pickering emulsion	Chen and Zhang (2019)
		Improved rheology of gluten-free dough	Villanueva et al. (2018)
		Used as emulsifier and fat analogue in cheese	Liu et al. (2008)
		Stabilized fish oil nanoemulsion	Yang et al. (2019a)
		Microencapsulation of bioactive compounds	Vaucher et al. (2019), Yan et al. (2020)
Whey protein	Milk	Stabilization of O/W emulsion and improved digestion of bioactive compound	Boonlao et al. (2020), Pan et al. (2019)
		Improved physico-chemical and sensory properties of ice cream	Danesh et al. (2017)
		Used as emulsifying salt analogue in cheese	Sołowiej et al. (2020)
		Enhanced oxidative stability of fortified mayonnaise	Rahmani-Manglano et al. (2020)
		Encapsulation of bioactive compounds	Liu et al. (2020), Rodrigues et al. (2020), Durán et al. (2020), Talón et al. (2019)
Gelatin	Animal collagen	Encapsulation of oregano essential oil	Hernández-Nava et al. (2020)
		Improved the stability of double emulsion (W/O/W)	Zhu et al. (2018)
		Increased lubrication, whiteness and physical stability of coffee whitener	Huang et al. (2020)
		Improved properties of frozen dough	Yu et al. (2020)
		Inhibited bread stalling	Yu et al. (2019)
		Improved stability of beer and ice cream	Duan et al. (2018)
		Encapsulation of *L. reuteri*	Zhao et al. (2020)
		Used as film forming material	Nazmi and Sarbon (2019), Stejskal et al. (2020), Tian et al. (2020), Xiong et al. (2020)

(continued)

Table 2.2 (continued)

Emulsifier	Source	Application	References
Ovalbumin	Egg white	Enhanced the rheological attributes of steamed dough	Sang et al. (2018)
		Encapsulation of curcumin	Feng et al. (2016)
		Used as stabilizer in biosensor	Li et al. (2019)
		Encapsulation of vitamin D_3	Xiang et al. (2020)
		Stabilized the emulsion	Li et al. (2020b), Niu et al. (2016), Zhang et al. (2020)
		Improved bread texture	Nozawa et al. (2016)
Plant proteins			
Gluten	Wheat	Stabilized O/W emulsion and replaced the mayonnaise	Liu et al. (2018)
		Improved mechanical and barrier properties of film	Deepika et al. (2020)
Zein	Corn	Used as fat replacer in mayonnaise	Gu et al. (2016)
		Improved rheological and sensory properties	
		Stabilization of oil-in-glycerol emulsion gel	Chen et al. (2016)
		Stabilization of foam and emulsions	Nephomnyshy et al. (2020)
Rice protein	Rice	Encapsulation of flaxseed	Perrechil et al. (2021)
		Stabilized O/W emulsion	Mun et al. (2016)
		Enhanced nutritional, rheological and sensory properties	Phongthai et al. (2016)
Soy protein	Soy bean	Stabilized ice cream structure	Chen et al. (2019b)
		Encapsulation of anthocyanin	Patel et al. (2020)
Pea protein	Pea	Increased the emulsion stability	Chen et al. (2019a)
Chickpea protein	Chickpea	Stabilization of O/W emulsion	Glusac et al. (2020), Felix et al. (2020)

(continued)

Table 2.2 (continued)

Emulsifier	Source	Application	References
Polysaccharides			
Pectin	Citrus peel, apple peel, sugarbeet	Stabilization of nanoemulsions	Mungure et al. (2018)
		As emulsifier in orange juice	Rivadeneira et al. (2020)
		Stabilization in multilayer emulsion	Fan et al. (2020)
		Encapsulant in probiotic beverages	Santos et al. (2017)
		As lecithin replacer in chocolate	Datsomor et al. (2019)
		Increased crumb compressibility and volume of gluten-free bread	Dubrovskaya et al. (2017)
Gum Arabic		Encapsulation of TEO	Cai et al. (2019)
		Development of biodegradable films	Xu et al. (2019a), Yang et al. (2019c)
		Enhanced crispiness and overall acceptability of biscuits	China et al. (2020)
		Nanoencapsulation of fish oil	Vahidmoghadam et al. (2019)
Alginate	Seaweeds	Pickering emulsion stabilization	Dai et al. (2018), Zhao et al. (2018a), Yang et al. (2019b)
		Stabilization of O/W	Su et al. (2018), Zhao et al. (2018b)
		Enhanced mechanical and antimicrobial potential of film	Yang et al. (2017)
		As emulsifier in peanut beverages	Gama et al. (2019)
		Improved the nutrition and quality of white bread	Supartini and Mushollaeni (2017)

(continued)

Table 2.2 (continued)

Emulsifier	Source	Application	References
Xanthan gum	*Xanthomonas campestris*	Used as coating material for lotus root	Lara et al. (2020)
		Inhibited enzymatic browning and *B. subtilis* growth	
		Extent the shelf life of lotus root	
		Stabilized the cinnamon-cocoa beverage	Muhammad et al. (2020a)
		Increased the viscosity	
		Prevented the sedimentation of cinnamon	
		Encapsulation of cinnamon extract	Muhammad et al. (2020b)
		Improved quality of barley bread	Jalayeri et al. (2020)
		Prevented bread stalling	
Phospholipids			
Lecithin	Milk, egg, soy bean, cotton seed etc.	Development and stabilization of nanoemulsion for delivery of vit E	Ozturk et al. (2014)
		Stabilizing dairy beverages	Nyuydze and Martínez-Monteagudo (2021)
		As emulsifier in tea whitener and salad dressing	Chung et al. (2017), Gavahian et al. (2018)
		Enhanced nutrition, rheology and shelf stability of dough	Ahmed et al. (2020)
Bio-emulsifiers and bio-surfactants			
Mannoprotein	Yeast cell wall	Enhanced alcohol level and taste of wine	Wu et al. (2015)
		Stabilization of salad dressing and mayonnaise	de Melo et al. (2015)
Saponin	*Quillaja saponaria* tree	Development and stabilization of nanoemulsion for bioactive compounds	Salminen et al. (2014), Ozturk et al. (2015), Lv et al. (2019)
		As emulsifier in liquid coffee whitener	Chung et al. (2017)
		Development of lipid nanocarriers	Kharat and McClements (2019)
		As emulsifier and antioxidant in nanoemulsions	Bai et al. (2016)
Rhamnolipids	Microbes	Emulsion stabilization	Li et al. (2018)
		Increase stability of whey protein-based emulsion	Zhang et al. (2019)
		As emulsifier in sesame-based beverage	Abbasi et al. (2020)

(continued)

Table 2.2 (continued)

Emulsifier	Source	Application	References
Sophorolipid	Yeast	Antimicrobial agent	Chen et al. (2020), Fontoura et al. (2020), Hipólito et al. (2020)
		Stabilized nanoparticle and encapsulation of curcumin	Yuan et al. (2021)

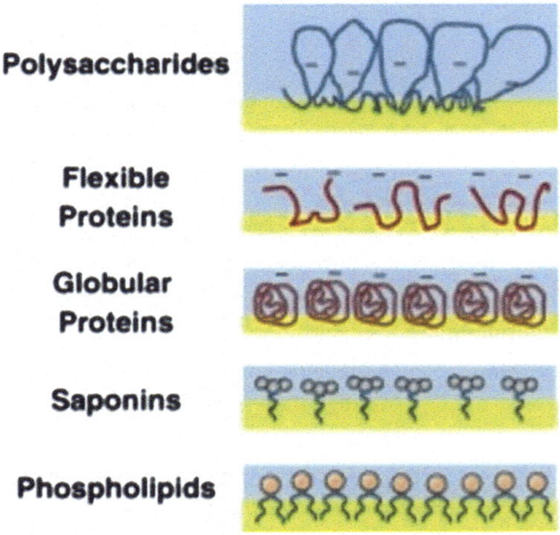

Fig. 2.7 Interactions of natural emulsifiers with continuous and dispersed phase. (Reprint with permission from Ozturk and McClements 2016)

2.3 Behavior of Different Emulsifiers Towards Interfacial Layer

Depending upon the point of interaction, these can be divided into two types; (a) in solution interaction, (b) interaction at the interface (McClements and Jafari 2018). Different types of interactions such as steric, electrostatic, covalent bonding, etc. were observed between emulsifiers and interfacial layer. Different emulsifiers interact differently with interface as explained in sections below and reflected in Fig. 2.7.

2.3.1 Proteins

Proteins are generally comparatively small molecules (≈10–50 kDa) that can quickly adsorb to droplet surfaces and make electrically charged thin interfacial layers. Proteins can develop several conformations with interface depending on type of interaction and molecular structure (Sarkar and Singh 2016). Flexible proteins (e.g. casein, gelatin) quickly go through conformational alteration so that the hydrophobic groups attach with oil while the hydrophilic groups attach to water. Whereas, rigid globular proteins (like pea, egg, whey or soy proteins) may undergo partial unfolding after adsorption and make viscoelastic cohesive layers (Bouyer et al. 2012). Presence of hydrophobic and hydrophilic groups on protein and phospholipids make them suitable for electrostatic stabilization. Moreover, steric repulsion also helps the proteins during emulsion stabilization (Griffin and Khouryieh 2020).

2.3.2 *Polysaccharides*

Polysaccharides owing to their high MW adsorb comparatively slowly to droplet and make thick hydrophilic interfacial layers. The large molecules of polysaccharide refers that they also have high surface saturation as compared to proteins, hence, require more emulsifier to occupy the entire droplet surfaces. Polysaccharides usually stabilize the emulsions through steric interactions. They adsorb at the interface of aqueous and lipid phase and stabilize the emulsions (Bouyer et al. 2012). Moreover, carbohydrates bound the proteinaceous material which adsorbs on the water and oil interface, form the hydrated layer around the oil droplets to prevent the flocculation (Nakauma et al. 2008). Vélez-Erazo et al. (2020) reported the effect of steric interactions of eight different polysaccharides (carrageenan, xanthan gum, gum Arabic, sodium alginate, pectin, gellan gum, locust bean gum and tara gum) in sunflower oil-based emulsions. The authors explained that these polysaccharides were efficient enough to reduce the interfacial tension at lipid-water interface.

2.3.3 *Phospholipids*

Phospholipids seem to have interaction behavior somewhere between polysaccharides and saponins in spite of their low molecular weight (≈0.760 kDa). This could be due to their ability of forming large complex molecular structures, like vesicles or bilayers in solution instead of prevailing as single molecules or small micelles.

2.3.4 *Saponins*

Saponins are generally more effective for making small droplets and thin interfacial layer with low concentrations as compared to biopolymers. Moreover, the high non-polar to polar ratio makes them more surface-active compounds compared to biopolymers. They tend to reduce interfacial tension more efficiently due to their efficient oil–water interface packing ability. So, saponins generally form smaller droplets than biopolymers during homogenization (Ozturk and McClements 2016).

2.4 Factor Affecting the Performance and Interaction of Emulsifiers

Emulsifiers play a major role not only in the formation of emulsions but also to make them stable throughout their intended shelf life. Clean label emulsifiers work in the similar manner as emulsifiers from other sources. The emulsions could destabilize as a result of various mechanisms including coalescence, gravitational separation (sedimentation, creaming) Ostwald ripening, flocculation and phase inversion (Fig. 2.8) (Hu et al. 2016). The following factors affect the properties of emulsifiers that ultimately reduce the emulsion stability.

2.4.1 Temperature

Stability of emulsions and performance of emulsifiers are greatly influenced by the storage temperature. Proteins processed at high temperatures are subjected to denaturation and minimize the stability through steric interactions. However, emulsions containing proteins as an emulsifier are stable at room temperature. Saponins containing emulsions are stable even at high temperature due to the presence of strong electrostatic interaction between the molecules of dispersed and continuous phase. Phospholipids (such as soybean phospholipids) exhibit a reduction in their polarity at elevated temperatures. This, in turn, leads to the higher mobility of lipid phase and increased lipid chain disorder (Tai et al. 2020).

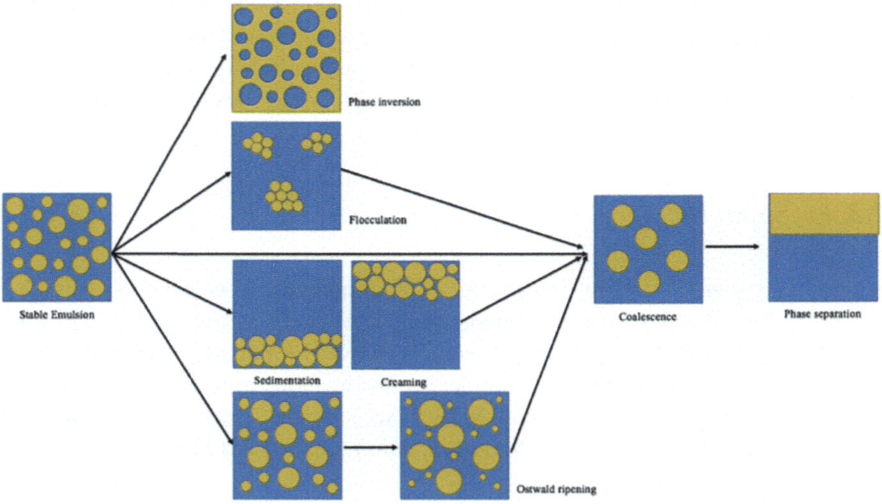

Fig. 2.8 Mechanisms of emulsion destabilization. (Reprint with permission from Hu et al. 2016)

2.4.2 *pH*

pH of the solution is also a determining factor in the stability of emulsions. For instance, adsorbed proteins at the interface start to flocculate near isoelectric point and disturb the stability of emulsions. However, proteins are stable below the isoelectric point of proteins in the emulsion. Phospholipids and saponins are considered to be stable against flocculation due to their ability to form charged particles at neutral pH. However, in case of saponins, high ionic strength under neutral conditions and acidic conditions for phospholipids promote flocculation. Ionic strength and pH of the solution also affect the electrostatic interactions between the saponins and phospholipid coated droplets.

2.4.3 *Ionic Strength*

The stability of emulsions containing proteins and polysaccharides as an emulsifier is highly dependent on the ionic strength of the solution and significantly reduce with increase in the ionic strength of the solution (Shao et al. 2017; Taha et al. 2019). Moreover, high ionic strength changes the binding behavior and structure of proteins. It also reduces the electrostatic repulsion, thus reducing the stability of emulsions (Taha et al. 2019). Saponins are stable even at high salt concentrations but droplet aggregation starts after further increase in ionic strength of emulsions (Xu et al. 2019b).

2.5 Conclusion

Advancement in technology, food safety awareness and labeling requirements have made consumers more informed about their diet. This has necessitated the use of green label ingredients, (including emulsifiers) in processed foods. The natural emulsifiers have the capability to produce emulsions with stability over a wide range of environmental conditions. An individual or combination of natural emulsifiers may be used to form a stable emulsion over a wide range of conditions of ionic strength, storage temperature, pH etc. A number of applications of natural emulsifiers have been reported in foods, pharmaceuticals and cosmetics for emulsion/foam stabilization, texture improvement, lubrication, fat replacement and microencapsulation. The natural emulsifiers behave differently with interface through various interactions. Polysaccharides adsorb slowly to the interface compared to proteins, whereas; phospholipids show intermediate behavior. Saponins reduce interfacial tensions more efficiently even at low concentrations and can form smaller droplets.

References

Abbasi H, Karimi S, Gharibzahedi SMT (2020) Rhamnolipid as a unique emulsifier to stabilize sesame oil-in-water beverage emulsions formed by ultrasound-induced cavitation: optimizing the formulation and physical properties. J Food Process Preserv 44(10):e14810. https://doi.org/10.1111/jfpp.14810

Abdullah MSP, Noordin MI, Mohd Ismail SI, Mustapha NM, Jasamai M, Danik MF, Wan Ismail W, Shamsuddin AF (2018) Recent advances in the use of animal-sourced gelatine as natural polymers for food, cosmetics and pharmaceutical applications. Sains Malaysiana 47(2):323–336

Abu Elella MH, Goda ES, Gab-Allah MA, Hong SE, Pandit B, Lee S, Gamal H, Rehman A, Yoon KR (2020) Xanthan gum-derived materials for applications in environment and eco-friendly materials: a review. J Environ Chem Eng 9(1):104702. https://doi.org/10.1016/j.jece.2020.104702

Adjonu R, Doran G, Torley P, Agboola S (2014) Whey protein peptides as components of nanoemulsions: a review of emulsifying and biological functionalities. J Food Eng 122:15–27. https://doi.org/10.1016/j.jfoodeng.2013.08.034

Ahmed IBH, Hannachi A, Haros CM (2020) Combined effect of chia flour and soy lecithin incorporation on nutritional and technological quality of fresh bread and during staling. Foods 9(4):446

Alting A, Van De Velde F (2012) Proteins as clean label ingredients in foods and beverages. In: Natural food additives, ingredients and flavourings. Elsevier, Amsterdam, pp 197–211

Amagliani L, O'Regan J, Kelly AL, O'Mahony JA (2017) The composition, extraction, functionality and applications of rice proteins: a review. Trends Food Sci Technol 64:1–12

Anvari M, Joyner HS (2017) Effect of fish gelatin-gum arabic interactions on structural and functional properties of concentrated emulsions. Food Res Int 102:1–7. https://doi.org/10.1016/j.foodres.2017.09.085

Bai L, Huan S, Gu J, McClements DJ (2016) Fabrication of oil-in-water nanoemulsions by dual-channel microfluidization using natural emulsifiers: saponins, phospholipids, proteins, and polysaccharides. Food Hydrocoll 61:703–711. https://doi.org/10.1016/j.foodhyd.2016.06.035

BeMiller JN (2019) 15—Pectins. In: BeMiller JN (ed) Carbohydrate chemistry for food scientists (third edition). AACC International Press, Washington, DC, pp 303–312. https://doi.org/10.1016/B978-0-12-812069-9.00015-7

Berta M, Koelewijn I, Öhgren C, Stading M (2019) Effect of zein protein and hydroxypropyl methylcellulose on the texture of model gluten-free bread. J Texture Stud 50(4):341–349

Boland M (2011) Whey proteins. In: Handbook of food proteins. Elsevier, Amsterdam, pp 30–55

Boonlao N, Shrestha S, Sadiq MB, Anal AK (2020) Influence of whey protein-xanthan gum stabilized emulsion on stability and in vitro digestibility of encapsulated astaxanthin. J Food Eng 272:109859. https://doi.org/10.1016/j.jfoodeng.2019.109859

Bouyer E, Mekhloufi G, Rosilio V, Grossiord J-L, Agnely F (2012) Proteins, polysaccharides, and their complexes used as stabilizers for emulsions: alternatives to synthetic surfactants in the pharmaceutical field? Int J Pharm 436(1–2):359–378

Cai C, Ma R, Duan M, Lu D (2019) Preparation and antimicrobial activity of thyme essential oil microcapsules prepared with gum arabic. RSC Adv 9(34):19740–19747

Casado V, Martín D, Torres C, Reglero G (2012) Phospholipases in food industry: a review. In: Lipases and phospholipases. Springer, Berlin, pp 495–523

Chakraborty SK, Kotwaliwale N, Navale SA (2020) Selection and incorporation of hydrocolloid for gluten-free leavened millet breads and optimization of the baking process thereof. LWT 119:108878. https://doi.org/10.1016/j.lwt.2019.108878

Chang C, Wang T, Hu Q, Luo Y (2017) Zein/caseinate/pectin complex nanoparticles: formation and characterization. Int J Biol Macromol 104:117–124. https://doi.org/10.1016/j.ijbiomac.2017.05.178

Chen L (2015) Emulsifiers as food texture modifiers. In: Modifying food texture. Elsevier, Amsterdam, pp 27–49

Chen S, Zhang L-M (2019) Casein nanogels as effective stabilizers for pickering high internal phase emulsions. Colloids Surf A Physicochem Eng Aspects 579:123662. https://doi.org/10.1016/j.colsurfa.2019.123662

Chen X-W, Fu S-Y, Hou J-J, Guo J, Wang J-M, Yang X-Q (2016) Zein based oil-in-glycerol emulgels enriched with β-carotene as margarine alternatives. Food Chem 211 836–844

Chen M, Lu J, Liu F, Nsor-Atindana J, Xu F, Goff HD, Ma J, Zhong F (2019a) Study on the emulsifying stability and interfacial adsorption of pea proteins. Food Hydrocoll 88:247–255. https://doi.org/10.1016/j.foodhyd.2018.09.003

Chen W, Liang G, Li X, He Z, Zeng M, Gao D, Qin F, Goff HD, Chen J (2019b) Effects of soy proteins and hydrolysates on fat globule coalescence and meltdown properties of ice cream. Food Hydrocoll 94:279–286. https://doi.org/10.1016/j.foodhyd.2019.02.045

Chen J, Lü Z, An Z, Ji P, Liu X (2020) Antibacterial activities of sophorolipids and nisin and their combination against foodborne pathogen Staphylococcus aureus. Eur J Lipid Sci Technol 122(3):1900333

Cheng Y-S, Lu P-M, Huang C-Y, Wu J-J (2017) Encapsulation of lycopene with lecithin and α-tocopherol by supercritical antisolvent process for stability enhancement. J Supercrit Fluids 130:246–252

Cheng W, Wang B, Zhang C, Dong Q, Qian J, Zha L, Chen W, Hong L (2019) Preparation and preliminary pharmacokinetics study of GNA-loaded zein nanoparticles J Pharm Pharmacol 71(11):1626–1634

China M, Oguzor U, Ujong A (2020) Effect of gum arabic incorporation on the proximate composition and sensory properties of biscuits produced from flour blends of wheat and water yam. Asian Food Sci J 18:1–11

Chung C, Sher A, Rousset P, Decker EA, McClements DJ (2017) Formulation of food emulsions using natural emulsifiers: utilization of quillaja saponin and soy lecithin to fabricate liquid coffee whiteners. J Food Eng 209:1–11. https://doi.org/10.1016/j.jfoodeng.2017.04.011

Clare DA, Daubert CR (2011) Expanded functionality of modified whey protein dispersions after transglutaminase catalysis. J Food Sci 76(4):C576–C584

Clements T, Ndlovu T, Khan S, Khan W (2019) Biosurfactants produced by Serratia species: classification, biosynthesis, production and application. Appl Microbiol Biotechnol 103(2):589–602

Dai L, Zhan X, Wei Y, Sun C, Mao L, McClements DJ, Gao Y (2018) Composite zein-propylene glycol alginate particles prepared using solvent evaporation: characterization and application as pickering emulsion stabilizers. Food Hydrocoll 85:281–290

Dammak I, Sobral PJA, Aquino A, Neves MA, Conte-Junior CA (2020) Nanoemulsions: using emulsifiers from natural sources replacing synthetic ones—a review. Compr Rev Food Sci Food Saf 19(5):2721–2746. https://doi.org/10.1111/1541-4337.12606

Damodaran S, Parkin KL, Fennema OR (2007) Fennema's food chemistry. CRC Press, Boca Raton

Danesh E, Goudarzi M, Jooyandeh H (2017) Short communication: effect of whey protein addition and transglutaminase treatment on the physical and sensory properties of reduced-fat ice cream. J Dairy Sci 100(7):5206–5211. https://doi.org/10.3168/jds.2016-12537

Datsomor DN, Agbenorhevi JK, Kpodo FM, Oduro IN (2019) Okra pectin as lecithin substitute in chocolate. Sci Afr 3:e00070. https://doi.org/10.1016/j.sciaf.2019.e00070

Day L (2011) Wheat gluten: production, properties and application. In: Handbook of food proteins. Elsevier, Amsterdam, pp 267–288

De Iseppi A, Curioni A, Marangon M, Vincenzi S, Kantureeva G, Lomolino G (2019) Characterization and emulsifying properties of extracts obtained by physical and enzymatic methods from an oenological yeast strain. J Sci Food Agric 99(13):5702–5710

De Kruif CG, Huppertz T, Urban VS, Petukhov AV (2012) Casein micelles and their internal structure. Adv Colloid Interface Sci 171:36–52

de Melo ANF, de Souza EL, da Silva Araujo VB, Magnani M (2015) Stability, nutritional and sensory characteristics of French salad dressing made with mannoprotein from spent brewer's yeast. LWT Food Sci Technol 62(1, Part 2):771–774. https://doi.org/10.1016/j.lwt.2014.06.050

Deepika K, Praveena PL, Srisugamathi G, Nilofar Nisha J (2020) Development and evaluation study of polyvinyl alcohol with gluten film. Mater Today 45:597. https://doi.org/10.1016/j.matpr.2020.02.389

Deeth H, Bansal N (2019) Chapter 1—Whey proteins: an overview. In: Deeth HC, Bansal N (eds) Whey proteins. Academic Press, New York, pp 1–50. https://doi.org/10.1016/B978-0-12-812124-5.00001-1

Dickinson E (2010) Flocculation of protein-stabilized oil-in-water emulsions. Colloids Surf B Biointerfaces 81(1):130–140

Dissanayake M, Vasiljevic T (2009) Functional properties of whey proteins affected by heat treatment and hydrodynamic high-pressure shearing. J Dairy Sci 92(4):1387–1397

Duan R, Zhang J, Liu L, Cui W, Regenstein JM (2018) The functional properties and application of gelatin derived from the skin of channel catfish (Ictalurus punctatus). Food Chem 239:464–469. https://doi.org/10.1016/j.foodchem.2017.06.145

Dubrovskaya N, Savkina O, Kuznetsova L, Parakhina O (2017) Development of glutenfree bread with unconventional raw ingredients of high nutritional value and antimicrobial activity. Agron Res 15(S2):1276–1286

Durán E, Churio O, Arias JL, Neira-Carrillo A, Valenzuela C (2020) Preparation and characterization of novel edible matrices based on alginate and whey for oral delivery of iron. Food Hydrocoll 98:105277

Elzoghby AO, Abo El-Fotoh WS, Elgindy NA (2011) Casein-based formulations as promising controlled release drug delivery systems. J Control Release 153(3):206–216. https://doi.org/10.1016/j.jconrel.2011.02.010

Erickson MC (2008) Chemistry and function of phospholipids. In: Food lipids: chemistry, nutrition and biotechnology. CRC Press, Boca Raton, pp 39–62

Evans M, Ratcliffe I, Williams PA (2013) Emulsion stabilisation using polysaccharide–protein complexes. Curr Opin Colloid Interface Sci 18(4):272–282

Fan C, Chen X, He J (2020) Effect of calcium chloride on emulsion stability of methyl-esterified citrus pectin. Food Chem 332:127366

Felix M, Cermeño M, FitzGerald RJ (2020) Influence of hydrolysis on the bioactive properties and stability of chickpea-protein-based O/W emulsions. J Agric Food Chem 68(37):10118–10127

Feng J, Wu S, Wang H, Liu S (2016) Improved bioavailability of curcumin in ovalbumin-dextran nanogels prepared by Maillard reaction. J Funct Foods 27:55–68. https://doi.org/10.1016/j.jff.2016.09.002

Fontoura ICC, Saikawa GIA, Silveira VAI, Pan NC, Amador IR, Baldo C, Rocha SPD, Celligoi MAPC (2020) Antibacterial activity of sophorolipids from Candida bombicola against human pathogens. Braz Arch Biol Technol 63:e20180568

Frede K, Henze A, Khalil M, Baldermann S, Schweigert FJ, Rawel H (2014) Stability and cellular uptake of lutein-loaded emulsions. J Funct Foods 8:118–127

Gagliardi A, Bonacci S, Paolino D, Celia C, Procopio A, Fresta M, Cosco D (2019) Paclitaxel-loaded sodium deoxycholate-stabilized zein nanoparticles: characterization and in vitro cytotoxicity. Heliyon 5(9):e02422

Gama AP, Hung Y-C, Adhikari K (2019) Optimization of emulsifier and stabilizer concentrations in a model peanut-based beverage system: a mixture design approach. Foods 8(4):116

Gavahian M, Chen Y-M, Mousavi Khaneghah A, Barba FJ, Yang BB (2018) In-pack sonication technique for edible emulsions: understanding the impact of acacia gum and lecithin emulsifiers and ultrasound homogenization on salad dressing emulsions stability. Food Hydrocoll 83:79–87. https://doi.org/10.1016/j.foodhyd.2018.04.039

Glusac J, Isaschar-Ovdat S, Fishman A (2020) Transglutaminase modifies the physical stability and digestibility of chickpea protein-stabilized oil-in-water emulsions. Food Chem 315:126301. https://doi.org/10.1016/j.foodchem.2020.126301

Griffin K, Khouryieh H (2020) Influence of electrostatic interactions on the formation and stability of multilayer fish oil-in-water emulsions stabilized by whey protein-xanthan-locust bean complexes. J Food Eng 277:109893

Gu J, Xin Z, Meng X, Sun S, Qiao Q, Deng H (2016) A "reduced-pressure distillation" method to prepare zein-based fat analogue for application in mayonnaise formulation. J Food Eng 182:1–8
Guedes Silva KC, Feltre G, Dupas Hubinger M, Kawazoe Sato AC (2021) Protection and targeted delivery of β-carotene by starch-alginate-gelatin emulsion-filled hydrogels. J Food Eng 290:110205. https://doi.org/10.1016/j.jfoodeng.2020.110205
Guiotto EN, Cabezas DM, Diehl BW, Tomás MC (2013) Characterization and emulsifying properties of different sunflower phosphatidylcholine enriched fractions. Eur J Lipid Sci Technol 115(8):865–873
Habibi H, Khosravi-Darani K (2017) Effective variables on production and structure of xanthan gum and its food applications: a review. Biocatal Agric Biotechnol 10:130–140
Hariyadi DM, Islam N (2020) Current status of alginate in drug delivery. Adv Pharmacol Pharm Sci 2020:8886095
Hasenhuettl GL, Hartel RW (2008) Food emulsifiers and their applications, vol 40 and 6. Springer, Berlin
Haug IJ, Draget KI (2011) 5—Gelatin. In: Phillips GO, Williams PA (eds) Handbook of food proteins. Woodhead Publishing, New Delhi, pp 92–115. https://doi.org/10.1533/9780857093639.92
Hernández-Nava R, López-Malo A, Palou E, Ramírez-Corona N, Jiménez-Munguía MT (2020) Encapsulation of oregano essential oil (Origanum vulgare) by complex coacervation between gelatin and chia mucilage and its properties after spray drying. Food Hydrocoll 109:106077
Hipólito A, Alves da Silva RA, Caretta TO, Silveira VAI, Amador IR, Panagio LA, Borsato D, Celligoi MAPC (2020) Evaluation of the antifungal activity of sophorolipids from Starmerella bombicola against food spoilage fungi. Biocatal Agric Biotechnol 29:101797. https://doi.org/10.1016/j.bcab.2020.101797
Holt C (2016) Casein and casein micelle structures, functions and diversity in 20 species. Int Dairy J 60:2–13. https://doi.org/10.1016/j.idairyj.2016.01.004
Horne DS (2020) Chapter 6—Casein micelle structure and stability. In: Boland M, Singh H (eds) Milk proteins (third edition). Academic Press, New York, pp 213–250. https://doi.org/10.1016/B978-0-12-815251-5.00006-2
Hu Y-T, Ting Y, Hu J-Y, Hsieh S-C (2016) Techniques and methods to study functional characteristics of emulsion systems. J Food Drug Anal 25(1):16–26. https://doi.org/10.1016/j.jfda.2016.10.021
Huang T, Tu Z, Zou Z, Shangguan X, Wang H, Bansal N (2020) Glycosylated fish gelatin emulsion: rheological, tribological properties and its application as model coffee creamers. Food Hydrocoll 102:105552. https://doi.org/10.1016/j.foodhyd.2019.105552
Huppertz T, Fox P, Kelly A (2018a) The caseins: structure, stability, and functionality. In: Proteins in food processing. Elsevier, Amsterdam, pp 49–92
Huppertz T, Fox PF, Kelly AL (2018b) 3—The caseins: structure, stability, and functionality. In: Yada RY (ed) Proteins in food processing (second edition). Woodhead Publishing, New Delhi, pp 49–92. https://doi.org/10.1016/B978-0-08-100722-8.00004-8
Hussain A, Hasan A, Babadaei MMN, Bloukh SH, Edis Z, Rasti B, Sharifi M, Falahati M (2020) Application of gelatin nanoconjugates as potential internal stimuli-responsive platforms for cancer drug delivery. J Mol Liquids 318:114053. https://doi.org/10.1016/j.molliq.2020.114053
Jalayeri H, Karimi M, Abdollahzadeh A (2020) Effect of xanthan gum and transglutaminase on improving the quality and reducing staling of barley bread. J Innov Food Sci Technol 12(1):125–133
Jambrak AR, Mason TJ, Lelas V, Paniwnyk L, Herceg Z (2014) Effect of ultrasound treatment on particle size and molecular weight of whey proteins. J Food Eng 121:15–23
Jiménez-Peñalver P, Koh A, Gross R, Gea T, Font X (2020) Biosurfactants from waste: structures and interfacial properties of sophorolipids produced from a residual oil cake. J Surfactants Deterg 23(2):481–486
Jurado Gonzalez P, Sörensen PM (2020) Characterization of saponin foam from Saponaria officinalis for food applications. Food Hydrocoll 101:105541. https://doi.org/10.1016/j.foodhyd.2019.105541

Karaca AC, Low N, Nickerson M (2015) Potential use of plant proteins in the microencapsulation of lipophilic materials in foods. Trends Food Sci Technol 42(1):5–12

Kasaai MR (2018) Zein and zein-based nano-materials for food and nutrition applications: a review. Trends Food Sci Technol 79:184–197

Kharat M, McClements DJ (2019) Fabrication and characterization of nanostructured lipid carriers (NLC) using a plant-based emulsifier: Quillaja saponin. Food Res Int 126:108601. https://doi.org/10.1016/j.foodres.2019.108601

Khotimchenko M (2020) Pectin polymers for colon-targeted antitumor drug delivery. Int J Biol Macromol 158:1110–1124. https://doi.org/10.1016/j.ijbiomac.2020.05.002

Kilara A, Vaghela M (2018) Whey proteins. In: Proteins in food processing. Elsevier, Amsterdam, pp 93–126

Klang V, Valenta C (2011) Lecithin-based nanoemulsions. J Drug Deliv Sci Technol 21(1):55–76

Koo CK, Chung C, Fu J-TR, Sher A, Rousset P, McClements DJ (2019) Impact of sodium caseinate, soy lecithin and carrageenan on functionality of oil-in-water emulsions. Food Res Int 123:779–789

Kralova I, Sjöblom J (2009) Surfactants used in food industry: a review. J Dispers Sci Technol 30(9):1363–1383

Kumar A, Rao KM, Han SS (2018) Application of xanthan gum as polysaccharide in tissue engineering: a review. Carbohydr Polym 180:128–144

Lam RSH, Nickerson MT (2013) Food proteins: a review on their emulsifying properties using a structure–function approach. Food Chem 141(2):975–984. https://doi.org/10.1016/j.foodchem.2013.04.038

Lam RS, Nickerson MT (2015) The effect of pH and temperature pre-treatments on the structure, surface characteristics and emulsifying properties of alpha-lactalbumin. Food Chem 173:163–170

Lara G, Yakoubi S, Villacorta CM, Uemura K, Kobayashi I, Takahashi C, Nakajima M, Neves MA (2020) Spray technology applications of xanthan gum-based edible coatings for fresh-cut lotus root (Nelumbo nucifera). Food Res Int 137:109723. https://doi.org/10.1016/j.foodres.2020.109723

Li J, Karboune S (2019) Characterization of the composition and the techno-functional properties of mannoproteins from Saccharomyces cerevisiae yeast cell walls. Food Chem 297:124867. https://doi.org/10.1016/j.foodchem.2019.05.141

Li Z, Dai L, Wang D, Mao L, Gao Y (2018) Stabilization and rheology of concentrated emulsions using the natural emulsifiers quillaja saponins and rhamnolipids. J Agric Food Chem 66(15):3922–3929

Li X, Wu X, Zhang F, Zhao B, Li Y (2019) Label-free detection of folic acid using a sensitive fluorescent probe based on ovalbumin stabilized copper nanoclusters. Talanta 195:372–380. https://doi.org/10.1016/j.talanta.2018.11.067

Li Y, Li Y, Yuan D, Wang Y, Li M, Zhang L (2020a) The effect of caseins on the stability and whipping properties of recombined dairy creams. Int Dairy J 105:104658. https://doi.org/10.1016/j.idairyj.2020.104658

Li Z, Kuang H, Yang J, Hu J, Ding B, Sun W, Luo Y (2020b) Improving emulsion stability based on ovalbumin-carboxymethyl cellulose complexes with thermal treatment near ovalbumin isoelectric point. Sci Rep 10(1):1–9

List GR (2015) 1—Soybean lecithin: food, industrial uses, and other applications. In: Ahmad MU, Xu X (eds) Polar lipids. Elsevier, Amsterdam, pp 1–33. https://doi.org/10.1016/B978-1-63067-044-3.50005-4

Liu H, Xu XM, Guo SD (2008) Comparison of full-fat and low-fat cheese analogues with or without pectin gel through microstructure, texture, rheology, thermal and sensory analysis. Int J Food Sci Technol 43(9):1581–1592

Liu D, Nikoo M, Boran G, Zhou P, Regenstein JM (2015) Collagen and gelatin. Annu Rev Food Sci Technol 6:527–557

Liu Z, Li Z, Zhong H, Zeng G, Liang Y, Chen M, Wu Z, Zhou Y, Yu M, Shao B (2017) Recent advances in the environmental applications of biosurfactant saponins: a review. J Environ Chem Eng 5(6):6030–6038. https://doi.org/10.1016/j.jece.2017.11.021

Liu X, Guo J, Wan Z-L, Liu Y-Y, Ruan Q-J, Yang X-Q (2018) Wheat gluten-stabilized high internal phase emulsions as mayonnaise replacers. Food Hydrocoll 77 168–175. https://doi.org/10.1016/j.foodhyd.2017.09.032

Liu K, Kong X-L, Li Q-M, Zhang H-L, Zha X-Q, Luo J-P (2020) Stability and bioavailability of vitamin D3 encapsulated in composite gels of whey protein isolate and lotus root amylopectin. Carbohydr Polym 227:115337

Lv S, Zhang Y, Tan H, Zhang R, McClements DJ (2019) Vitamin E encapsulation within oil-in-water emulsions: impact of emulsifier type on physicochemical stability and bioaccessibility. J Agric Food Chem 67(5):1521–1529

Lv P, Wang D, Dai L, Wu X, Gao Y, Yuan F (2020) Pickering emulsion gels stabilized by high hydrostatic pressure-induced whey protein isolate gel particles: characterization and encapsulation of curcumin. Food Res Int 132 109032

Ma Z, Boye JI, Simpson BK (2016) Preparation of salad dressing emulsions using lentil, chickpea and pea protein isolates: a response surface methodology study. J Food Qual 39(4):274–291. https://doi.org/10.1111/jfq.12190

McClements DJ (2015) Food emulsions: principles, practices, and techniques. CRC Press, Boca Raton

McClements DJ, Jafari SM (2018) Improving emulsion formation, stability and performance using mixed emulsifiers: a review. Adv Colloid Interface Sci 251:55–79

McMahon DJ, Oommen BS (2013) Casein micelle structure, functions, and interactions. In: Advanced dairy chemistry. Springer, Berlin, pp 185–209

Meng Y, Cloutier S (2014) Gelatin and other proteins for microencapsulation. In: Microencapsulation in the food industry. Elsevier, Amsterdam, pp 227–239

Mirmoghtadaie L, Shojaee Aliabadi S, Hosseini SM (2016) Recent approaches in physical modification of protein functionality. Food Chem 199:619–627. https://doi.org/10.1016/j.foodchem.2015.12.067

Mohan A, McClements DJ, Udenigwe CC (2016) Encapsulation of bioactive whey peptides in soy lecithin-derived nanoliposomes: influence of peptide molecular weight. Food Chem 213:143–148

Moldes A, Vecino X, Rodríguez-López L, Rincón-Fontán M, Cruz JM (2020) Chapter 8—Biosurfactants: the use of biomolecules in cosmetics and detergents. In: Rodrigues AG (ed) New and future developments in microbial biotechnology and bioengineering. Elsevier, Amsterdam, pp 163–185. https://doi.org/10.1016/B978-0-444-64301-8.00008-1

Muhammad DRA, Kongor JE, Dewettinck K (2020a) Investigating the effect of different types of cocoa powder and stabilizers on suspension stability of cinnamon-cocoa drink. J Food Sci Technol 58(10):3933–3941

Muhammad DRA, Sedaghat Doost A, Gupta V, bin Sintang MD, Van de Walle D, Van der Meeren P, Dewettinck K (2020b) Stability and functionality of xanthan gum–shellac nanoparticles for the encapsulation of cinnamon bark extract. Food Hydrocoll 100:105377. https://doi.org/10.1016/j.foodhyd.2019.105377

Mulyadewi KA, Balia RL, Suryaningsih L (2019) Biochemical activity of *saccharomyces cerevisiae* yeast in the formation of mannoprotein on physical properties of sausage products. Scientific Papers: Management, Economic Engineering in Agriculture & Rural Development, vol 19(4)

Mun S, Shin M, Kim Y-R (2016) Emulsifying properties of proteins isolated from various rice cultivars. Food Bioprocess Technol 9(5):813–821

Mungure TE, Roohinejad S, Bekhit AE-D, Greiner R, Mallikarjunan K (2018) Potential application of pectin for the stabilization of nanoemulsions. Curr Opin Food Sci 19:72–76. https://doi.org/10.1016/j.cofs.2018.01.011

Nakauma M, Funami T, Noda S, Ishihara S, Al-Assaf S, Nishinari K, Phillips GO (2008) Comparison of sugar beet pectin, soybean soluble polysaccharide, and gum arabic as food emulsifiers. 1. Effect of concentration, pH, and salts on the emulsifying properties. Food Hydrocoll 22(7):1254–1267

Nazmi N, Sarbon N (2019) Response surface methodology on development and formulation optimisation of chicken skin gelatine film blended with carboxymethyl cellulose as affected by varying plasticiser concentrations. Int Food Res J 26(1):47–57

Nephomnyshy I, Rosen-Kligvasser J, Davidovich-Pinhas M (2020) The development of a direct approach to formulate high oil content zein-based emulsion gels using moderate temperatures. Food Hydrocoll 101:105528. https://doi.org/10.1016/j.foodhyd.2019.105528

Nesic AR, Seslija SI (2017) The influence of nanofillers on physical–chemical properties of polysaccharide-based film intended for food packaging. In: Food packaging. Elsevier, Amsterdam, pp 637–697

Ngouémazong ED, Christiaens S, Shpigelman A, Van Loey A, Hendrickx M (2015) The emulsifying and emulsion-stabilizing properties of pectin: a review. Compr Rev Food Sci Food Saf 14(6):705–718

Nishinari K, Fang Y, Nagano T, Guo S, Wang R (2018) Soy as a food ingredient. In: Proteins in food processing. Elsevier, Amsterdam, pp 149–186

Nitschke M, Costa S (2007) Biosurfactants in food industry. Trends Food Sci Technol 18(5):252–259

Niu F, Pan W, Su Y, Yang Y (2016) Physical and antimicrobial properties of thyme oil emulsions stabilized by ovalbumin and gum arabic. Food Chem 212:138–145

Nozawa M, Ito S, Arai E (2016) Effect of ovalbumin on the quality of gluten-free rice flour bread made with soymilk. LWT Food Sci Technol 66:598–605. https://doi.org/10.1016/j.lwt.2015.11.010

Nyuydze C, Martínez-Monteagudo SI (2021) Role of soy lecithin on emulsion stability of dairy beverages treated by ultrasound. Int J Dairy Technol 74(1):84–94

Orthoefer F, Kim D (2019) Applications of emulsifiers in baked foods. In: Food emulsifiers and their applications. Springer, Berlin, pp 299–321

Otoni CG, Avena-Bustillos RJ, Chiou BS, Bilbao-Sainz C, Bechtel PJ, McHugh TH (2012) Ultraviolet-B radiation induced cross-linking improves physical properties of cold- and warm-water fish gelatin gels and films. J Food Sci 77(9):E215–E223

Ozturk B, McClements DJ (2016) Progress in natural emulsifiers for utilization in food emulsions. Curr Opin Food Sci 7:1–6. https://doi.org/10.1016/j.cofs.2015.07.008

Ozturk B, Argin S, Ozilgen M, McClements DJ (2014) Formation and stabilization of nanoemulsion-based vitamin E delivery systems using natural surfactants: Quillaja saponin and lecithin. J Food Eng 142:57–63. https://doi.org/10.1016/j.jfoodeng.2014.06.015

Ozturk B, Argin S, Ozilgen M, McClements DJ (2015) Nanoemulsion delivery systems for oil-soluble vitamins: influence of carrier oil type on lipid digestion and vitamin D3 bioaccessibility. Food Chem 187:499–506

Pan Y, Xie Q-T, Zhu J, Li X-M, Meng R, Zhang B, Chen H-Q, Jin Z-Y (2019) Study on the fabrication and in vitro digestion behavior of curcumin-loaded emulsions stabilized by succinylated whey protein hydrolysates. Food Chem 287:76–84. https://doi.org/10.1016/j.foodchem.2019.02.047

Paniagua-Michel JJ, Olmos-Soto J, Morales-Guerrero ER (2014) Chapter 11—Algal and microbial exopolysaccharides: new insights as biosurfactants and bioemulsifiers. In: Kim S-K (ed) Advances in food and nutrition research, vol 73. Academic Press, New York, pp 221–257. https://doi.org/10.1016/B978-0-12-800268-1.00011-1

Papalamprou EM, Doxastakis GI, Kiosseoglou V (2010) Chickpea protein isolates obtained by wet extraction as emulsifying agents. J Sci Food Agric 90(2):304–313

Patel AS, Kar A, Mohapatra D (2020) Development of microencapsulated anthocyanin-rich powder using soy protein isolate, jackfruit seed starch and an emulsifier (NBRE-15) as encapsulating materials. Sci Rep 10(1):1–12

Perrechil F, Louzi VC, da Silva A, Paiva L, Valentin Natal GS, Braga MB (2021) Evaluation of modified starch and rice protein concentrate as wall materials on the microencapsulation of flaxseed oil by freeze-drying. LWT 140:110760. https://doi.org/10.1016/j.lwt.2020.110760

Phongthai S, D'Amico S, Schoenlechner R, Rawdkuen S (2016) Comparative study of rice bran protein concentrate and egg albumin on gluten-free bread properties. J Cereal Sci 72:38–45. https://doi.org/10.1016/j.jcs.2016.09.015

Phongthai S, Homthawornchoo W, Rawdkuen S (2017) Preparation, properties and application of rice bran protein: a review. Int Food Res J 24(1):25–34

Pichot R, Watson RL, Norton IT (2013) Phospholipids at the interface: current trends and challenges. Int J Mol Sci 14(6):11767–11794

Piorkowski DT, McClements DJ (2014) Beverage emulsions: recent developments in formulation, production, and applications. Food Hydrocoll 42:5–41

Prasad C, Bhatia E, Banerjee R (2020) curcumin encapsulated lecithin nanoemulsions: an oral platform for ultrasound mediated spatiotemporal delivery of curcumin to the tumor. Sci Rep 10(1):1–15

Rahmani-Manglano NE, González-Sánchez I, García-Moreno PJ, Espejo-Carpio FJ, Jacobsen C, Guadix EM (2020) Development of fish oil-loaded microcapsules containing whey protein hydrolysate as film-forming material for fortification of low-fat mayonnaise. Foods 9(5):545

Ranadheera CS, Liyanaarachchi WS, Chandrapala J, Dissanayake M, Vasiljevic T (2016) Utilizing unique properties of caseins and the casein micelle for delivery of sensitive food ingredients and bioactives. Trends Food Sci Technol 57:178–187. https://doi.org/10.1016/j.tifs.2016.10.005

Rivadeneira JP, Wu T, Gaban PJV, Tamolang KA (2020) Rheological behavior of purified banana peel pectin from 'saba' banana [Musa BBB saba (Musa acuminata x Musa balbisiana)] peel applied to beverage

Rodrigues RM, Ramos PE, Cerqueira M, Teixeira JA, Vicente AA, Pastrana LM, Pereira RN, Cerqueira MA (2020) Electrosprayed whey protein-based nanocapsules for β-carotene encapsulation. Food Chem 314:126157

Salminen H, Aulbach S, Leuenberger BH, Tedeschi C, Weiss J (2014) Influence of surfactant composition on physical and oxidative stability of Quillaja saponin-stabilized lipid particles with encapsulated ω-3 fish oil. Colloids Surf B Biointerfaces 122:46–55

Sang S, Zhang H, Xu L, Chen Y, Xu X, Jin Z, Yang N, Wu F, Li D (2018) Functionality of ovalbumin during Chinese steamed bread-making processing. Food Chem 253:203–210. https://doi.org/10.1016/j.foodchem.2018.01.150

Santos E, Andrade R, Gouveia E (2017) Utilization of the pectin and pulp of the passion fruit from Caatinga as probiotic food carriers. Food Biosci 20:56–61. https://doi.org/10.1016/j.fbio.2017.08.005

Sarkar A, Singh H (2016) Emulsions and foams stabilised by milk proteins. In: Advanced dairy chemistry. Springer, Berlin, pp 133–153

Scalone GLL, Ioannidis AG, Lamichhane P, Devlieghere F, De Kimpe N, Cadwallader K, De Meulenaer B (2020) Impact of whey protein hydrolysates on the formation of 2,5-dimethylpyrazine in baked food products. Food Res Int 132:109089. https://doi.org/10.1016/j.foodres.2020.109089

Schmidt U, Schmidt K, Kurz T, Endreß H-U, Schuchmann H (2015) Pectins of different origin and their performance in forming and stabilizing oil-in-water-emulsions. Food Hydrocoll 46:59–66

Schrieber R, Gareis H (2007) Gelatine handbook: theory and industrial practice. Wiley, New York

Shao P, Ma H, Zhu J, Qiu Q (2017) Impact of ionic strength on physicochemical stability of o/w emulsions stabilized by Ulva fasciata polysaccharide. Food Hydrocoll 69:202–209

Silva Araújo VB, Melo ANF, Costa AG, Castro-Gomez RH, Madruga MS, Souza EL, Magnani M (2014) Followed extraction of β-glucan and mannoprotein from spent brewer's yeast (Saccharomyces uvarum) and application of the obtained mannoprotein as a stabilizer in mayonnaise. Innov Food Sci Emerg Technol 23:164–170. https://doi.org/10.1016/j.ifset.2013.12.013

Sołowiej BG, Nastaj M, Szafrańska JO, Muszyński S, Gustaw W, Tomczyńska-Mleko M, Mleko S (2020) Effect of emulsifying salts replacement with polymerised whey protein isolate on textural, rheological and melting properties of acid casein model processed cheeses. Int Dairy J 105:104694. https://doi.org/10.1016/j.idairyj.2020.104694

Sosa M, Califano A, Lorenzo G (2019) Influence of quinoa and zein content on the structural, rheological, and textural properties of gluten-free pasta. Eur Food Res Technol 245(2):343–353

Stejskal N, Miranda JM, Martucci JF, Ruseckaite RA, Aubourg SP, Barros-Velázquez J (2020) The effect of gelatine packaging film containing a spirulina platensis protein concentrate on atlantic mackerel shelf life. Molecules 25(14):3209

Su C, Feng Y, Ye J, Zhang Y, Gao Z, Zhao M, Yang N, Nishinari K, Fang Y (2018) Effect of sodium alginate on the stability of natural soybean oil body emulsions. RSC Adv 8(9):4731–4741

Sugiah S, Balia RL, Utama GL (2019) The potential of mannoprotein extracted from *candida apicola* cell wall as emulsification agent. Scientific Papers: Management, Economic Engineering in Agriculture & Rural Development, vol 19(4)

Sun Y, Liu Z, Zhang L, Wang X, Li L (2020) Effects of plasticizer type and concentration on rheological, physico-mechanical and structural properties of chitosan/zein film. Int J Biol Macromol 143:334–340

Supartini N, Mushollaeni W (2017) Local alginate as a food additive and nutritional improvement for white bread. Am J Res Commun 5(1):1–22

Szafrańska JO, Muszyński S, Sołowiej BG (2020) Effect of whey protein concentrate on physicochemical properties of acid casein processed cheese sauces obtained with coconut oil or anhydrous milk fat. LWT 127:109434. https://doi.org/10.1016/j.lwt.2020.109434

Taha A, Ahmed E, Hu T, Xu X, Pan S, Hu H (2019) Effects of different ionic strengths on the physicochemical properties of plant and animal proteins-stabilized emulsions fabricated using ultrasound emulsification. Ultrason Sonochem 58:104627

Tai K, Rappolt M, Mao L, Gao Y, Yuan F (2020) Stability and release performance of curcumin-loaded liposomes with varying content of hydrogenated phospholipids. Food Chem 326:126973

Takeshita S, Oda T (2016) Usefulness of alginate lyases derived from marine organisms for the preparation of alginate oligomers with various bioactivities. In: Advances in food and nutrition research, vol 79. Elsevier, Amsterdam, pp 137–160

Talón E, Lampi A-M, Vargas M, Chiralt A, Jouppila K, González-Martínez C (2019) Encapsulation of eugenol by spray-drying using whey protein isolate or lecithin: release kinetics, antioxidant and antimicrobial properties. Food Chem 295:588–598

Tian L, Liang C, Fu C, Qiang T, Liu Y, Ju X, Shi Z, Xia J, Li H (2020) Esculin and ferric citrate-incorporated sturgeon skin gelatine as an antioxidant film for food packaging to prevent *Enterococcus faecalis* contamination. Food Funct 11(10):9129–9143

Troncoso Recio R (2019) Development of biscuits enriched with phytosterols and fortified with iron and calcium using casein hydrolysates as carriers. Química analítica e alimentaria

Uzoigwe C, Burgess JG, Ennis CJ, Rahman PK (2015) Bioemulsifiers are not biosurfactants and require different screening approaches. Front Microbiol 6:245

Vahidmoghadam F, Pourahmad R, Mortazavi A, Davoodi D, Azizinezhad R (2019) Characteristics of freeze-dried nanoencapsulated fish oil with whey protein concentrate and gum arabic as wall materials. Food Sci Technol 39(4)

Vaucher ACS, Dias PCM, Coimbra PT, Costa ISM, Marreto RN, Dellamora-Ortiz GM, De Freitas O, Ramos MFS (2019) Microencapsulation of fish oil by casein-pectin complexes and gum arabic microparticles: oxidative stabilisation. J Microencapsul 36(5):459–473. https://doi.org/10.1080/02652048.2019.1646335

Vélez-Erazo EM, Bosqui K, Rabelo RS, Kurozawa LE, Hubinger MD (2020) High internal phase emulsions (HIPE) using pea protein and different polysaccharides as stabilizers. Food Hydrocoll 105:105775

Villanueva M, Pérez-Quirce S, Collar C, Ronda F (2018) Impact of acidification and protein fortification on rheological and thermal properties of wheat, corn, potato and tapioca starch-based gluten-free bread doughs. LWT 96:446–454. https://doi.org/10.1016/j.lwt.2018.05.069

Wagoner TB, Foegeding EA (2017) Whey protein–pectin soluble complexes for beverage applications. Food Hydrocoll 63:130–138

Wang P, Jin Z, Xu X (2015) Physicochemical alterations of wheat gluten proteins upon dough formation and frozen storage—a review from gluten, glutenin and gliadin perspectives. Trends Food Sci Technol 46(2):189–198
Wei Y, Tong Z, Dai L, Wang D, Lv P, Liu J, Mao L, Yuan F, Gao Y (2020) Influence of interfacial compositions on the microstructure, physiochemical stability, lipid digestion and β-carotene bioaccessibility of pickering emulsions. Food Hydrocoll 104:105738
Wouters AGB, Delcour JA (2019) Cereal protein-based nanoparticles as agents stabilizing air–water and oil–water interfaces in food systems. Curr Opin Food Sci 25:19–27. https://doi.org/10.1016/j.cofs.2019.02.002
Wu J, Guan Y, Zhong Q (2015) Yeast mannoproteins improve thermal stability of anthocyanins at pH 7.0. Food Chem 172:121–128. https://doi.org/10.1016/j.foodchem.2014.09.059
Wusigale, Liang L, Luo Y (2020) Casein and pectin: structures, interactions, and applications. Trends Food Sci Technol 97:391–403. https://doi.org/10.1016/j.tifs.2020.01.027
Xiang C, Gao J, Ye H, Ren G, Ma X, Xie H, Fang S, Lei Q, Fang W (2020) Development of ovalbumin-pectin nanocomplexes for vitamin D3 encapsulation: enhanced storage stability and sustained release in simulated gastrointestinal digestion. Food Hydrocoll 106:105926. https://doi.org/10.1016/j.foodhyd.2020.105926
Xie M, Dunford NT (2017) Lipid composition and emulsifying properties of canola lecithin from enzymatic degumming. Food Chem 218:159–164
Xiong W, Wang Y, Zhang C, Wan J, Shah BR, Pei Y, Zhou B, Li J, Li B (2016) High intensity ultrasound modified ovalbumin: structure, interface and gelation properties. Ultrason Sonochem 31:302–309
Xiong Y, Chen M, Warner RD, Fang Z (2020) Incorporating nisin and grape seed extract in chitosan-gelatine edible coating and its effect on cold storage of fresh pork. Food Control 110:107018
Xu T, Gao C, Feng X, Yang Y, Shen X, Tang X (2019a) Structure, physical and antioxidant properties of chitosan-gum arabic edible films incorporated with cinnamon essential oil. Int J Biol Macromol 134:230–236. https://doi.org/10.1016/j.ijbiomac.2019.04.189
Xu X, Sun Q, McClements DJ (2019b) Enhancing the formation and stability of emulsions using mixed natural emulsifiers: hydrolyzed rice glutelin and quillaja saponin. Food Hydrocoll 89:396–405
Xu Y-T, Wang Y-H, Chen F-P, Tang C-H (2020) Whether ovalbumin performs as a particulate or polymeric emulsifier is largely determined by pH. Food Hydrocoll 103:105694. https://doi.org/10.1016/j.foodhyd.2020.105694
Xue J, Zhong Q (2014) Thyme oil nanoemulsions coemulsified by sodium caseinate and lecithin. J Agric Food Chem 62(40):9900–9907
Yamashita Y, Miyahara R, Sakamoto K (2017) Chapter 28—Emulsion and emulsification technology. In: Sakamoto K, Lochhead RY, Maibach HI, Yamashita Y (eds) Cosmetic science and technology. Elsevier, Amsterdam, pp 489–506. https://doi.org/10.1016/B978-0-12-802005-0.00028-8
Yan B, Davachi SM, Ravanfar R, Dadmohammadi Y, Deisenroth TW, Pho TV, Odorisio PA, Darji RH, Abbaspourrad A (2020) Improvement of vitamin C stability in vitamin gummies by encapsulation in casein gel. Food Hydrocoll 113:106414. https://doi.org/10.1016/j.foodhyd.2020.106414
Yang Y, Leser ME, Sher AA, McClements DJ (2013) Formation and stability of emulsions using a natural small molecule surfactant: Quillaja saponin (Q-Naturale®). Food Hydrocoll 30(2):589–596
Yang G, Yao Y, Wang C (2017) Green synthesis of silver nanoparticles impregnated bacterial cellulose-alginate composite film with improved properties. Mater Lett 209:11–14. https://doi.org/10.1016/j.matlet.2017.07.097
Yang L, Qin X, Kan J, Liu X, Zhong J (2019a) Improving the physical and oxidative stability of emulsions using mixed emulsifiers: casein-octenyl succinic anhydride modified starch combinations. Nanomaterials 9(7):1018

Yang X, Wang Y, Bai R, Ma H, Wang W, Sun H, Dong Y, Qu F, Tang Q, Guo T (2019b) Pickering emulsion-enhanced interfacial biocatalysis: tailored alginate microparticles act as particulate emulsifier and enzyme carrier. Green Chem 21(9):2229–2233

Yang Z, Zou X, Li Z, Huang X, Zhai X, Zhang W, Shi J, Tahir HE (2019c) Improved postharvest quality of cold stored blueberry by edible coating based on composite gum arabic/roselle extract. Food Bioprocess Technol 12(9):1537–1547

Yu W, Xu D, Li D, Guo L, Su X, Zhang Y, Wu F, Xu X (2019) Effect of pigskin-originated gelatin on properties of wheat flour dough and bread. Food Hydrocoll 94:183–190. https://doi.org/10.1016/j.foodhyd.2019.03.016

Yu W, Xu D, Zhang H, Guo L, Hong T, Zhang W, Jin Y, Xu X (2020) Effect of pigskin gelatin on baking, structural and thermal properties of frozen dough: comprehensive studies on alteration of gluten network. Food Hydrocoll 102:105591. https://doi.org/10.1016/j.foodhyd.2019.105591

Yuan Y, Ma M, Zhang S, Liu C, Chen P, Li H, Wang D, Xu Y (2021) Effect of sophorolipid on the curcumin-loaded ternary composite nanoparticles self-assembled from zein and chondroitin sulfate. Food Hydrocoll 113:106493. https://doi.org/10.1016/j.foodhyd.2020.106493

Zhang G, Li Y, Song T, Bao M, Li Y, Li X (2019) Improvement in emulsifying properties of whey protein–rhamnolipid conjugates through short-time heat treatment. Colloids Surf B Biointerfaces 181:688–695. https://doi.org/10.1016/j.colsurfb.2019.06.015

Zhang Z-K, Xiao J-X, Huang G-Q (2020) Pickering emulsions stabilized by ovalbumin-sodium alginate coacervates. Colloids Surf A Physicochem Eng Aspects 595:124712

Zhao X, Yu G, Li J, Feng Y, Zhang L, Peng Y, Tang Y, Wang L (2018a) Eco-friendly pickering emulsion stabilized by silica nanoparticles dispersed with high-molecular-weight amphiphilic alginate derivatives. ACS Sustain Chem Eng 6(3):4105–4114

Zhao Y, Chen Z, Wu T (2018b) Cryogelation of alginate improved the freeze-thaw stability of oil-in-water emulsions. Carbohydr Polym 198:26–33

Zhao M, Huang X, Zhang H, Zhang Y, Gänzle M, Yang N, Nishinari K, Fang Y (2020) Probiotic encapsulation in water-in-water emulsion via heteroprotein complex coacervation of type-A gelatin/sodium caseinate. Food Hydrocoll 105:105790. https://doi.org/10.1016/j.foodhyd.2020.105790

Zhu Q, Qiu S, Zhang H, Cheng Y, Yin L (2018) Physical stability, microstructure and micro-rheological properties of water-in-oil-in-water (W/O/W) emulsions stabilized by porcine gelatin. Food Chem 253:63–70. https://doi.org/10.1016/j.foodchem.2018.01.119

Ziaeifar L, Labbafi Mazrae Shahi M, Salami M, Askari GR (2018) Effect of casein and inulin addition on physico-chemical characteristics of low fat camel dairy cream. Int J Biol Macromol 117:858–862. https://doi.org/10.1016/j.ijbiomac.2018.05.135

Zou L, Akoh CC (2013) Characterisation and optimisation of physical and oxidative stability of structured lipid-based infant formula emulsion: effects of emulsifiers and biopolymer thickeners. Food Chem 141(3):2486–2494

Chapter 3
Natural Antioxidants and Flavorings for Clean Label Foods

Mehdi Nikoo and Hassan Ahmadi Gavlighi

3.1 Introduction

Oxidative rancidity in food products refers to the perception of undesirable flavors and odors caused by lipid oxidation. Lipid oxidation results from attack of molecular oxygen to unsaturated lipid molecules and can give rise to potentially toxic oxidation products and deterioration in the quality of food lipids (Velasco et al. 2010). Food processing also may degrade lipids and form harmful compounds such as cholesterol-oxidation products (Dantas et al. 2021) and high concentrations of cytotoxic and genotoxic lipid oxidation products (LOPs) such as aldehydes and epoxy-fatty acids. These oxidation products may pose health hazards such as enhanced risks of chronic non-communicable diseases (NCDs), e.g. cancer, along with cardiovascular and neurological diseases (Grootveld et al. 2020; Vieira et al. 2017). Food proteins are preferential targets of radical species, and protein oxidation has relevant consequences on protein functionality and food quality. Protein oxidation is the cause of the loss of nutritional value due to the depletion of essential amino acids and the impaired digestibility of oxidized proteins. Gastrointestinal tract and internal organs are exposed to the cytotoxic and mutagenic potential of these species (Estévez and Luna 2017). Therefore, antioxidants are the necessary additives for foods that contain a high amount of unsaturated lipids. In the food industry, addition of chain-breaking antioxidants is widely used to increase oxidative stability in lipids and oils. Currently, synthetic antioxidants are widely used for inhibition

M. Nikoo (✉)
Department of Pathobiology and Quality Control, Artemia and Aquaculture Research Institute, Urmia University, Urmia, West Azerbaijan, Iran
e-mail: m.nikoo@urmia.ac.ir

H. A. Gavlighi
Department of Food Science and Technology, Faculty of Agriculture, Tarbiat Modares University, Tehran, Iran
e-mail: ahmadi_ha@modares.ac.ir

C. M. Galanakis (ed.), *The Age of Clean Label Foods*,
https://doi.org/10.1007/978-3-030-96698-0_3

or reduction of quality deterioration in foods. However, in recent years, research results indicated some health and safety issues resulting from the use of synthetic antioxidants. Thus, the application of natural antioxidants in various food products has been considered by food manufacturers and consumers as well as regulatory authorities (Shahidi and Zhong 2015). This chapter addresses the recent findings on natural antioxidant and flavoring compounds as natural alternative additives in foods.

3.1.1 Mechanisms of Lipid Oxidation

Lipid oxidation is a multifunctional process and thus it is difficult to evaluate the effect of a specific factor in the overall oxidation process. Some factors affect the rate of lipid oxidation include external factors such as concentration of oxygen, temperature, light, and intrinsic factors such as lipid composition and degree of unsaturation, free fatty acids, metal ions, antioxidants, etc.; the composition of food matrix and the physical properties of the food. Lipid oxidation not only affects the quality of foods with deteriorated flavors and odors, but also with loss of essential nutrients such as fatty acids and vitamins, and with changes in texture and color associated with reactions of lipid oxidation products with other food components such as proteins (Velasco et al. 2010). The main reactants of lipid oxidation are oxygen and unsaturated fatty acids (PUFA). Initiation begins with abstraction of a hydrogen atom from fatty acid acyl chain (LH). The lipid radicals generated ($L^•$) reacts with oxygen and peroxyl radicals ($LOO^•$) are produced. Propagation phase starts when peroxyl radicals abstract hydrogen from close lipid molecules, leading to generation of more lipid radicals and lipid hydroperoxides (LOOH) molecules. LOOH cannot be detected via sensory analysis which is the reason why lipid peroxides do not correlate with off-flavor. The degradation of these hydroperoxides, owing to metal catalysis, results in unfavorable secondary intermediates of shorter chain length such as ketones, aldehydes, alcohols, and small alkanes which causes off-odors, off-flavors and color changes. Termination occurs when a radical reaction is quenched by a reaction with another radical or by an antioxidant (AH). All these reactions occur simultaneously (Karlsdottir et al. 2014).

3.1.2 Mechanisms of Protein Oxidation

Food proteins are preferential targets of radical species, and protein oxidation has relevant consequences on protein functionality and food quality (Lorido et al. 2016) such as reduced meat tenderness and juiciness, development of off-flavor in milk, and overall eating quality of frozen and processed fish (Lund and Baron 2010). Protein oxidation is defined as modification of proteins by reactive oxygen species (ROS) including free radicals (such as hydroxyl and peroxyl radicals) and non-radical species such as hydrogen peroxide. Protein oxidation is seen as large modifications on the protein side chains, or on protein backbone including cleavage of peptide bonds and protein crosslinking. Protein oxidation generates alkyl, peroxyl, and alkoxyl radicals, hydroperoxides, and carbonyls and this is also the case for lipid oxidation. But due to the heterogeneous nature of proteins, the oxidation mechanism is more complicated than that for lipids as it also includes formation of backbone and side-chains radicals (Lund and Baron 2010). Protein oxidation is the cause of the loss of nutritional value due to the depletion of essential amino acids and the impaired digestibility of oxidized proteins. Gastrointestinal tract and internal organs are exposed to the cytotoxic and mutagenic potential of these species (Estévez and Luna 2017). Carbonylation is the most notable expressions of the oxidative damage to proteins. Primary protein carbonyls (i.e. α-aminoadipic and γ-glutamic semialdehydes) have been linked to radical-mediated oxidative stress (Estévez et al. 2021). Both α-aminoadipic (AAS, oxidation product from lysine) and γ-glutamic semialdehydes (GGS, oxidation product from arginine and proline) are good markers of protein oxidation in meat products. Besides their role as oxidation markers, their interaction with other food components could lead to relevant consequences on food quality (Estévez et al. 2011). Secondary protein carbonyls are introduced in proteins via covalent linkage of lipid carbonyls (i.e. protein-bound malondialdehyde (MDA)) (Estévez et al. 2021). The incubation of different proteins (human serum albumin, human hemoglobin, and β-lactoglobulin) with MDA increased protein hydrazones (DNPH-derivatized carbonyls). When lipid oxidation is intense and MDA is largely formed, the nucleophilic addition of MDA to proteins by reacting with ε-amino groups will hamper the primary oxidation of alkaline amino acids such as lysine, and thus the subsequent formation of allysine (primary protein carbonyls) is declined. However, MDA contributes to the formation of hydrazones upon reaction with the DNPH reagent. As a result, the attachment of MDA to proteins could interfere in the assessment of protein carbonyls using DNPH method (Estévez et al. 2019). More detailed information on amino acids side chain modifications and their respective oxidation products can be found in the review chapter by Lund and Baron (2010), Jacobsen et al. (2010), and Medina and Pazos (2010).

3.2 Prevention of Lipid and Protein Oxidation

In food industry, an effective way of preventing oxidative deteriorations of food lipids and proteins is the use of antioxidants. However, antioxidants that are able to prevent lipid oxidation are not always able to prevent oxidation of protein. Inhibition of lipid oxidation is expected to prevent protein oxidation to some extent by minimizing the formation of secondary lipid oxidation products and thereby preventing their interactions with proteins (Lund and Baron 2010). In food industry, addition of chain-breaking antioxidants is widely used to increase oxidative stability in lipids and oils. Synthetic phenolic antioxidants such as *tert*-butylhydroxytoluene (BHT), *tert*-butylhydroxyanisole (BHA), *tert*-butylhydroquinone (TBHQ), and propyl gallate (PG) are very effective to inhibit or reduce lipid oxidation and to retard the development of rancidity in most foods (Velasco et al. 2010). However, these synthetic antioxidants have fallen out of favor due to the consumers demand for "clean label", all natural ingredients (Shahidi and Zhong 2015). In frozen foods such mince and surimi, cryoprotectants have long been used to maintain the stability of myofibrillar proteins and water-holding capacity. Recent studies have shown the potential of protein hydrolysates and peptides as natural alternatives to commercial additives such as sucrose and sorbitol (Nikoo et al. 2016; Zhang et al. 2020). Polysaccharides and low-molecular weight oligosaccharides (Walayat et al. 2020, 2022) along with ice-structuring proteins (Du et al. 2020; Wang et al. 2021) also effectively inhibited or reduced freeze-induced protein denaturation and inhibited water mobility in muscle foods.

3.3 Antioxidants

Antioxidants are compounds that are capable of inhibiting or delaying onset of oxidation and are classified as synthetic or natural. Due to the safety concerns for synthetic antioxidants and the increasing desire of consumers towards natural additives in foods or "clean-label" foods, there is an increasing demand for natural antioxidants. Herbs and spices consisted of various bioactive compounds that act as antioxidant in different foods. Those compounds possess additional advantageous that are their health benefits. Herbs and spices also may be used as natural food flavors while synthetic antioxidants may not contribute additional nutritional benefits (Embuscado 2015). Antioxidants are classified into free radical scavengers (such as BHT, BHA, TBHQ, PG, tocopherols, and extracts of herbs and spices (rosemary, oregano, green tea, seaweeds, clove)), oxygen scavengers (such as ascorbic acid, ascorbates, sulphites, and bisulphites), and chelating agents (such as citric acid, EDTA, and phosphates). As stated, antioxidants protect against oxidation. Factors that affect lipid and protein oxidation are the presence of oxygen and transitional metal ions, light, moisture, hemoglobin, and heat. To prevent or slow down the oxidative deterioration of foods during storage, oxygen and metal catalysts must be removed or sequestered to render them unreactive (Embuscado 2015; Medina and Pazos 2010).

3.3.1 Natural Antioxidants

Several studies aimed to protect muscle foods against lipid and protein oxidation by means of natural antioxidants. The increasing interests of food scientists in phenolic-rich extracts as natural antioxidants provided critical insights into the application of those natural antioxidants as functional ingredients in meat products (Estévez and

Xiong 2021). Higher plants and their constituents are rich source of natural antioxidants such as tocopherols and polyphenols which are found in spices, herbs, fruits, vegetables, cereals, grains, seeds. Also, antioxidants are present in agriculture and food wastes such as hulls, shells, and peels of nuts and cereals, pistachio, apple, potatoes, pomegranate, fish by-products, and marine algae such as sulfated polysaccharides, fucoxanthin, sterols, peptides, carotenoids, and mycosporine-like amino acids (Alipour et al. 2018; Basiri et al. 2015; Rakariyatham et al. 2021; Shahidi and Zhong 2015; Turgut et al. 2017; Vallverdú-Queralt et al. 2014; Wang et al. 2014). Natural antioxidants effectively reduced oxidation in different foods (Cheng et al. 2021; del Pilar Garcia-Mendoza et al. 2021; Medina et al. 2012; Nirmal and Benjakul 2010). The public awareness of consumers towards safety and health risks related to synthetic food antioxidants has led to the growing interest in clean label foods. An example was the use of Camu-camu (*Myrciaria dubia*) seed extract (LCE) as a natural antioxidant in yogurt. The main phenolic compounds of LCE were determined as vescalagin, castalagin, gallic acid, procyanidin A2, and (−)-epicatechin. LCE showed high radical scavenging activity, reducing power, and Cu^{2+}-chelating capacity. LCE at 0.25–1 g/100 g increased antioxidant and reducing capacity and at 0.25 g/100 g led to 84% acceptance index of yogurt (Fidelis et al. 2020). Morcuende et al. (2020) reported the reduction of protein oxidation by sprayed fruit extracts in lamb cutlets subjected to high-oxygen packaging atmosphere (High-Ox-MAP) for 10 days at 4 °C followed by roasting at 180 °C for 10 min. Extracts reduced protein oxidation as evidenced by higher thiols content. Extract from oaknut (QI) was most effective in protecting lamb cutlets against protein carbonylation and lower instrumental hardness in cooked lamb cutlets more likely due to the action on ellagitannins in this extract. This antioxidant strategy was suggested to counteract the negative impact of protein oxidation that occurred during cold storage of high-oxygen modified atmosphere packaging lamb cutlets. Furthermore, Neto et al. (2021) used yellow mombin (*Spondias mombin* L.) bagasse phenolic-rich extract (YMBE) as a source of natural antioxidant in ready-to-eat chicken patties. The extract was able to maintain the red color and to inhibit myoglobin degradation. YMBE reduced the formation of carbonyl compounds and also lipid oxidation products (peroxides, thiobarbituric-acid reactive substances (TBARS), conjugated dienes, and *p*-anisidine) during storage compared to the control samples, being potential preservatives to replace synthetic antioxidants in meat products. Larsson and Undeland (2010) investigated the potential of caffeic acid as a natural antioxidant to decrease lipid and protein oxidation in washed cod mince during ice or frozen storage in the presence of hemoglobin (Hb), which is a pro-oxidant. Hemoglobin (20 μmol/L) added into cod mince increased peroxides, rancid odor, protein carbonyls while decreased redness, protein solubility and α-tocopherol. These reactions were inhibited by ≥50 mg/kg caffeic acid in washed cod mince during ice storage. During frozen storage, caffeic acid was less effective antioxidant due to the high viscosity of the frozen food matrix.

3.3.1.1 Green Tea Extract

Green tea is a source of polyphenols of natural antioxidants that can be used as an alternative. These substances are usually harmless, and oxidation inhibits no less active. These antioxidants, preventing the oxidation of organic molecules, are essential not only for preserving food products but also for protecting living systems from oxidative stress. Green tea extracts and its bioactive compounds (EGCG), as well as other catechins, have been tested for their antioxidant activity in different foods. GTE and its bioactive catechins including EGCG could inhibit oxidative modifications in food products using electron transfer, hydrogen atom transfer, and chelation of metal ions as oxidation catalysts. In EGCG, the structure of trihydroxy B-ring and gallate group in the 3-position contributed to its free radicals scavenging activity. Particularly in muscle-based foods and some other foods containing unstable unsaturated fatty acids, iron and copper are the main catalysts of oxidation. It was found that EGCG, due to its high affinity for some metal ions, could prevent lipid oxidation (Nikoo et al. 2018). Acylation of epigallocatechin (EGC) with different fatty acids such as propionic acid, caprylic acid, lauric acid, stearic acid, and docosahexaenoic acid significantly increased lipophilicity of EGC. The esters and EGC showed high free radical scavenging activity and suggested as natural antioxidants for application in both hydrophilic and lipophilic food matrix (Ambigaipalan et al. 2020).

Green tea extract or its polyphenols have been shown to work as antioxidant. They preserved or improved quality and safety of different food products such as seafood (Chaijan et al. 2020; Cao et al. 2020; Firdous et al. 2021; Nirmal and Benjakul 2011; Roy et al. 2021), dairy products (Najgebauer-Lejko et al. 2020; Pluta-Kubica et al. 2020; Rashidinejad et al. 2016), meat products (Alirezalu et al. 2021; Demirok and Kolsarıcı 2014; Schilling et al. 2018; Siripatrawan and Noipha 2012), and plant food products such as edible oils, bread, biscuits, cakes, and noodles (Gómez-Mascaraque et al. 2017; Gramza-Michałowska et al. 2016; Han et al. 2020; Jia et al. 2021; Von Staszewski et al. 2014; Xu et al. 2019; Yu et al. 2019).

Quality of vacuum-packed tilapia fillets stored at 4 °C for 21 days was preserved using antioxidative EGCG-gelatin biofilm (EGT) was determined (Cao et al. 2020). TBARS increased during 21 days of storage, and fillets with EGT showed the lowest value. However, TBARS of all fillets were lower than the acceptable limit for fish (5–8 mg MDA meq/kg). Slow dissociation of EGCG from the biofilm allowed EGCG to play an antioxidant role in fillets for a long time. Nirmal and Benjakul (2011) showed that green tea extract with and without chlorophyll removal at 0.1–1 g/L showed high polyphenoloxidase (PPO) inhibitory activity in Pacific white shrimp during iced storage. EGCG in gelatin biofilm (EGT) was able to reduce quality deterioration of tilapia fillets during refrigerated storage (Cao et al. 2020). Compared with the control fillets, EGT reduced the rate of lipid and protein oxidation and had lower K-value and biogenic amines. Based on sensory evaluation, the shelf life of fillets extended by 6 days in the EGT samples. Green tea extracts (100–300 mg/kg) maintained quality of fresh pork sausages (Schilling et al. 2018).

Green tea extracts improved oxidative stability in sausages as shown by lower TBARS formation. Consumer acceptability scores were greater (98%) in sausages containing extract than the control. EGCG at 3.3, 6.6, and 9.9 g/kg was able to reduce the formation of acrylamide in white bread. It reduced acrylamide formation by 37% compared to the control while decreasing its moisture content by 6%. EGCG did not affect bread texture but increased lightness and yellowness and decreased the redness of bread crust, porosity and granular size (Fu et al. 2018). Antioxidant activity of EGCG on the physicochemical and oxidative stability of myofibrillar protein–soybean oil emulsions was studied by Cao et al. (2018). EGCG at all levels tested (50–1000 mg/L) decreased lipid oxidation in myofibrillar protein emulsion composite gels during 7 days of chill storage (4 °C). At >100 mg/L, EGCG increased sulfhydryl loss, reduced surface hydrophobicity and aggregation and cross linking of proteins, resulting in hampered emulsification and gel formation of myofibrillar proteins. However, at lower levels (50–200 mg/L) EGCG improved oxidative stability of meat emulsions without affecting texture. More information on application of green tea extract and EGCG as natural antioxidants in foods and their effects on sensory properties and health benefits of foods can be found in the review by Nikoo et al. (2018).

3.3.1.2 Oregano

In foods, effect of oregano extract with antioxidant properties has been documented. Rosmarinic acid, caffeic acid, *P*-coumaric acid, and caffeoyl derivatives, flavones (apigenin and luteolin), and flavanols (myricetin and quercetin) were the main active components (Jimenez-Alvarez et al. 2008). Remarkable inhibition or reduction of lipid oxidation is foods such as bulk oils, salad dressing, fish, and frozen meat products with added oregano extract of the isolated active components has been studied. Jimenez-Alvarez et al. (2008) investigated antioxidant activity and influence on the physical properties of oregano in bulk oil or oil-in-water emulsions enriched with 5% tuna oil and 3.5 mg/kg Fe^{2+}. The antioxidant effect was tested compared with other synthetic antioxidants such as EDTA and Trolox (a water-soluble α-tocopherol analog). It was found that antioxidant activity of oregano was similar to that of trolox in both bulk oil and o/w emulsion, and EDTA in bulk oil. Olmedo et al. (2008) studied the effect of oregano essential oil on the oxidative stability of fried salted peanuts during storage at 23 °C for 112 days and results were compared with those added with BHT. There was no significant effect of the essential oil on sensory properties of the peanuts. It improved the stability of the products in terms of lipid oxidation and inhibition of rancid flavor in fried salted peanuts. Daniells (2010) conducted a study on the antioxidant effect of oregano oil at 0.5% in processed meat and found that the addition of oregano oil decreased the rate of oxidation and the loss in red color. However, the addition of oregano oil created an aftertaste; probably affecting acceptability of the product.

3.3.1.3 Rosemary

Rosemary extract is a natural antioxidant that is the primary phenolic diterpene compound (carnosic acid, carnosol, and rosmarinic acid) discovered in rosemary leaves, and carnosic acid is considered as the primary index component. There are three grade of rosemary antioxidants based on carnosic acid content (30%, 60%, and 90%) in China. The antioxidant activity of carnosic acid is related to two ortho phenolic hydroxyl groups that removes ${}^{\bullet}OH$, $ROO^{\bullet}$ and singlet oxygen. Rosemary extract as natural additive and antioxidant (Shi et al. 2019; Qiu et al. 2018) in food and health industry has been approved and can reduce consumer concern related to safety of foods. Al-Hijazeen and Al-Rawashdeh (2019) studied the effect of different levels (300 and 350 ppm) of rosemary extract (RE) (*Rosmarinus officinalis* Linn.) on the quality and stability of ground chicken meat. The results have shown that TBARS and total carbonyl values of RE and sodium nitrite after 7 days were the lowest value compared to control. The RE and sodium nitrite also showed the highest significant effect delaying aldehydes formation, and positively affect meat sensory attributes. Therefore, RE (350 ppm) was beneficial antioxidant and comparable to the other commercial antioxidants and could be an excellent substitution to many synthetic antioxidants used in the meat industry. Saini et al. (2019) developed chicken meat powder (CMP) incorporated ready-to-eat shelf-stable fried chicken snacks and evaluated the effect of rosemary leaves extract (RE) incorporation on oxidative stability and sensory properties of the product. The study applied 3% of RE during ambient storage up to 60 days. Results indicated that RE could reduce lipid oxidation while improving sensory score (texture, flavor, and overall acceptability) of the product during the storage period. However, the study also revealed that the appearance of samples incorporated with RE was not acceptable. Martínez et al. (2019) studied effect different natural antioxidant (citric (C), rosemary (R), paprika, garlic, oregano, beet (B), lettuce (L), arugula (A), spinach (S), chard (Ch), celery (Ce), and watercress (W)) obtained from fruit and vegetables on cured meat product (chorizo). They found that rosemary was the highest antioxidant activity among all antioxidants. Therefore, they offered excellent antioxidants and could be a perfect strategy to produce clean label dry-cured meat products.

3.3.1.4 Extracts of Seaweeds

Phenolic compounds, carotenoids, tocopherols, polysaccharides, and peptides are natural antioxidants in seaweeds. Phlorotannins are one of the major polyphenolic compounds composed of oligo- or polymers of phloroglucinol (1,3,5-trihydroxy benzene). The observed molecular weight of Phlorotannin varies in the range of 10–100 kDa. The other compound in algae is carotenoids that are lipophilic. Tocopherols are fat-soluble antioxidants that protect lipids by their efficient radical scavenging activity. There are four isomers of tocopherols are α-, β-, γ-, and δ- in which antioxidant activity of α is the highest (Steevensz et al. 2012).

Various biological property of polysaccharides from marine sources (such as blood thinner, antitumor, antineoplastic, microbial control, anti-inflammatory, and antioxidant activity) have been well documented, suggesting their potential application if food industry. Polysaccharides or hydrocolloids can used in food or cosmetic products as emulsion stabilizer. The composition, rheological, and physicochemical properties of algae polysaccharides affect emulsion stability. The main polysaccharides in marine algae are fucoidan and laminarans of brown algae, carrageenan of red algae, and ulvan of green algae. Hermund et al. (2015) investigated the bioactive compound with antioxidant activity from Icelandic brown algae (*Fucus vesiculosus*) using ethyl acetate and water. Both extracts were supplemented in fish-oil-enriched milk or mayonnaise. The *in vitro* antioxidant properties test has shown that ethyl acetate extract had week metal chelating ability, but showed high radical scavenging activity. Ethyl acetate fraction at 2 g/kg was the most efficient antioxidant in fish-oil-enriched milk emulsions. The high antioxidant activity in the foods has been attributed to high radical scavenging, high or moderate metal chelating ability, as well as high phenolic and carotenoid contents. Water extract at 2 g/kg in mayonnaise was more efficient in decreasing the formation of oxidation products than the lower concentrations while, ethyl acetate fraction was most efficient in reducing the degradation of healthy n-3 PUFA. In general, both water extract and ethyl acetate fraction worked as antioxidants in fish-oil-enriched mayonnaise.

O'Sullivan et al. (2016) enriched yogurt with extracts from *Ascophyllum nodosum* (100% H_2O (AN_{100}), 80% ethanol:20% H_2O (AN_{80e})) and *Fucus vesiculosus* (60% ethanol:40% H_2O (FV_{60e})) at 0.25% and 0.50% (w/w) and evaluated antioxidant, physicochemical, and textural properties of yogurts. Yogurts with FV_{60e} and AN_{80e} had lower lipid oxidation and higher yellowness (b*-value). The pH, microorganisms, and syneresis were not influenced by seaweed extract addition. Antioxidant activity (DPPH) was stable duding 28 days of refrigerated storage. Extract-enriched yogurt or digestates did not alter cellular antioxidant status. Among all yogurts, AN_{100} at both levels were highly accepted by sensory panelists.

In another study on fish oil-enriched mayonnaise, Honold et al. (2016) compared the efficacy of different extract (acetone extract, ethanol extract, and two water extracts) obtained from brown seaweed *F. vesiculosus* as natural source of natural antioxidants. The results showed that amount of phenolic compounds and carotenoids from acetone and ethanol were the highest. The primary oxidation products of enriched mayonnaise formulation with different extract showed that acetone and ethanol were more efficient than water extract. A similar result was observed for secondary oxidation products. The efficacy of the extracts was related to high phenolic compounds and high radical scavenging activity and probably to the location of phenolic compounds at the interface. Based on this, especially the acetone and ethanol extracts from *F. vesiculosus* might be used as the potential replacement of synthetic antioxidants in food formulation.

Karadağ et al. (2017) investigated oxidative stability and microstructure of 5% fish oil-in-enriched granola bars added natural antioxidants from *F. vesiculosus*. The oxidative stability of the bars at 20 °C was evaluated during 10 days storage by measuring peroxides and volatile compounds. Ethanol (EE) and acetone (AE) extracts showed the highest antioxidant activity compared to water extract. All extracts, except water at low concentrations, reduced lipid oxidation during 10 week of storage at 0.5 or 1 g extract/100 g emulsion. The antioxidant efficacy of EE and AE extract was among other related high total phenolic content, high radical scavenging activity of the extracts, together with a high interfacial affinity of phenolic compounds, and probably regeneration of tocopherol.

The water extract of *Ascophyllum nodosum*, *Bifurcaria bifurcata*, and *F. vesiculosus* as an antioxidant were added into canola oil. Lipid oxidation parameters (peroxide value (PV), *p*-anisidine value (AV), TBARS value, conjugated dienes (CD), and TOTOX index) were evaluated for 16 days at 60 °C. The addition of seaweed extracts to canola oil can delay primary and secondary compounds during storage and can be used as a natural antioxidant as efficient BHT added at 50 mg/kg (Agregán et al. 2017).

Jónsdóttir et al. (2016) evaluated cod protein hydrolysate and ethyl acetate extracts of *F. vesiculosus* in the hemoglobin-fortified washed cod mince and iron-containing cod liver oil emulsion. The results revealed that ethyl acetate extract was more efficient in cod liver oil emulsion because of the highest in ferrous chelating capacity. In another model (hemoglobin-fortified washed cod mince), *F. vesiculosus* had more antioxidant properties as evidenced by effective reduction of lipid and protein oxidation and rancid odor development during 8 days. The hemoglobin activity inhibition by seaweed extract was the reason for lower oxidation in washed cod mince.

Babakhani et al. (2016) assessed the antioxidant activity of ethanol and water extract from *Fucus serratus* and *Polysiphonia fucoides* on mackerel (*Scomber scombrus*) mince. The 50% ethanolic extract (500 ppm) resulted in low levels of peroxide value, volatiles, and carbonyl compounds and protected against the loss of α-tocopherol and tryptophan residues similar to synthetic antioxidants such as BHT (200 ppm). Water extract with the lowest phenolic content showed no antioxidant effect, whereas extracts obtained with absolute ethanol from both species showed a prooxidative tendency. In conclusion, the 50% ethanolic extracts of *P. fucoides* was useful as natural antioxidants in fish mince.

Antioxidant activity of crude fucoidan extracted from *Laminaria japonica* on surimi products was evaluated by Zheng et al. (2021). The main oxidation in this product is protein oxidation and the loss in gelation. The results showed that at 0.125% addition of fucoidan, DPPH radical scavenging activity was the highest dose to inhibit free radicals. ABTS cation radical scavenging and reducing power activity were dose dependence with concentration of fucoidan and 1% showed the highest effect on inhibition of oxidation.

3.3.1.5 Apple Peel Extract, Pistachio Green Haul Extract, and Pomegranate Peel Extract

Polyphenolic compounds from agricultural waste are considered as one of the valuable sources with natural antioxidant and antimicrobial properties. Apple peels are a great source of natural antioxidants especially phytochemicals like flavonoids (quercetin), phenolic acids (caffeic acid, ferulic acid, chlorogenic acid, isoferulic acid), dihydrochalcones (phloridzin and phloretin), anthocyanins, and flavan-3-ols ((+)-catechin, epigalocatechin) (Sekhon-Loodu et al. 2013). Apple skin has three- to sixfold higher flavonoid content than apple flesh and has unique flavonoids, such as quercetin glycosides, not found in the flesh (Vinson et al. 2001; Wolfe and Liu 2003; Bitalebi et al. 2019). Several studies tested the antioxidant activity of apple peel skin extract in food matrices. The polyphenolic fractions from apple peel (died or frozen) containing quercetin glycosides and epicatechin in combination with other polyphenols, such as phloridzin and cyanidin-3-*O*-galactoside, showed the most significant antioxidant capacity in stabilization of omega-3 polyunsaturated fatty acid-enriched fish oil (Sekhon-Loodu et al. 2013). Antioxidant capacity of apple skin extracts from the "Northern Spy" cultivar was examined for inhibition of lipid oxidation in aqueous EPA emulsions and bulk fish oil. Apple peel extract was effective in reducing the oxidation induced by heat, UV light, and peroxyl radicals. Additionally, the removal of sugars and organic acids (ASE 2) from the crude ethanol extract of apple skins (ASE 1) enhanced the antioxidant properties in both the emulsion and bulk fish oil systems. The average induction times of accelerated oxidation at 50–80 °C of fish oil incorporated with ASE 2 (400 μg/mL) was similar to that of α-tocopherol (400 μg/mL) and BHT (200 μg/mL) (Rupasinghe et al. 2010). Apple peel phenolic was also effective in reducing the oxidative damage to myofibrillar proteins induced by metal-catalyzed oxidation systems (Rysman et al. 2016). Three pure phenolic compounds (chlorogenic acid, (−)-epicatechin and phloridzin) and a crude apple peel extract at 50, 100, and 200 μM were added to myofibrillar proteins. Samples were then subjected to Fe^{3+}/H_2O_2 oxidation at 37 °C for 10 days, and protein oxidation was evaluated as carbonylation (α-amino adipic and γ-glutamic semialdehydes) and Schiff base cross-links. Significant inhibition by apple phenolics was found as compared to the control treatment, with (−)-epicatechin being the most efficient antioxidant and phloridzin showing the weakest antioxidant effect. Results demonstrated the potential of apple peel phenolics as natural inhibitors of protein oxidation in meat products. Another study that investigated the antioxidant potential of apple peel extract was the apple peel powder/CMC solution as an edible coating. The coating solution is also added with tartaric acid as an antimicrobial agent. Results showed that active coating treatment completely inhibited lipid oxidation, and efficiently suppressed the growth of microbial (mesophilic aerobic bacteria, molds and yeasts, and *Salmonella enterica*) on raw beef patties. The active coating formulation did not affect the sensory characteristics of raw and cooked beef patties (Shin et al. 2017).

Pistachio green hull (PGH), which constitutes the most considerable amount of pistachio industry waste (34–45%), contains many polyphenolic compounds such as gallic acid, catechin, 4-hydroxybenzoic acid, protocatechuic acid, naringin, isorhamnetin-7-O-glucoside, eriodictyol-7-*O*-glucoside, quercetin-3-*O*-rutinoside, and isorhamnetin-3-*O*-glucoside (Barreca et al. 2016). Hence, the extraction of bioactive compounds (polyphenols) and application in food products were considered as an antioxidant, antimicrobial and antidiabetic activity. Different concentration of lyophilized PGH (0–1000 mg/kg) was evaluated as natural antioxidant and preservative in beef patty during chill storage (Sadeghinejad et al. 2019). Although, lyophilized PGH higher than 500 mg/kg controlled lipid oxidation, lyophilized PGH had more robust protein carbonyl formation than lipid oxidation during 8 days of storage. Interestingly, redness and oxymyoglobin percentage that indicator of freshness color of beef patty was controlled during storage. The growth of lactic acid bacteria, *Staphylococcus aureus*, yeast, and mold were inhibited by Lyophilized PGH except for *Enterobacteriaceae*. Lashgari et al. (2020) investigated the effect of PGH addition of oxidative stability and shelf life of dry fermented sausage. PGH supplemented (500 and 750 mg/kg) was decreased significantly TBARS value compared to control. This could be due to polyphenols that can inhibit radical during the storage of sausage. Moreover, the sensory properties of samples were not changed by the addition of PGH up to 750 mg/kg. Several types of research showed that flavonoids and terpenoids could be used as nitrite replacer with antioxidant activity in processed meat (de Oliveira et al. 2012; Šojić et al. 2015). Lipid oxidation is one of the main problems in food formulation that can affect safety and development off-flavor. Green technology without using a solvent is one of safe approach to extract of bioactive compounds. The effect of essential oil containing monoterpene phenols, thymol, and carvacrol, and supercritical extract (0.075 and 0.150 μL/g) from winter savory as a natural additive on lipid oxidation, microbial growth and sensory quality of fresh pork sausages during storage was investigated. The results showed that TBARS value was decreased by all concentrations of essential oil and extract, and 0.150 μL/g of an extract with lower pressure (100 bar) had the highest antioxidant activity. This activity could be related to the main bioactive compound of carvacrol and thymol. However, antimicrobial activity of extract at 0.150 μL/g in high pressure (300 bar) was more substantial than other treatments (Enterobacteriaceae count to under 3 log CFU/g). In addition, the sensory properties of enriched sausages with winter savory were improved (Šojić et al. 2015). Aliyari et al. (2020) replaced different concentration of PGH (250–1250 mg/kg) in cooked beef sausage with nitrite (0–100 mg/kg). The peroxide and TBARS values were reduced during 1 month of storage, and nitrite reduction up to 50% had any effect on shelf life, oxidation, and sensory property. The addition of 0–30% of PGH in formulation ice cream on the antioxidant activity of the product was investigated. The ice cream with 20% PGH has eight-time antioxidant activity (DPPH$^•$) more than control. In addition, enrichment of ice cream up to 20% had no significant effect on sensory properties. Hence, it can be used as a natural antioxidant in iced dairy products (Mohammadian et al. 2020).

Pomegranate peel is a byproduct of pomegranate juice industry (40–50% of total weight fruit) that discarded in an environment without any valorization but have an excellent source of polyphenol compounds including, phenolic acids (hydroxycinnamic and hydroxybenzoic acids), flavonoids (anthocyanins, catechins, and other complex flavonoids) and hydrolyzable tannins (ellagic and gallic acids, pedunculagin, punicalin, and punicalagin) (flavonoids, phenolic acids, and tannins) (Pirzadeh et al. 2020). Turgut et al. (2017) evaluated the incorporated of 0.5% and 1% of pomegranate peel extract (PPE) on lipid and protein oxidation in beef meatballs. The beef meatballs supplemented with 1% PPE showed better lipid stability than BHT (0.01%) due to the high phenolic content of PPE that inhibits the formation of lipid peroxides and malondialdehydes. The TBARS values for control, 0.5% PPE, BHT, and 1% PPE were 1.19 mg MDA/kg, 0.71 mg MDA/kg, 0.60 mg MDA/kg and 0.56 mg MDA/kg, respectively after 6 month storage. In addition, PPE 1% had the highest reduction (25%) of the protein carbonyl content compared to control by delaying carbonyl formation. The sulfhydryl content after 6 month storage for control, BHT, 0.5% PPE and 1.0% PPE were ~24, 31, 34 and 40 (nmol/mg protein). Another study conducted by Firuzi et al. (2019) has shown that pomegranate rind powder extract (PRPE) reduced oxidation of frankfurter during 60 days storage at 4 °C. The peroxide value of frankfurters supplemented with pomegranate rind powder extract was decreased 59.22% compared to control throughout the storage period. Regarding TBARS value, samples with PRPE had lower values than BHT and nitrite. It could be due to the higher polyphenol content of PRPE.

3.3.1.6 Protein Hydrolysates and Peptides

Fish industry has been growing fast in the past decades and generates huge amounts of by-products which are about 60% of fish weight after industrial processing, although it differs according to species (Marti-Quijal et al. 2020; Stevens et al. 2018). Protein hydrolysates and bioactive peptides from fish by-products have received increasing attention due to their benefits in human food or animal feed (Desai et al. 2022). Peptides with various biological activities (antioxidant, anticancer, anti-diabetic, antihypertensive, anti-obesity, and immunomodulatory activities) have been produced from marine or shellfish proteins (Heffernan et al. 2021; Sarteshnizi et al. 2021). Enzymes with different cleavage specificities break down the proteins to peptides of varying size (i.e., polypeptides, oligopeptides, dipeptides and tripeptides, and free amino acids). There is a close relationship between antioxidant properties of peptides and the peptide's structure such as peptide size and amino acid residues with particular side chains as active site of antioxidant peptides (Wu et al. 2021; Zheng et al. 2018). Peptides and protein hydrolysates with antioxidant activity have been isolated from several fish and shellfish (Wang et al. 2021; Zhang et al. 2021; Camargo et al. 2021; Rivero-Pino et al. 2020; Rodrigues et al. 2021; Sarteshnizi et al. 2019; Rabiei et al. 2021). The majority of the antioxidative peptides have small chain length (e.g. 3–16 amino acids) with MW less than 3 kDa. Peptides generally exert their antioxidant activity through direct scavenging of free radicals or by increasing the expression of antioxidant defense proteins or enzymes (Pangestuti and Kim 2017). The sequence and structure of peptides play significant roles in their antioxidant functions. It is believed that antioxidant activity of peptides is associated with the presence of specific amino acid residues that possess side chain groups with antioxidant potential such as thiol in cysteine, phenolic hydroxyl group in tyrosine, imidazole group in histidine, thioether in methionine, and indole group in tryptophan (Wu et al. 2020). Besides, different sequences of amino acids might affect the structure and physicochemical properties of peptides, thereby changing their antioxidant activity (Wu et al. 2021).

Freezing is the most widely way to preserve the quality of seafoods. Nevertheless, the eating quality of fresh fish and shellfish is superior compared to frozen seafoods. Quality deterioration of frozen seafoods has been attributed to the formation of ice crystals, lipid oxidation, protein denaturation, leading to the loss of texture and the development of off-odors and off-flavors (Nielsen and Jessen 2007). Fish protein hydrolysates have been used as antioxidants and cryoprotectants to maintain the stability of myofibrillar proteins and to reduce the rate of lipid oxidation during frozen storage or repeated freeze-thaw cycles (Nikoo et al. 2016). Nikoo et al. (2019) indicated that protein hydrolysates from rainbow trout by-products were able to reduce the formation of TBARS and protein carbonyls in raw fish emulsion during refrigerated storage. The tetrapeptide isolated from Amur sturgeon skin gelatin reduced lipid oxidation and maintained thermal properties and the water populations associated with myofibrillar proteins in Japanese seabass (*Lateolabrax japonicas*) mince subjected to six freeze-thaw cycles (Nikoo et al. 2014). Recently, Gheshlaghi et al. (2021) reported that sturgeon gelatin hydrolysates (SGH) increased the fermentation rate, and enhanced rheological properties and elastic gel structure of a fat-free set-type yogurt, especially at 0.1 g/dL. These yogurts showed lower syneresis and higher water holding capacity. Zhang et al. (2020) used surimi processing by-products (SPB) protein hydrolysates to inhibit protein oxidation and maintain gel properties of silver carp (*Hypophthalmichthys molitrix*) surimi during frozen storage. Partial replacement of sucrose with 2% of hydrolysates produced using trypsin or Alcalase effectively delayed cysteine oxidation, the formation of total carbonyls, the loss in Ca^{2+}-ATPase activity while reduced the loss in gelation and water holding capacity. Results showed that SPB hydrolysate showed the potential as natural antioxidant and cryoprotectant to maintain frozen surimi properties. Natural additives in frozen fish muscle and surimi and effects on quality and gelation of myofibrillar proteins are shown in Table 3.1.

Plant-based proteins might be considered a potential alternative to animal proteins. However, grassy and bean odor/flavor and inferior functional properties are the main drawback that influence their food application. Modification of proteins by enzyme leads to change in the molecular structure, physicochemical properties, as well as improvement sensory properties (Sharif et al. 2018). Hydrolyzed vegetable proteins (HVPs) have been used to create meaty flavor for over 100 years. In general, HVP is described as a flavoring, flavor enhancer, or combination that can be produced by enzymatic or acidic hydrolysis of plant proteins. It should be noted that these protein hydrolysates have natural antioxidant properties that can be used in food formulations. Hu et al. (2020) used corn gluten meal (CGM) hydrolysates prepared using three plant enzymes (papain, ficin, and bromelain) with different molecular weight as natural antioxidant in fish mince. The results showed that lower

Table 3.1 Natural additives in frozen fish muscle and surimi and effects on quality and gelation of myofibrillar proteins (MP)

Additives	Properties and preparation of additives	Surimi or MP preparation and storage	Findings	References
Kappa-carrageenan	Commercial Kappa-carrageenan (KC) $[C_6H_9O_8SNa]n$	Grass carp (*Ctenopharyngodon idella*) muscle surimi was first prepared and then MP was extracted using 0.05 mmol/L KCl, 20 mmol/L Tris-maleate, pH: 7.0. The MP (60 mg/mL) was dissolved into the PBS buffer solution (pH = 7.2) and added with KC at 1% or 2% powder. MP gel was obtained by heating at 40 °C (20 min) and then transferred to 90 °C (30 min).	KC showed a significant control on the reduction of functional properties due to the enhanced inter/intra-molecular interactions and less formation of ice crystals. Addition of KC (2%) significantly restricted the decline in WHC from 95.45% to 75.47%, which is well associated with high hydrogen proton density. Besides that, low field nuclear magnetic resonance (LF-NMR) analysis exhibited a restricted increase in T22 relaxation time in samples added with KC (2%).	Walayat et al. (2022)

(continued)

Table 3.1 (continued)

Additives	Properties and preparation of additives	Surimi or MP preparation and storage	Findings	References
Egg white proteins + xylooligosaccharides (EWP-XO)	Commercial xylooligosaccharides ($[C_5H_{10}O_5]n$, 86,017,200) was used. EWP was prepared	Different concentrations of EWP and XO mixture (0%, 2%, 4% and 6%) were added to the top mouth culter (*Culter alburnus*) MPs during frozen storage (−18 °C) of 60 days. MP was first extracted using low salt buffer solution (0.05 M NaCl, 20 mM Tris-maleate, pH 7.2) and then with high salt buffer solution (0.6 M NaCl, 20 mM Tris-maleate, pH 7.2)	EWP-XO reduced the Ca-ATPase activity, and samples treated with 6% EWP-XO showed less increase in carbonyl content. EWP-XO inhibited the decline in the sulfhydryl contents. EWP-XO increased the secondary structural stability by preventing the reduction in α-helix content and led to well-structured protein network that reduced the porosity and protein aggregation of MPs gel.	Walayat et al. (2020)
Inulin	Inulin with different degrees of polymerization (DP), including short-chain inulin (S-INU, DP: 2–6), nature inulin (N-INU, DP: 10–23) and long-chain inulin (L-INU, DP: 23–46) was used.	Silver carp surimi was mixed with inulin (1%, 4%, 8%, and 10% w/w) or with commercial cryoprotectant (CC, 4% w/w sucrose + 4% w/w sorbitol), and surimi without adding any cryoprotectant was used as a control. Surimi were subjected to 1, 2, 3, and 5 freeze-thaw cycles and gels were produced by chopping surimi with 1.5% NaCl for 3 min. Surimi gels were heated at 40 °C for 60 min and then at 90 °C for 30 min.	8% S-INU better maintained the surimi gel properties in terms of breaking force and deformation, water holding capacity, texture and whiteness of surimi gel while 8% L-INU better restrained the thawing water loss.	Cao et al. (2022)

Additives	Properties and preparation of additives	Surimi or MP preparation and storage	Findings	References
Ice structuring protein (ISP)	Commercial ISP from Ryegrass (*Lolium perenne*) with purity >95% was used.	Mirror carp (Cyprinus carpio L.) small muscle pieces (5 × 3 × 2 cm^3) and vacuum impregnated (0.08 MPa) at 4 °C in 0.0, 0.5, 1.0, 2.0 and 3.0 (g/L) of ISP for 45 min. The samples were frozen at −25 °C for 7 days and then thawed at 4 °C for 12 h. The first F-T cycle was completed when the core temperature was 0–4 °C. The third and fifth F-T cycles were frozen and thawed according to the above condition.	ISP could restrain moisture migration and destruction of microstructure of mirror carp during F-T cycles. More importantly, inhibitory effect of 2.0 ISP on oxidation was verified from the decrease in carbonyl contents (16.7%) and TBARS value (21.5%) of mirror carp compared with control.	Du et al. (2020)
L-arginine (Arg) + oxidized caffeic acid (OCF)	CF solution (1.0 g/100 mL) was adjusted to pH 8.0. The solution was placed heated to 40 °C and subjected to oxygenation for 1 h by bubbling the solution with oxygen to convert the phenolic compounds to quinones. Thereafter, the solution was adjusted to pH 7.0 by using 6.0 mol/L HCl and was referred to as OCF compound	Frozen bighead carp (*Aristichthys nobilis*) surimi with 76.56% moisture and 17.89% protein was added with 0.5 or 3.0 g/100 g salt. Arg (0.1 g/100 g) and OCF (0.1, 0.2, 0.3, 0.4, 0.5, 0.6, 0.7, 0.8, and 0.9 g/100 g; based on protein content) were added. Samples were preheated at 40 °C for 30 min followed by 90 °C for 20 min. The gel was obtained and cooled with iced water for 1 h, and then stored at 4 °C in a refrigerator at least 3 h prior to analysis.	Improvements in low-salt surimi gels were achieved when 0.1% Arg was combined with various levels of OCF, especially in combination with 0.5% OCF (LC-A-O). The water distribution in LC-A-O was markedly improved and its β-sheet structure content was significantly increased. Denser and more uniform network structure of LC-A-O was obtained due to much stronger and more intra-molecular cross linking of proteins, via the formation of new adducts between OCF and surimi proteins.	Xiong et al. (2021)

(continued)

Table 3.1 (continued)

Additives	Properties and preparation of additives	Surimi or MP preparation and storage	Findings	References
Sulfated polysaccharides	Sulfated polysaccharides from green alga *Ulva intestinalis (UIP) were extracted. UIP consisted of* carbohydrate (65.3%), protein (8.4%), sulfate (18%) and uronic acid (5.6%) with the average molecular weight of 64.2×10^3 g/mol	UIP was added into surimi paste in four levels (0.25–1.0 g/100 g)	Emulsifying stability of surimi pastes were notably improved by UIP addition while emulsification capacity showed no changes. The addition of UIP up to 0.25 g/100 g maintained the textural characteristics and water holding capacity of surimi gels. The gel whiteness was not adversely changed by UIP addition up to 0.75 g/100 g. The SEM images revealed the presence of compact microstructures with dense surfaces due to UIP addition.	Alipour et al. (2018)

molecular weight of hydrolysates had better antioxidant activities. Considering its antioxidant function and activity, both bromelain and ficin were suitable for the production of CGM hydrolysates with high antioxidant activity. Papain-produced hydrolysates (CH-P4) with a molecular weight of 5–10 kDa, followed by ficin-produced hydrolysates (CH-F1) with a molecular weight of less than 1 kDa, and bromelain-produced hydrolysates with a molecular weight of 3-5.5 kDa (CH-B3) showed the strongest antioxidant activity. The rate of lipid oxidation in minced meat containing CH-P4, CH-B3 and CH-F1 hydrolysates decreased by 30.45%, 27.2% and 13.83%, respectively. This study showed that CGM can be a good source for the production of hydrolysate with antioxidant properties and ficin and bromelain can act as effective enzymes to hydrolyze CGM proteins and improve its antioxidant function. In other study, Karimi et al. (2020) hydrolyzed corn germ protein by three enzymes, Alcalase, Flavourzyme, and trypsin and hydrolysates were separated by ultrafiltration membrane with different MWCOs (2, 10, 100 kDa). The degree of hydrolysis after 2 h was observed to be 17.5%, 11.1% and 2.55% for Alcalase, trypsin and Flavourzyme, respectively. Trypsin hydrolysates showed the highest inhibitory activity of DPPH, ABTS radicals and the highest chelating activity of Fe^{2+}. This could be due to higher hydrophobic amino acid contents. In addition, molecular weight less than 2 kDa showed the highest radical scavenging activity, while molecular weight 2–10 kDa showed the highest Fe^{2+} chelating activity.

3.4 Natural Food Flavorings

Natural flavor enhancers can produce from animal or vegetable protein by Maillard reaction (MR). The MR is one of the most important chemical modifications that can be considered for proteins. This non-enzymatic reaction occurs between the carbonyl groups of reducing sugars and amines and forms brown pigments, intermediates, melanoidins and other Millard reaction products (Hong et al. 2010). Due to the role of MR products in the food industry, such as flavor enhancing, antimicrobial and antioxidant properties, various studies have been performed on MR (Eric et al. 2014; Methven 2012). Protein hydrolysates and peptides have been found to contribute to the taste and aroma in foods through direct contribution, heat degradation, Maillard reaction, or synergy with other food components. Peptides are recognized as essential flavor potentiates and precursors of the Maillard reaction. Recently, an increasing number of studies have indicated that Maillard reaction products (MRPs) prepared from protein hydrolysates/peptides and carbohydrates can influence flavor properties of food products (Fu et al. 2019a). In addition, MPRs generated from various protein hydrolysates/peptides, and carbohydrate sources showed antioxidant activity, emphasizing their biological activities in addition to their flavor imparting to foods. Some protein hydrolysates and peptides with desirable flavor generated using Maillard reaction include smoothhound viscera protein hydrolysates (Abdelhedi et al. 2017), chicken bone hydrolysates (Nie et al. 2017), whey protein hydrolysates (Nooshkam and Madadlou 2016), shrimp byproduct protein hydrolysates (Zha et al. 2015), sunflower protein hydrolysates (Eric et al. 2014) and porcine plasma protein hydrolysates (Fu et al. 2019b).

A study conducted by Bagnasco et al. (2013) showed that hydrolysis using food-grade exo- plus endo-proteases (i.e., Umamizyme and Flavourzyme) led to the production of protein hydrolysates possessing umami taste as the most intense attribute, mainly in Umamizyme hydrolysates. Hydrolysates contained medium-low size polypeptides with a degree of hydrolysis (DH) of ~25% and 67% when Flavourzyme and Umamyzyme were used. Consumption of Cheddar cheese containing ~6 g sodium/kg cheese may increase the risk of hypertension and cardiovascular disease. Scalone et al. (2019) hydrolyzed vegetable protein, and adenosine-5′-monophosphate (AMP) gave the highest saltiness and lowest bitterness scores. Low sodium cheese with 75% NaCl substitution, 2 g/L hydrolyzed vegetable protein, and 300 mg/L AMP gave the best flavor without sensory loss. The effect of temperature (80–140 °C) was investigated on production of MRPs by reaction between sunflower peptide and D-xylose with or without L-cysteine (PXC or PX) for 2.0 h and a pH of 7.4 (Eric et al. 2014). Results shown those MRPs at higher temperature (PXC-120) with cysteine addition was the highest meat-like flavor, kokumi flavor, and continuity taste compared to other MRPs. The presence of cysteine inhibited peptide cross-linking and increased molecular weight degradation during higher temperature. In addition, results showed that peptides with higher molecular weight (>5 kDa) in the presence of cysteine had negative sensory properties but peptides between 1 and 5 kDa and lower than 1 kDa were positive effect of sensory attributes. On the other hand, peptide with molecular weight higher than 5 or below 1 kDa without cysteine had negative effect on sensory but peptides between 1 and 5 kDa had positive influence on sensory. This could be due to presence of glutamic acid, glycine and threonine in the structure of PX. In summary, sunflower MRPs prepared at higher temperature (120 °C) in with peptides 1–5 kDa and lower than 1 kDa could be used as a flavor enhancer in the food industry related flavor such as meat or soups. In other study, heating (120 °C) of soy protein hydrolysate with cysteine and xylose showed higher kokumi flavor formation (Habinshuti et al. 2021). In addition, enhance the intensity of kokumi properties of peptides can be by synergistic between oligosaccharide (stachyose and raffinose) and γ-glutamyl-peptides (Shibata et al. 2017).

Peony seed meal is produced as a major environmental problem during peony seed oil extraction. In a study by Shang et al. (2020), the effect of different sugars on the structural and aromatic properties of MRPs was investigated. It was observed that the effect of ribose, xylose and glucose on the structure is greater than that of maltose and xylooligosaccharides. Ribose and xylose sugars had a stronger meaty and umami flavor to MRPs. In addition, MRPs prepared by ribose and xylose showed higher browning intensity, degree of substitution (DS), and fluorescence intensity. The molecular weight distributions of 500–1000, 1000–3000 and more than 3000 Da also increased significantly in all Millard products. In general, the results showed that the distribution of molecular weight, free amino acids and volatile compounds significantly affect the sensory properties of MRPs, which can increase their industrial potential as flavor enhancers (Shang et al. 2020). The

Table 3.2 Natural flavors produced by Millard reaction products (MRPs) from different source of peptides

Source	Reaction condition	Flavors	References
Cod (*Gadus morhua* L.) skin collagen peptides	Xylose (at 80, 100, and 120 °C for up to 150 min)	Twenty-six volatile compounds, including 2,5-dimethyl-pyrazine and 2-ethyl-3,5-dimethylpyrazine	Chen et al. (2020a, b)
Blue Round Scad (*Decapterus maruadsi*) hydrolysates	Xylose and glucose with cysteine (100 °C for 30 min)	Fifty-nine volatile compounds e.g., furans, pyrazines, and sulphur-containing compounds	Zhong et al. (2021)
Gelatin hydrolysates from grass carp (*Ctenopharyngodon idellus*) scales	Glucose, xylose, and ribose (75–95 °C for 5 h)	Thirty-one volatile compounds (butanal, benzaldehyde, 2-methyl-3-furanthiol, and maltol)	Chen et al. (2019)
Chicken liver protein hydrolysate	Xylose (120 °C for 90 min)	Eighty-nine volatile compounds including 17 aldehydes, 8 alcohols, 10 esters, 18 ketones and 12 other	Chen et al. (2020a, b)
Sweet potato protein hydrolysates	Arabinose, xylose, galactose and glucose (120 °C for 5 h)	112 volatile compounds including 16 aldehydes, 29 ketones, 10 esters, 17 alcohols, 8 acids, 3 furans, 3 thiophenes, 4 pyrazines, 2 disulfides, 4 alkanes, 2 alkenes, 1 imine and 6 others	Habinshuti et al. (2021)
Hydrolyzed whey protein isolate	Glucose (180 °C for 90 min)	Pyrazines and pyrazine	Scalone et al. (2019)

natural flavors produced by Maillard reaction products (MRPs) from different source of peptides are presented in Table 3.2.

3.5 Challenges When "Chemicals" Are Replacing with "Natural" Ingredient

Polyphenols can be degraded by changing pH, light, temperature, and during the intestinal digestion (Rafiee et al. 2017). Verbeyst et al. (2010) examined the degradation of anthocyanins in strawberry paste across a temperature range of 80–130 °C. The anthocyanins decreased with higher time temperatures. Hartmann et al. (2008) evaluated the effect of processing steps on polyphenol and anthocyanin contents of strawberry. The pressing and pasteurization had the most impact in the decrease of these compounds. Moreover, the application of polyphenols in food formulation can change sensory properties such as flavor and taste. This can be a significant drawback when applying polyphenolic compounds in a particular food.

Toxicity and dosage of each natural compound should be taken into account. The hydrophilic nature of natural additive made lower solubility and decreased efficiency in the hydrophobic matrix of food such as sauces. The interaction of polyphenols with protein in dairy products can reduce antioxidant activity (Xiao et al. 2011). The stability and bioaccessibility of phenolic compounds and carotenoids in yogurt formulation with strawberry and peach were evaluated (Oliveira and Pintado 2015). In strawberry following intestinal digestion, pelargonidin-3-glucoside decreased 65%, and pelargonidin-3-rutinoside decreased by 58%. In the peach formula, (+)-catechin content showed the highest reduction (80%), whereas neo chlorogenic, chlorogenic acid, rutin, and the carotenoid zeaxanthin decreased to a lower level, between 39% and 45%. Caffeic acid, a hydroxycinnamic acid common in different vegetable sources, has been used as a natural antioxidant for inhibiting the oxidation of fish lipids and proteins present in various food matrices including bulk fish oils, liposomes, fish oil emulsions, washed cod mince, and regular horse mackerel mince (Medina et al. 2012). The effect of caffeic acid was dependent on the storage temperature, showing higher effectiveness above than below 0 °C. Caffeic acid was not able to inhibit the oxidation of bulk fish oils, the fish oil in water emulsions, and the fish-oil enriched fitness bar. In the liposome system, caffeic acid inhibited hemoglobin (Hb)-promoted oxidation but strongly mediated Fe^{2+} mediated oxidation. Besides such limitations, the higher price of caffeic acid compared with synthetic antioxidants might be a drawback influencing the decision of whether the food industry prefers to use this natural antioxidant. The better diffusion of caffeic acid into frozen mince in the presence of a fish peptide that reduces the viscosity of a rigid food matrix positively influenced the antioxidant activity of caffeic acid in fish mince subjected to different freeze-thaw cycles. However, the high price of pure peptides for food applications might limit its potential application in foods. The major drawback can be the sensory properties of the food added with hydrolysates. Although protein hydrolysates maintained the thermal properties of major muscle structural proteins (i.e., myosin and actin) and increased the amount of unfrozen water in frozen myofibrillar proteins, foods with hydrolysates, particularly those derived from processing byproducts, always showed inferior sensory perception. EGCG is categorized as GRAS by the U.S. FDA. However, EGCG or green eta extract have been shown to impart toxicity at higher doses. Lambert et al. (2010) found the cytotoxicity effect of EGCG fed to mice at 1500 mg/kg. James et al. (2018) found a similar results in which EGCG at concentrations ranging from 500 to 750 mg/kg for 3 days caused hepatic inflammation, necrosis, and hemorrhage in C57B/6L mice. Therefore, dose-dependent response studies are needed to assure the safety of EGCG or GTE in foods.

3.6 Conclusion

Natural antioxidants and flavorings are of interest because of their effectiveness in enhancing food safety and quality. As documented by many researches, lipid and protein oxidation was largely inhibited in foods aged with natural antioxidants.

However, it should be considered that some natural antioxidant compounds, especially at higher concentrations, caused cytotoxicity and may be linked to adverse health effects. Thus, when formulating novel functional foods that use the benefits of natural antioxidants, the food safety concerns must be further investigated in terms of dose-response studies to assure minimum adverse effects, if any. The stability and function of the natural antioxidants and the flavor and taste of foods with natural antioxidants and flavorings after processing and during storage with regards to consumer acceptability must be considered.

References

Abdelhedi O, Mora L, Jemil I, Jridi M, Toldrá F, Nasri M, Nasri R (2017) Effect of ultrasound pretreatment and Maillard reaction on structure and antioxidant properties of ultrafiltrated smooth-hound viscera proteins-sucrose conjugates. Food Chem 230:507–515

Agregán R, Munekata PE, Domínguez R, Carballo J, Franco D, Lorenzo JM (2017) Proximate composition, phenolic content and in vitro antioxidant activity of aqueous extracts of the seaweeds *Ascophyllum nodosum*, *Bifurcaria bifurcata* and *Fucus vesiculosus*. Effect of addition of the extracts on the oxidative stability of canola oil under accelerated storage conditions. Food Res Int 99:986–994

Al-Hijazeen M, Al-Rawashdeh M (2019) Preservative effects of rosemary extract (*Rosmarinus officinalis* L.) on quality and storage stability of chicken meat patties. Food Sci Technol 39(1):27–34

Alipour HJ, Rezaei M, Shabanpour B, Tabarsa M (2018) Effects of sulfated polysaccharides from green alga *Ulva intestinalis* on physicochemical properties and microstructure of silver carp surimi. Food Hydrocoll 74:87–96

Alirezalu K, Hesari J, Yaghoubi M, Khaneghah AM, Alirezalu A, Pateiro M, Lorenzo JM (2021) Combined effects of ε-polylysine and ε-polylysine nanoparticles with plant extracts on the shelf life and quality characteristics of nitrite-free frankfurter-type sausages. Meat Sci 172:108318

Aliyari P, Bakhshi Kazaj F, Barzegar M, Ahmadi Gavlighi H (2020) Production of functional sausage using pomegranate peel and pistachio green hull extracts as natural preservatives. J Agric Sci Technol 22(1):159–172

Ambigaipalan P, Oh WY, Shahidi F (2020) Epigallocatechin (EGC) esters as potential sources of antioxidants. Food Chem 309:125609

Babakhani A, Farvin KS, Jacobsen C (2016) Antioxidative effect of seaweed extracts in chilled storage of minced Atlantic mackerel (*Scomber scombrus*): effect on lipid and protein oxidation. Food Bioprocess Technol 9(2):352–364

Bagnasco L, Pappalardo VM, Meregaglia A, Kaewmanee T, Ubiali D, Speranza G, Cosulich ME (2013) Use of food-grade proteases to recover umami protein–peptide mixtures from rice middlings. Food Res Int 50(1):420–427

Barreca D, Laganà G, Leuzzi U, Smeriglio A, Trombetta D, Bellocco E (2016) Evaluation of the nutraceutical, antioxidant and cytoprotective properties of ripe pistachio (*Pistacia vera* L., variety Bronte) hulls. Food Chem 196:493–502

Basiri S, Shekarforoush SS, Aminlari M, Akbari S (2015) The effect of pomegranate peel extract (PPE) on the polyphenol oxidase (PPO) and quality of Pacific white shrimp (*Litopenaeus vannamei*) during refrigerated storage. LWT-Food Sci Technol 60(2):1025–1033

Bitalebi S, Nikoo M, Rahmanifarah K, Noori F, Gavlighi HA (2019) Effect of apple peel extract as natural antioxidant on lipid and protein oxidation of rainbow trout (*Oncorhynchus mykiss*) mince. Int Aqua Res 11(2):135–146

Camargo TR, Ramos P, Monserrat JM, Prentice C, Fernandes CJ, Zambuzzi WF, Valenti WC (2021) Biological activities of the protein hydrolysate obtained from two fishes common in the fisheries bycatch. Food Chem 342:128361

Cao Y, Ai N, True AD, Xiong YL (2018) Effects of (−)-epigallocatechin-3-gallate incorporation on the physicochemical and oxidative stability of myofibrillar protein–soybean oil emulsions. Food Chem 245:439–445

Cao J, Wang Q, Ma T, Bao K, Yu X, Duan Z et al (2020) Effect of EGCG-gelatin biofilm on the quality and microbial composition of tilapia fillets during chilled storage. Food Chem 305:125454

Cao Y, Zhao L, Huang Q, Xiong S, Yin T, Liu Z (2022) Water migration, ice crystal formation, and freeze-thaw stability of silver carp surimi as affected by inulin under different additive amounts and polymerization degrees. Food Hydrocoll 124:107267

Chaijan S, Panpipat W, Panya A, Cheong LZ, Chaijan M (2020) Preservation of chilled Asian sea bass (*Lates calcarifer*) steak by whey protein isolate coating containing polyphenol extract from ginger, lemongrass, or green tea. Food Control 118:107400

Chen K, Yang X, Huang Z, Jia S, Zhang Y, Shi J et al (2019) Modification of gelatin hydrolysates from grass carp (*Ctenopharyngodon idellus*) scales by Maillard reaction: antioxidant activity and volatile compounds. Food Chem 295:569–578

Chen K, Yang Q, Hong H, Feng L, Liu J, Luo Y (2020a) Physicochemical and functional properties of Maillard reaction products derived from cod (*Gadus morhua* L.) skin collagen peptides and xylose. Food Chem 333:127489

Chen X, Zou Y, Wang D, Xiong G, Xu W (2020b) Effects of ultrasound pretreatment on the extent of Maillard reaction and the structure, taste and volatile compounds of chicken liver protein. Food Chem 331:127369

Cheng J, Xiang R, Tang D, Zhu M, Liu X (2021) Regulation of protein oxidation in Cantonese sausages by rutin, quercetin and caffeic acid. Meat Sci 175:108422

Daniells D (2010) Oregano oil shows promise for processed meat preservative. J Food Sci

Dantas NM, de Oliveira VS, Sampaio GR, Chrysostomo YSK, Chávez DWH, Gamallo OD et al (2021) Lipid profile and high contents of cholesterol oxidation products (COPs) in different commercial brands of canned tuna. Food Chem 352:129334

Demirok E, Kolsarıcı N (2014) Effect of green tea extract and microwave pre-cooking on the formation of acrylamide in fried chicken drumsticks and chicken wings. Food Res Int 63:290–298

Desai AS, Brennan M, Gangan SS, Brennan C (2022) Utilization of fish waste as a value-added ingredient: sources and bioactive properties of fish protein hydrolysate. In: Sustainable fish production and processing. Academic Press, London, pp 203–225

Du X, Chang P, Tian J, Kong B, Sun F, Xia X (2020) Effect of ice structuring protein on the quality, thermal stability and oxidation of mirror carp (*Cyprinus carpio* L.) induced by freeze-thaw cycles. LWT Food Sci Technol 124:109140

Embuscado ME (2015) Spices and herbs: natural sources of antioxidants–a mini review. J Funct Foods 18:811–819

Eric K, Raymond LV, Abbas S, Song S, Zhang Y, Masamba K, Zhang X (2014) Temperature and cysteine addition effect on formation of sunflower hydrolysate Maillard reaction products and corresponding influence on sensory characteristics assessed by partial least square regression. Food Res Int 57:242–258

Estévez M, Luna C (2017) Dietary protein oxidation: a silent threat to human health? Crit Rev Food Sci Nutr 57(17):3781–3793

Estévez M, Xiong YL (2021) Protein oxidation in foods: mechanisms, consequences, and antioxidant solutions. *Foods 10*(10):2346

Estévez M, Ventanas S, Heinonen M (2011) Formation of Strecker aldehydes between protein carbonyls–α-aminoadipic and γ-glutamic semialdehydes–and leucine and isoleucine. Food Chem 128(4):1051–1057

Estévez M, Padilla P, Carvalho L, Martín L, Carrapiso A, Delgado J (2019) Malondialdehyde interferes with the formation and detection of primary carbonyls in oxidized proteins. Redox Biol 26:101277

Estévez M, Díaz-Velasco S, Martínez R (2021) Protein carbonylation in food and nutrition: a concise update. Amino Acids:1–15

Fidelis M, de Oliveira SM, Santos JS, Escher GB, Rocha RS, Cruz AG et al (2020) From byproduct to a functional ingredient: Camu-camu (*Myrciaria dubia*) seed extract as an antioxidant agent in a yogurt model. J Dairy Sci 103(2):1131–1140

Firdous A, Ringø E, Elumalai P (2021) Effects of green tea-and amla extracts on quality and melanosis of Indian white prawn (*Fenneropenaeus indicus*, Milne Edwards, 1837) during chilled storage. Aquacult Fish 6(6):617–627

Firuzi MR, Niakousari M, Eskandari MH, Keramat M, Gahruie HH, Mousavi KA (2019) Incorporation of pomegranate juice concentrate and pomegranate rind powder extract to improve the oxidative stability of frankfurter during refrigerated storage. LWT-Food Sci Technol 102:237–245

Fu Z, Yoo MJ, Zhou W, Zhang L, Chen Y, Lu J (2018) Effect of (−)-epigallocatechin gallate (EGCG) extracted from green tea in reducing the formation of acrylamide during the bread baking process. Food Chem 242:162–168

Fu Y, Zhang Y, Soladoye OP, Aluko RE (2019a) Maillard reaction products derived from food protein-derived peptides: insights into flavor and bioactivity. Crit Rev Food Sci Nutr 60:3429–3442

Fu Y, Bak KH, Liu J, De Gobba C, Tøstesen M, Hansen ET et al (2019b) Protein hydrolysates of porcine hemoglobin and blood: peptide characteristics in relation to taste attributes and formation of volatile compounds. Food Res Int 121:28–38

Gheshlaghi SP, Khaledabad MA, Nikoo M, Regenstein JM, Gavlighi HA (2021) Impact of sturgeon gelatin hydrolysates (SGH) on physicochemical and microbiological properties of fat-free set-type yogurt. LWT Food Sci Technol 148:111665

Gómez-Mascaraque LG, Hernández-Rojas M, Tarancón P, Tenon M, Feuillère N, Ruiz JFV et al (2017) Impact of microencapsulation within electrosprayed proteins on the formulation of green tea extract-enriched biscuits. LWT-Food Sci Technol 81:77–86

Gramza-Michałowska A, Kobus-Cisowska J, Kmiecik D, Korczak J, Helak B, Dziedzic K, Górecka D (2016) Antioxidative potential, nutritional value and sensory profiles of confectionery fortified with green and yellow tea leaves (*Camellia sinensis*). Food Chem 211:448–454

Grootveld M, Percival BC, Leenders J, Wilson PB (2020) Potential adverse public health effects afforded by the ingestion of dietary lipid oxidation product toxins: significance of fried food sources. Nutrients 12(4):974

Habinshuti I, Zhang M, Sun HN, Mu TH (2021) Effects of ultrasound-assisted enzymatic hydrolysis and monosaccharides on structural, antioxidant and flavour characteristics of Maillard reaction products from sweet potato protein hydrolysates. Int J Food Sci Technol. https://doi.org/10.1111/ijfs.15249

Han CW, Ma M, Zhang HH, Li M, Sun QJ (2020) Progressive study of the effect of superfine green tea, soluble tea, and tea polyphenols on the physico-chemical and structural properties of wheat gluten in noodle system. Food Chem 308:125676

Hartmann A, Patz CD, Andlauer W, Dietrich H, Ludwig M (2008) Influence of processing on quality parameters of strawberries. J Agric Food Chem 56(20):9484–9489

Heffernan S, Giblin L, O'Brien N (2021) Assessment of the biological activity of fish muscle protein hydrolysates using in vitro model systems. Food Chem 359:129852

Hermund DB, Yeşiltaş B, Honold P, Jónsdóttir R, Kristinsson HG, Jacobsen C (2015) Characterisation and antioxidant evaluation of Icelandic F. vesiculosus extracts in vitro and in fish-oil-enriched milk and mayonnaise. J Funct Foods 19:828–841

Hong JH, Jung DW, Kim YS, Lee SM, Kim KO (2010) Impacts of glutathione Maillard reaction products on sensory characteristics and consumer acceptability of beef soup. J Food Sci 75(8):S427–S434

Honold PJ, Jacobsen C, Jónsdóttir R, Kristinsson HG, Hermund DB (2016) Potential seaweed-based food ingredients to inhibit lipid oxidation in fish-oil-enriched mayonnaise. Eur Food Res Technol 242(4):571–584

Hu R, Chen G, Li Y (2020) Production and characterization of antioxidative hydrolysates and peptides from corn gluten meal using papain, ficin, and bromelain. Molecules 25(18):4091

Jacobsen C, Nielsen HH, Jørgensen B, Nielsen J (2010) Chemical processes responsible for quality deterioration in fish. In: Chemical deterioration and physical instability of food and beverages. Woodhead Publishing, Cambridge, pp 439–465

James KD, Kennett MJ, Lambert JD (2018) Potential role of the mitochondria as a target for the hepatotoxic effects of (-)-epigallocatechin-3-gallate in mice. Food Chem Toxicol 111:302–309

Jia C, Li J, Zhang M, Ma W, Zhao S, Liu R et al (2021) Antioxidant properties of the extracts of vine tea (*Ampelopsis grossedentata*) with the different color characteristics and inhibition of rapeseed and sunflower oil oxidation. LWT Food Sci Technol 136:110292

Jimenez-Alvarez D, Giuffrida F, Golay PA, Cotting C, Lardeau A, Keely BJ (2008) Antioxidant activity of oregano, parsley, and olive mill wastewaters in bulk oils and oil-in-water emulsions enriched in fish oil. J Agric Food Chem 56(16):7151–7159

Jónsdóttir R, Geirsdóttir M, Hamaguchi PY, Jamnik P, Kristinsson HG, Undeland I (2016) The ability of in vitro antioxidant assays to predict the efficiency of a cod protein hydrolysate and brown seaweed extract to prevent oxidation in marine food model systems. J Sci Food Agric 96(6):2125–2135

Karadağ A, Hermund DB, Jensen LHS, Andersen U, Jónsdóttir R, Kristinsson HG et al (2017) Oxidative stability and microstructure of 5% fish-oil-enriched granola bars added natural antioxidants derived from brown alga *Fucus vesiculosus*. Eur J Lipid Sci Technol 119(4):1500578

Karimi A, Azizi MH, Ahmadi Gavlighi H (2020) Fractionation of hydrolysate from corn germ protein by ultrafiltration: *in vitro* antidiabetic and antioxidant activity. Food Sci Nutr 8:5192

Karlsdottir MG, Petty HT, Kristinsson HG (2014) Oxidation in aquatic foods and analysis methods. In: Antioxidants and functional components in aquatic foods. John Wiley & Sons, Ltd, Chichester, pp 1–22

Lambert JD, Kennett MJ, Sang D, Reuhl KR, Ju J, Yang CS (2010) Hepatotoxicity of high oral dose (-)-epigallocatechin-3-gallate in mice. Food Chem Toxicol 48(1): 409–416

Larsson KJ, Undeland IK (2010) Effect of caffeic acid on haemoglobin-mediated lipid and protein oxidation in washed cod mince during ice and frozen storage. J Sci Food Agric 90(14):2531–2540

Lashgari SS, Noorolahi Z, Sahari MA, Ahmadi Gavlighi H (2020) Improvement of oxidative stability and textural properties of fermented sausage via addition of pistachio hull extract. Food Sci Nutr 8(6):2920–2928

Lorido L, Ventanas S, Akcan T, Estévez M (2016) Effect of protein oxidation on the impaired quality of dry-cured loins produced from frozen pork meat. Food Chem 196:1310–1314

Lund MN, Baron CP (2010) Protein oxidation in foods and food quality. In: Chemical deterioration and physical instability of food and beverages. Woodhead Publishing, Cambridge, pp 33–69

Martínez L, Bastida P, Castillo J, Ros G, Nieto G (2019) Green alternatives to synthetic antioxidants, antimicrobials, nitrates, and nitrites in clean label Spanish chorizo. Antioxidants 8(6):184

Marti-Quijal FJ, Remize F, Meca G, Ferrer E, Ruiz MJ, Barba FJ (2020) Fermentation in fish and by-products processing: an overview of current research and future prospects. Curr Opin Food Sci 31:9–16

Medina I, Pazos M (2010) Oxidation and protection of fish. In: Oxidation in foods and beverages and antioxidant applications. Volume 2: Management in different industry sectors. Woodhead Publishing Ltd, Cambridge, pp 91–120

Medina I, Undeland I, Larsson K, Storrø I, Rustad T, Jacobsen C et al (2012) Activity of caffeic acid in different fish lipid matrices: a review. Food Chem 131(3):730–740

Methven L (2012) Natural food and beverage flavour enhancer. In: Natural food additives, ingredients and flavourings. Woodhead Publishing, Cambridge, pp 76–99

Mohammadian AH, Ghandehari Yazdi AP, Barzegar M (2020) Inhibition activity on alpha-glucosidase of experimental, functional ice cream with the addition of pistachio peel extract. J Food Bioprocess Eng 3(1):53–60

Morcuende D, Vallejo-Torres C, Ventanas S, Martínez SL, Ruiz SC, Estévez M (2020) Effectiveness of sprayed bioactive fruit extracts in counteracting protein oxidation in lamb cutlets subjected to a high-oxygen MAP. Foods 9(11):1715

Najgebauer-Lejko D, Witek M, Żmudziński D, Ptaszek A (2020) Changes in the viscosity, textural properties, and water status in yogurt gel upon supplementation with green and Pu-erh teas. J Dairy Sci 103(12):11039–11049

Neto DC, Cordeiro ÂM, Meireles BR, Araújo Í, Estévez M, Ferreira V, Silva FA (2021) Inhibition of protein and lipid oxidation in ready-to-eat chicken patties by a *Spondias mombin* L. bagasse phenolic-rich extract. Foods 10(6):1338

Nie X, Xu D, Zhao L, Meng X (2017) Antioxidant activities of chicken bone peptide fractions and their Maillard reaction products: effects of different molecular weight distributions. Int J Food Prop 20(sup1):S457–S466

Nielsen J, Jessen F (2007) Quality of frozen fish. In: Nollet LML (ed) Handbook of meat, poultry, and seafood quality. Blackwell Publishing, Ames, IA, pp 577–586

Nikoo M, Benjakul S, Ehsani A, Jing L, Wu FF, Yang N, Xu B, Jin Z, Xu X (2014) Antioxidant and cryoprotective effects of a tetrapeptide isolated from Amur sturgeon skin gelatin. J Funct Foods 7:609–620

Nikoo M, Benjakul S, Rahmanifarah K (2016) Hydrolysates from marine sources as cryoprotective substances in seafoods and seafood products. Trends Food Sci Technol 57:40–51

Nikoo M, Regenstein JM, Ahmadi Gavlighi H (2018) Antioxidant and antimicrobial activities of (-)-epigallocatechin-3-gallate (EGCG) and its potential to preserve the quality and safety of foods. Compr Rev Food Sci Food Saf 17:732–753

Nikoo M, Benjakul S, Yasemi M, Ahmadi Gavlighi H, Xu X (2019) Hydrolysates from rainbow trout (*Oncorhynchus mykiss*) processing by-product with different pretreatments: antioxidant activity and their effect on lipid and protein oxidation of raw fish emulsion. LWT Food Sci Technol 108:120–128

Nirmal NP, Benjakul S (2010) Effect of catechin and ferulic acid on melanosis and quality of Pacific white shrimp subjected to prior freeze–thawing during refrigerated storage. Food Control 21(9):1263–1271

Nirmal NP, Benjakul S (2011) Use of tea extracts for inhibition of polyphenoloxidase and retardation of quality loss of Pacific white shrimp during iced storage. LWT-Food Sci Technol 44(4):924–932

Nooshkam M, Madadlou A (2016) Maillard conjugation of lactulose with potentially bioactive peptides. Food Chem 192:831–836

O'sullivan AM, O'grady MN, O'callaghan YC, Smyth TJ, O'brien NM, Kerry JP (2016) Seaweed extracts as potential functional ingredients in yogurt. Innovative Food Sci Emerg Technol 37:293–299

Oliveira A, Pintado M (2015) Stability of polyphenols and carotenoids in strawberry and peach yoghurt throughout in vitro gastrointestinal digestion. Food Funct 6(5):1611–1619

de Oliveira TLC, de Carvalho SM, de Araújo Soares R, Andrade MA, das Graças Cardoso M, Ramos EM, Piccoli RH (2012) Antioxidant effects of *Satureja montana* L. essential oil on TBARS and color of mortadella-type sausages formulated with different levels of sodium nitrite. LWT Food Sci Technol 45(2):204–212

Olmedo R, Nepote V, Mestrallet MG, Grosso NR (2008) Effect of the essential oil addition on the oxidative stability of fried–salted peanuts. Int J Food Sci Technol 43(11) 1935–1944

Pangestuti R, Kim SK (2017) Bioactive peptide of marine origin for the prevention and treatment of non-communicable diseases. Marine Drugs 15(3):67

del Pilar Garcia-Mendoza M, Espinosa-Pardo FA, Savoire R, Harscoat-Schiavo C, Cansell M, Subra-Paternault P (2021) Improvement of the oxidative stability of camelina oil by enrichment with phospholipid-quercetin formulations. Food Chem 341:128234

Pirzadeh M, Caporaso N, Rauf A, Shariati MA, Yessimbekov Z, Khan MU et al (2020) Pomegranate as a source of bioactive constituents: a review on their characterization, properties and applications. Crit Rev Food Sci Nutr 61:982–999

Pluta-Kubica A, Jamróz E, Kawecka A, Juszczak L, Krzyściak P (2020) Active edible furcellaran/whey protein films with yerba mate and white tea extracts: preparation, characterization and its application to fresh soft rennet-curd cheese. Int J Biol Macromol 155:1307–1316

Qiu X, Jacobsen C, Sørensen ADM (2018) The effect of rosemary (*Rosmarinus officinalis* L.) extract on the oxidative stability of lipids in cow and soy milk enriched with fish oil. Food Chem 263:119–126

Rabiei S, Rezaei M, Nikoo M, Khezri M, Rafieian-Kopai M, Anjomshoaa M (2021) Antioxidant properties of Klunzinger's mullet (*Liza klunzingeri*) protein hydrolysates prepared with enzymatic hydrolysis using a commercial protease and microbial hydrolysis with *Bacillus licheniformis*. Food Sci Technol Int:10820132211005297

Rafiee Z, Barzegar M, Sahari MA, Maherani B (2017) Nanoliposomal carriers for improvement the bioavailability of high–valued phenolic compounds of pistachio green hull extract. Food Chem 220:115–122

Rakariyatham K, Zhou D, Lu T, Yin F, Yu Z, Li D et al (2021) Synergistic effects of longan (*Dimocarpus longan*) peel extracts and food additives on oxidative stability of tuna oil. LWT Food Sci Technol 152:112275

Rashidinejad A, Birch EJ, Everett DW (2016) Antioxidant activity and recovery of green tea catechins in full-fat cheese following gastrointestinal simulated digestion. J Food Compos Anal 48:13–24

Rivero-Pino F, Espejo-Carpio FJ, Guadix EM (2020) Bioactive fish hydrolysates resistance to food processing. LWT Food Sci Technol 117:108670

Rodrigues DP, Calado R, Ameixa OM, Valcarcel J, Vázquez JA (2021) Valorisation of Atlantic codfish (*Gadus morhua*) frames from the cure-salting industry as fish protein hydrolysates with in vitro bioactive properties. LWT Food Sci Technol 149:111840

Roy VC, Chamika WAS, Park JS, Ho TC, Khan F, Kim YM, Chun BS (2021) Preparation of bio-functional surimi gel incorporation of fish oil and green tea extracts: physico-chemical activities, in-vitro digestibility, and bacteriostatic properties. Food Control 130:108402

Rupasinghe HV, Erkan N, Yasmin A (2010) Antioxidant protection of eicosapentaenoic acid and fish oil oxidation by polyphenolic-enriched apple skin extract. J Agric Food Chem 58(2):1233–1239

Rysman T, Utrera M, Morcuende D, Van Royen G, Van Weyenberg S, De Smet S, Estévez M (2016) Apple phenolics as inhibitors of the carbonylation pathway during in vitro metal-catalyzed oxidation of myofibrillar proteins. Food Chem 211:784–790

Sadeghinejad N, Sarteshnizi RA, Gavlighi HA, Barzegar M (2019) Pistachio green hull extract as a natural antioxidant in beef patties: effect on lipid and protein oxidation, color deterioration, and microbial stability during chilled storage. LWT Food Sci Technol 102:393–402

Saini A, Pandey A, Sharma S, Ambedkar YR, Suradkar US, Meena P et al (2019) Effect of rosemary (*Rosmarinus officinalis* L.) leaves extract on quality attributes of chicken powder incorporated fried chicken snacks. J Anim Res 9(4):565–571

Sarteshnizi RA, Sahari MA, Gavlighi HA, Regenstein JM, Nikoo M (2019) Antioxidant activity of Sind sardine hydrolysates with pistachio green hull (PGH) extracts. Food Biosci 27:37–45

Sarteshnizi RA, Sahari MA, Gavlighi HA, Regenstein JM, Nikoo M, Udenigwe CC (2021) Influence of fish protein hydrolysate-pistachio green hull extract interactions on antioxidant activity and inhibition of α-glucosidase, α-amylase, and DPP-IV enzymes. LWT Food Sci Technol 142:111019

Scalone GLL, Lamichhane P, Cucu T, De Kimpe N, De Meulenaer B (2019) Impact of different enzymatic hydrolysates of whey protein on the formation of pyrazines in Maillard model systems. Food Chem 278:533–544

Schilling MW, Pham AJ, Williams JB, Xiong YL, Dhowlaghar N, Tolentino AC, Kin S (2018) Changes in the physiochemical, microbial, and sensory characteristics of fresh pork sausage containing rosemary and green tea extracts during retail display. Meat Sci 143:199–209

Sekhon-Loodu S, Warnakulasuriya SN, Rupasinghe HV, Shahidi F (2013) Antioxidant ability of fractionated apple peel phenolics to inhibit fish oil oxidation. Food Chem 140(1–2):189–196

Shahidi F, Zhong Y (2015) Measurement of antioxidant activity. J Funct Foods 18:757–781
Shang YF, Cao H, Wei CK, Thakur K, Liao AM, Huang JH, Wei ZJ (2020) Effect of sugar types on structural and flavor properties of peony seed derived Maillard reaction products. Food Process Preserv 44(3):e14341
Sharif HR, Williams PA, Sharif MK, Abbas S, Majeed H, Masamba KG, Safdar W, Zhong F (2018) Current progress in the utilization of native and modified legume proteins as emulsifiers and encapsulants – a review. Food Hydrocoll 76 2–16
Shi J, Lei Y, Shen H, Hong H, Yu X, Zhu B, Luo Y (2019) Effect of glazing and rosemary (*Rosmarinus officinalis*) extract on preservation of mud shrimp (*Solenocera melantho*) during frozen storage. Food Chem 272:604–612
Shibata M, Hirotsuka M, Mizutani Y, Takahashi H, Kawada T, Matsumiya K et al (2017) Isolation and characterization of key contributors to the "kokumi" taste in soybean seeds. Biosci Biotechnol Biochem 81(11):2168–2177
Shin SH, Chang Y, Lacroix M, Han J (2017) Control of microbial growth and lipid oxidation on beef product using an apple peel-based edible coating treatment. LWT Food Sci Technol 84:183–188
Siripatrawan U, Noipha S (2012) Active film from chitosan incorporating green tea extract for shelf life extension of pork sausages. Food Hydrocoll 27(1):102–108
Šojić B, Tomović V, Kocić-Tanackov S, Škaljac S, Ikonić P, Džinić N et al (2015) Effect of nutmeg (*Myristica fragrans*) essential oil on the oxidative and microbial stability of cooked sausage during refrigerated storage. Food Control 54:282–286
Steevensz AJ, MacKinnon SL, Hankinson R, Craft C, Connan S, Stengel DB, Melanson JE (2012) Profiling phlorotannins in brown macroalgae by liquid chromatography–high resolution mass spectrometry. Phytochem Anal 23(5):547–553
Stevens JR, Newton RW, Tlusty M, Little DC (2018) The rise of aquaculture by-products: increasing food production, value, and sustainability through strategic utilisation. Mar Policy 90:115–124
Turgut SS, Işıkçı F, Soyer A (2017) Antioxidant activity of pomegranate peel extract on lipid and protein oxidation in beef meatballs during frozen storage. Meat Sci 129:111–119
Vallverdú-Queralt A, Regueiro J, Martínez-Huélamo M, Alvarenga JFR, Leal LN, Lamuela-Raventos RM (2014) A comprehensive study on the phenolic profile of widely used culinary herbs and spices: rosemary, thyme, oregano, cinnamon, cumin and bay. Food Chem 154:299–307
Velasco J, Dobarganes C, Márquez-Ruiz G (2010) Oxidative rancidity in foods and food quality. In: Chemical deterioration and physical instability of food and beverages. Woodhead Publishing, Cambridge, pp 3–32
Verbeyst L, Oey I, Van der Plancken I, Hendrickx M, Van Loey A (2010) Kinetic study on the thermal and pressure degradation of anthocyanins in strawberries. Food Chem 123(2):269–274
Vieira SA, Zhang G, Decker EA (2017) Biological implications of lipid oxidation products. J Am Oil Chem Soc 94(3):339–351
Vinson JA, Su X, Zubik L, Bose P (2001) Phenol antioxidant quantity and quality in foods: fruits. J Agric Food Chem 49(11):5315–5321
Von Staszewski M, Ruiz-Henestrosa VMP, Pilosof AM (2014) Green tea polyphenols-β-lactoglobulin nanocomplexes: interfacial behavior, emulsification and oxidation stability of fish oil. Food Hydrocoll 35:505–511
Walayat N, Xiong Z, Xiong H, Moreno HM, Niaz N, Ahmad MN et al (2020) Cryoprotective effect of egg white proteins and xylooligosaccharides mixture on oxidative and structural changes in myofibrillar proteins of *Culter alburnus* during frozen storage. Int J Biol Macromol 158:865–874
Walayat N, Wang X, Liu J, Nawaz A, Zhang Z, Khalifa I et al (2022) Kappa-carrageenan as an effective cryoprotectant on water mobility and functional properties of grass carp myofibrillar protein gel during frozen storage. LWT Food Sci Technol 154:112675

Wang T, Jonsdottir R, Olafsdottir G, Kristinsson HG (2014) Antioxidant properties of marine macroalgae. In: Antioxidants and functional components in aquatic foods. John Wiley & Sons, Ltd, Chichester, p 283

Wang B, Li F, Pan N, Kong B, Xia X (2021) Effect of ice structuring protein on the quality of quick-frozen patties subjected to multiple freeze-thaw cycles. Meat Sci 172:108335

Wolfe KL, Liu RH (2003) Apple peels as a value-added food ingredient. J Agric Food Chem 51(6):1676–1683

Wu D, Li M, Ding J, Zheng J, Zhu B, Lin S (2020) Structure-activity relationship and pathway of antioxidant shrimp peptides in a PC12 cell model. J Funct Foods 70:103978

Wu R, Huang J, Huan R, Chen L, Yi C, Liu D et al (2021) New insights into the structure-activity relationships of antioxidative peptide PMRGGGGYHY. Food Chem 337:127678

Xiao J, Mao F, Yang F, Zhao Y, Zhang C, Yamamoto K (2011) Interaction of dietary polyphenols with bovine milk proteins: molecular structure–affinity relationship and influencing bioactivity aspects. Mol Nutr Food Res 55(11):1637–1645

Xiong Z, Shi T, Zhang W, Kong Y, Yuan L, Gao R (2021) Improvement of gel properties of low salt surimi using low-dose L-arginine combined with oxidized caffeic acid. LWT Food Sci Technol 145:111303

Xu M, Wu Y, Hou GG, Du X (2019) Evaluation of different tea extracts on dough, textural, and functional properties of dry Chinese white salted noodle. LWT Food Sci Technol 101:456–462

Yu K, Zhou HM, Zhu KX, Guo XN, Peng W (2019) Increasing the physicochemical stability of stored green tea noodles: analysis of the quality and chemical components. Food Chem 278:333–341

Zha F, Wei B, Chen S, Dong S, Zeng M, Liu Z (2015) The Maillard reaction of a shrimp by-product protein hydrolysate: chemical changes and inhibiting effects of reactive oxygen species in human HepG2 cells. Food Funct 6(6):1919–1927

Zhang L, Li Q, Hong H, Luo Y (2020) Prevention of protein oxidation and enhancement of gel properties of silver carp (*Hypophthalmichthys molitrix*) surimi by addition of protein hydrolysates derived from surimi processing by-products. Food Chem 316:126343

Zhang X, Noisa P, Yongsawatdigul J (2021) Identification and characterization of tilapia antioxidant peptides that protect AAPH-induced HepG2 cell oxidative stress. J Funct Foods 86:104662

Zheng L, Yu H, Wei H, Xing Q, Zou Y, Zhou Y, Peng J (2018) Antioxidative peptides of hydrolysate prepared from fish skin gelatin using ginger protease activate antioxidant response element-mediated gene transcription in IPEC-J2 cells. J Funct Foods 51:104–112

Zheng M, Liu X, Chuai P, Jiang Z, Zhu Y, Zhang B et al (2021) Effects of crude fucoidan on physicochemical properties, antioxidation and bacteriostasis of surimi products. Food Control 122:107806

Zhong R, Lu X, Zhong J, Chen L, Cheng W, Liang P (2021) Influence of Maillard reaction in volatile flavor compounds of Blue Round Scad (*Decapterus maruadsi*) enzymatic hydrolysate. J Aqua Food Prod Technol 30(5):615–629

Chapter 4
Clean Label Foods with Reduced Fat Content

Amparo Quiles, Empar Llorca, Gemma Moraga, and Isabel Hernando

4.1 Foods with Reduced Fat Content

Although lipids are an essential component of our daily food intake, their overconsumption can lead to serious medical consequences (Norton et al. 2013). The excess of saturated fats in the diet is related to health problems like the increase of cardiovascular diseases, type II diabetes, metabolic syndrome, and obesity. Reducing the fat content of foods is frequently proposed as a method for reducing the consumer's energy intake (Lett et al. 2016) and remains one of the most effective approaches to minimize the impact of obesity and its health-related issues (Norton et al. 2013).

However, fats are responsible for many high-quality attributes in the food we consume, mainly related to texture, aroma, taste, and stability. This positive fundamental contribution of fats to sensory properties and technological food manufacture is the key reason why the replacement of certain fats (e.g., palm fat), or the reduction of saturated fats, is so complicated. The challenge for the food industry thus lies in designing healthier foodstuffs, with a focus on reducing fat without compromising the organoleptic properties of the food. Foods with reduced fat levels must, therefore, be carefully formulated to ensure they maintain their desirable physicochemical, sensory, and nutritional properties; otherwise, they will not be acceptable to consumers (Espinal-Ruiz et al. 2014).

Products labeled fat free and low-fat must contain <0.5 g and ≤3 g of fat per serving, respectively. Reduced or less fat may be used on product labels that contain 25% less fat than regular (full fat) products. The fat labeling claims do not provide any indication of the caloric content of the food item. However, products containing one third fewer calories or one half the fat of the reference food may be labeled as light. Furthermore, if ≥50% of calories in food are derived from fat, the fat content

A. Quiles (✉) · E. Llorca · G. Moraga · I. Hernando
Departamento de Tecnología de Alimentos, Universitat Politècnica de València, Camino de Vera, Valencia, Spain
e-mail: mquichu@tal.upv.es

C. M. Galanakis (ed.), *The Age of Clean Label Foods*,
https://doi.org/10.1007/978-3-030-96698-0_4

of the reduced-fat version must be reduced by 50%. The terms calorie-free and low calorie can only be used on products with <5 cal and ≤40 cal per serving, respectively, and reduced or fewer calories can only be used on products that have ≤25% of the calories in the regular product (Wylie-Rosett 2002).

Trends of healthiness and sustainability have triggered consumers to consider which components are in the food products they eat every day (Euromonitor International 2016). Consumers in industrialized countries are nowadays much more interested in information on production methods than they were 50 years ago. Some production methods are perceived as less "natural" (e.g., conventional agriculture), while some food components are "unhealthy" and "unfamiliar" (e.g., artificial additives). This trend often referred to as "clean label," which has driven the food industry to communicate whether a particular ingredient or additive is present or if the food has been produced using a more "natural" production method (e.g., organic agriculture) (Asioli et al. 2017). A recent trend in food products has emerged, which is often summarized under the umbrella of the so-called "clean label" (Cheung et al. 2016; Varela and Fiszman 2013) and has been taken up by a multitude of food industry stakeholders (Osborn 2015).

To date, there is no established, objective, or standard definition of what a "clean label" is, but there are several definitions or interpretations, often provided by market trend reports but not backed up by consumer behavior research or theory (Osborn 2015). Choosing healthy foods is vital to work towards a more homogenous understanding and application of the term "clean label" (Asioli et al. 2017). Thus, clean label products are associated with (1) all-natural: no artificial additives or 'chemical-sounding' ingredients; (2) simplicity: minimally processed and comprising a short list of recognizable ingredients; and (3) transparency: sustainably sourced ingredients.

4.2 Clean Label Ingredients to Replace Fat in Reduced-Fat Foods

Replacement of the fat components in foods is often difficult because they have multifunctional roles in foods as they influence chemical, physical, and sensory attributes as well as their processing characteristics. Reducing the fat content causes quality losses that decrease the acceptance of foods (Peng and Yao 2017). Depending on the food type, where fat is being removed, numerous fat replacers can be used depending on the physical attributes that need to be replaced (Rogers 2019). For example, in emulsions, they provide enhanced viscosity, mouthfeel, and body to the food, whereas in solid fat matrices, they provide an elastic network essential for the food structure (Rogers 2019). Fat replacers are categorized into two groups, **fat substitutes** and **fat mimetics** (O'Connor and O'Brien 2016).

Fat substitutes are lipid-based molecules that possess the physical and functional characteristics of conventional fat molecules like triglycerides. They can directly replace traditional fat molecules in foods on a weight-for-weight basis. Fat substitutes are typically synthetic molecules which provide no energy calories, or structured lipid molecules which contain reduced energy calories. They yield <9 kcal/g, which can zero if they are not absorbed (Shaltout and Youssef 2007). Therefore, while fat substitutes do not allow the design of reduced-fat foods, they can formulate diets with low-fat digestibility and, thus, with reduced calorie intake.

Fat substitutes can successfully maintain the palatability of foods as they can reproduce the texture and mouthfeel of fat. They are mostly heat stable and suitable for high-temperature cooking and frying. Furthermore, they may not reproduce the taste properties of fat, because fat itself provides a flavor to foods and is a carrier of other fat-soluble flavor compounds in foods (O'Connor and O'Brien 2016).

Fat mimetics are protein or carbohydrate-based ingredients, and they imitate the sensory or physical properties of triglycerides (Agyei-Amponsah et al. 2019). However, they cannot replace fat molecules in food on a weight-for-weight basis. Their energy contribution to the diet ranges from 0 to 4 kcal/g. Fat mimetics are not suitable for high-temperature process like frying, as they are susceptible to denaturation or caramelization. In addition, they are mostly polar and water-soluble compounds; thus, they cannot replace the nonpolar functional characteristics of fat, such as lipid-soluble flavor carrying capacity. However, their polar nature facilitates water-binding, which helps generate a sense of creaminess and lubricity in foods like those found in full-fat products (O'Connor and O'Brien 2016).

Many components used to produce fat replacers are entirely natural, such as carbohydrates (e.g., starch and fibers), proteins, partial glycerides and plant waxes or beeswax. Other sources, such as locust bean gum and carrageenan, are sourced from seeds and seaweeds (natural materials). However, it is essential to remember that not all additives are harmful, and "natural" is not always right, even in the eyes of the consumer (Osborn 2015). Some chemical-sounding ingredients, such as ethylcellulose and hydroxypropyl methylcellulose, are derived from cellulose. Because it is a naturally occurring plant fiber and one of the most abundant materials in nature, cellulose is a renewable resource that has a long history of safe use. In fact, contrary to the message delivered by popular media, the whole broad class of hydrocolloids is ideal materials for clean label product development because of their natural and familiar origins (Patel et al. 2020).

4.2.1 Fat Substitutes

Only lipid-based fat substitutes can mimic all the attributes of natural fat (Kanjilal et al. 1999). Therefore, an excellent fat substitute must have the taste, mouthfeel, stability for frying and baking, can reduce calories, and provide health benefits to the consumer (Akoh 2007). Most fat substitutes are emulsifiers or structured lipids. They are neither hydrolyzed nor absorbed by the body in the same manner as healthy fat and thus contribute substantially fewer calories (Shaltout and Youssef 2007). Emulsifiers, like Olestra and Sorbestrin, are fatty acids esters with chain lengths from C12 to C22, polyvalent alcohols like glycerol, sorbitol/sorbitan, and sucrose or organic acids like lactic acid. Structured lipids, like Salatrim and Caprenin, are triacylglycerols that have been modified via interesterification on their glycerol backbone to change the fatty acid composition and their positional distribution. Those are either chemically synthesized or derived from conventional fats and oils by enzymatic modification (Akoh 2007; Ognean et al. 2006), thus they cannot be considered as a clean label ingredient.

Although numerous fat substitutes exist to create low-fat foods, few of them can be considered "clean labels." The principal objective of using "natural" or "clean label" fat substitutes is to formulate healthy foods and not to design low-calorie foods. The excess of saturated fats in the diet is related to health problems, such as the increase of cardiovascular diseases, type II diabetes, metabolic syndrome, and obesity. Various methods have been developed throughout history to obtain solid fats from liquid oils, with the majority based on economic profitability. Hydrogenation has been one of the chemical methods most used to transform liquid oils into solid fats, to achieve the required plasticity. The food industry is interested in alternatives to saturated fats and partially/fully hydrogenated fats due to being unnatural and chemical-sounding; plus the association of the word "hydrogenated" with "trans-fats" (Osborn 2015).

Including vegetable oils as fat substitutes have been related to an essential increase of unsaturated fatty acids and decreasing low-density cholesterol (Álvarez et al. 2011). Thus, Domínguez et al. (2016) showed that the replacement of meat fat using olive oil gave food with a nutritional value higher than traditional food. Olive oil provides prominent levels of monounsaturated fatty acids which contain natural antioxidants like α-tocopherol; and reduce cholesterol levels without considerably affecting the physicochemical, color, or texture parameters. Furthermore, Álvarez et al. (2011) showed that using canola and olive oils as healthy natural replacement ingredients in Frankfurt sausages were nutritionally satisfactory and gave a more robust product. Rodríguez-Carpena et al. (2012) studied the impact of using avocado, sunflower, and olive oils as fat replacers for meat fats; they concluded that these vegetable oils lead to products with enhanced nutritional properties. Moreover, the replacement of saturated fats using linseed oil is a viable alternative to offer healthy products to consumers without affecting the sensory and technological features. Câmara and Pollonio (2015) showed that the linseed oil replacement level should not be so high that the attributes, appreciated by the consumers, were not adversely affected. Lee (2017) suggested using shea butter as a low-calorie fat, due to its low hydrolysis rate, because of the stearic acid at the sn-1,3 positions of triacylglycerol molecules. Therefore, using shea butter could be promising when designing clean label foods.

Diacylglycerol (DAG) is formed because enzymatic hydrolysis of triacylglycerol (TAG) present in animal tissues; they are also generated in seed oils through the action of plant lipases. Several vegetable oils inherently contain DAG, ranging from 0.8% to 5.8% (Yanai et al. 2007). Furthermore, recent studies have revealed DAG-rich oils help reduce plasma triglyceride levels, body weight, waist circumference, and visceral and subcutaneous fat; and have been associated with many other health benefits (Prabhavathi Devi et al. 2018). Thus, DAG has been reported to demonstrate an anti-obesity effect for its ability to suppress abdominal and visceral fat accumulation. DAG can be potentially used as a saturated fat replacer in food processing, entirely or partially (Lee et al. 2020). In addition, DAG has become a popular topic as it is considered as a novel functional lipid. Prabhavathi Devi et al. (2018) prepared a model 1,3-DAG-rich oil using a mixture of vegetable oils, specifically rice bran oil and sunflower oil, and they studied the impact of consumption in rats. The DAG-rich oil exhibited a 30% lower calorific value than TAG-rich oil and reduced LDL cholesterol besides lowering triglyceride levels in serum and liver of experimental rats. Therefore, DAG-rich oil can be exploited as a reduced-calorie fat with anti-obesity and hypolipidemic potential for promoting cardiovascular health.

Waxes are derived from natural sources; thus, they may be acceptable for use in "natural" food formulations; this could help solve the universal problem of oil separation and textural inconsistency in these products while maintaining a clean label (Winkler-Moser et al. 2019). Waxes are mixtures of long-chain high-melting lipid compounds such as wax esters, hydrocarbons, or long-chain alcohols (Doan et al. 2017) derived from animals (beeswax) or plants. Some waxes, such as rice bran wax or sunflower wax, are byproducts of oil refining. They are poorly absorbed in the gastrointestinal tract due to their hydrophobicity and limited hydrolysis of the wax esters (Hargrove et al. 2004). Accordingly, waxes are a good option as low-calorie ingredients to design clean label foods.

Lipid modification of foods by substitution of meat fat with fish oil has been shown an excellent strategy to improve their nutritional quality (Domínguez et al. 2017). The abundance of long-chained ω-3 fatty acids in fish oil, such as eicosapentaenoic acid (EPA) and docosahexaenoic acid (DHA), also have a positive impact on human health (Pourashouri et al. 2014). According to the international recommendations, EPA and DHA intake should be between 250 mg/day (EFSA 2009) and 1 g/day (Kris-Etherton et al. 2012). Thus, it would be interesting to develop strategies to allow using fish oil in low-fat/calorie food formulations. Delgado-Pando et al. (2011) observed that using oil fish with different combinations of vegetable oils (olive and linseed) and konjac gel, like saturated fat replacements, allowed the manufacture of low-fat products with a healthier lipid profile.

4.2.2 *Fat Mimetics*

The most common fat mimetics belong to the groups of polysaccharide and protein hydrocolloids. Hydrocolloids are a group of water-soluble naturally occurring polymers found abundant in nature. Their essential functions in food products are to control texture and organoleptic properties by enhancing the viscosity and gel characteristics (Kilcast 2004).

4.2.2.1 Protein-Based Mimetics

Proteins are widely used as ingredients in foods for their nutritional value, physiological, and technological functionality. The technical features of proteins include thickening, foaming, emulsifying, stabilizing, and jellifying. Protein-based fat mimetics are natural and clean label functional food ingredients. They are generally regarded as natural as they are obtained by minimal processing from various vegetable and animal sources. Furthermore, they need not be identified by an E number on the label of the food or beverage, except for gelatin (E number: E441) (Alting and Van De Velde 2012).

A wide range of protein sources is available: leguminous proteins (e.g., soy and pea), cereal proteins (e.g., wheat, barley, and rice), oilseed proteins (e.g., sunflower and rapeseed), dairy proteins (e.g., sodium caseinate and whey proteins), meat proteins (e.g., gelatin, collagen-rich derivates, and blood plasma), and egg proteins (e.g., albumin). According to Hall (2012), the most commonly used are soya concentrate (70% protein), soya isolate (90% protein), pea protein (80–90% protein), wheat protein (80% protein), dairy proteins (whey 35% or 80% and Sodium caseinate 90%), and pork (80% protein).

Most proteins are small molecules (molecular weight of 10–50 kDa), which reduces the impact on the viscosity. To apply proteins as a general source of viscosifiers, their functionality must be tailored to increase the volume occupied by the protein molecules. Structuring globular or unstructured proteins into large fibrillar structures, using heat treatment, is a natural process that maintains the clean label status of the proteins while increasing their functionality (Alting and Van De Velde 2012). The proteins are heat-sheared to form microscopic coagulated particles that mimic the mouthfeel and texture of fat; they may also be processed to modify their water-binding and emulsification properties. The heating characteristics can additionally be resolved by incorporating emulsifying agents and more nonpolar or lipophilic flavoring compounds (O'Sullivan 2017).

4.2.2.2 Carbohydrate-Based Mimetics

Carbohydrate-based fat replacers mimic fat particles in a particulate form or by contributing other textural characteristics that would be found in full fatty foods. They function primarily to give body, increase viscosity, and provide a creamy, slippery mouthfeel like that of fat. Carbohydrate-based fat replacers also can stabilize substantial quantities of water in a gel-like matrix, resulting in lubricating and flow properties like those of fat (Agyei-Amponsah et al. 2019). They perform a significant replacement role but are not typically suitable for frying applications (O'Connor and O'Brien 2016). Clean label carbohydrate-based fat mimetics include digestible starches to non-digestible plant fibers.

4.2.2.2.1 Starch-Based

Starch is a form of energy reserve for vegetables, usually stored in the grain or seed, but also present in roots, tubers, and tree trunks (stems). It is in corn, wheat, rice, oats, potatoes, tapioca (cassava), and sago trees, with different amylose and amylopectin content. Each starch source can be identified microscopically by the shape, diameter range, and coloration of its granules when stained by iodine. For example, rice starch offers some "fatty behavior," such as a soft gel structure and a fat-like texture with a creamy mouthfeel. It has rheological properties like those of fat due to its small particle size, the smallest of all commonly used starches (2–8 μm) (Joly and Anderstain 2009).

Native starches are clean label ingredients, whereas chemically modified starches carry an E number designation (E14xx) and are not perceived as natural. The term native applies to extracted non-processed starches, rather than natural. In terms of starch, natural would mean the use in foods of the whole vegetable or botanic plant from which the starch is derived and reliance on the starch being extracted (Mcdonagh 2012). From an industrial perspective, native carbohydrate-based fat replacers may have the advantages associated with the clean label; however, their functional suitability could be a challenge with highly specified food systems (Peng and Yao 2017). Pre-gelatinization of starch is a physical method with high desirability because no chemical material is used in its production process. After pregelatinization, starch drying prevents amylose and amylopectin crystallization, leading to a glassy structure with high and rapid hydration capacity. This starch absorbs both warm and cold water rapidly, with no additional heating needed to develop viscosity (Rahmati et al. 2014).

Clean label starches have been available in the marketplace for a long time. Consumers require clean label starches with the same function as modified starches but cost the same. Techniques that starch companies are using include physical modification through agglomeration and granulation, selective crop breeding (e.g., high amylopectin waxy potato and wheat starches), and thermal treatment of native flours (Mcdonagh 2012).

Starch-based fat replacers give the food fat properties such as mouthfeel, creaminess, appearance, and thickness. However, they may mask the food flavor in which they are incorporated. They do not carry lipid-soluble flavor compounds, so they may require emulsifiers to incorporate lipophilic flavors into foods (Fellows 2017). Their freeze/thaw-, heat-, acid-, and shear-stabilities may not be sufficient for some applications. However, they perform well as fat mimetics in high moisture foods such as salad dressings, sauces, baked goods, frostings, and sausages (O Connor and O'Brien 2016).

4.2.2.2.2 Fibers

Dietary fiber comprises complex substances which are not hydrolyzed by endogenous enzymes in the human small intestine, showing beneficial physiological effects (Cruz-Requena et al. 2019). Based on water solubility, dietary fiber can be classified as soluble and insoluble, with different effects on the gastrointestinal tract and different beneficial effects on human organisms. Soluble dietary fibers (β-glucan, psyllium, pectins, guar gum, arabinoxylans, and inulin) have been associated with the reduction of cholesterol levels, a decrease of gastric emptying, small intestine transit time, prebiotic effect, and fecal bulk effect. While insoluble dietary fibers (cellulose, hemicellulose, chitosan, and lignin) have been associated with the reduction of fermentability and stool bulk-forming (Surampudi et al. 2016). Dietary fibers are becoming one of the most critical food ingredients for their irreplaceable roles of techno-functional and bio-functional purposes in food products (Cui et al. 2013).

Critical dietary fiber physicochemical properties include solubility, viscosity, water holding capacity, oil-binding ability, bulking, and fermentability (Mudgil and Barak 2019). The water holding capacity is affected by factors such as the source of the dietary fiber (chemical composition and microstructure), the food system in which the dietary fiber is incorporated (ionic nature, pH, concentration, and ingredients with high water-binding capacity like starch and sugars), and the processing conditions. Therefore, different dietary fiber technological ingredients can be obtained by altering these factors. Dietary fiber's chemical composition and porous structure also play a role in oil-binding ability. This is also significant in the formulation of low-fat products, as the fiber molecules bind the low amount of fat present in the formulation, improving the texture of the final product (Mudgil and Barak 2019).

A promising way for the food industry to supply healthier foods could, therefore, be to replace fat with dietary fiber as a clean label ingredient. Soluble and insoluble indigestible vegetable fibers have multifunctional properties due to their ability to absorb water, adding a smooth texture as a fat mimetic. Combinations of fibers and other ingredients with unique and complementary properties may take advantage of synergistic effects, in terms of water-binding, creaminess, and structure (Herbert et al. 1999).

Relevant, sustainable sources of dietary fiber include cereals (e.g., oat, rye, barley, wheat, and rice), pseudocereals (e.g., quinoa), legumes (e.g., lentils, soy, peas, lupins, beans, and sprouts), fruits (e.g., citrus, apple, kiwi fruit, and prunes), and other vegetables (e.g., carrot, potato, bamboo, sugar cane, psyllium, sugar beet, and chicory root).

Regarding cereals, wheat bran, which is an excellent source of insoluble dietary fiber, has been used to replace a fat part in the production of different products obtaining very stable emulsions throughout the product processing and storage (Kim and Paik 2012; Talukder 2015; Özboy Özbas and Ardiç 2016). Bodner and Sieg (2009) compared wheat fiber made from wheat straw to fully extract oat fiber to find similar characteristics. They can be used interchangeably with little formula alteration because wheat fiber is bright and with little taste. There is a great interest in increasing the consumption of oat-based products that contain both soluble dietary fiber and insoluble dietary fiber (Talukder 2015). It is essential to differentiate between milled oat bran, refined insoluble oat fiber extracted from oat hulls, and purified β-glucan (Bodner and Sieg 2009). β-glucan is commonly found in oats, barley, mushrooms, and some microorganisms, whereas in cereals, it is generally present in the bran portion of the grains. β-glucan obtained from oats is reported to have a higher molecular weight than barley β-glucan (Lazaridou and Biliaderis 2004; Kouzounis et al. 2017). β-Glucan can hydrate well in water and forms a very viscous solution. In addition, barley is another excellent source of insoluble/soluble dietary fiber, including β-glucan, mainly found in the cell walls of the aleurone layer and endosperm of barley kernels (Andersson et al. 2004). The β-glucan content of barley ranges from 2% to 11%, which is higher than the β-glucan content of oats (2.2–7.8%) and wheat (0.4–1.4%). The health benefits associated with consuming β-glucan-rich foods are lowered blood glucose levels and lowering of serum total and LDL cholesterol levels. The beneficial health properties of β-glucan result from an increase in viscosity of intestinal content, which depends on the molecular weight and solubility of β-glucan (Mudgil and Barak 2019).

The legume family includes dry beans, broad beans, dry peas, chickpeas, lentils, soybeans, lupins, mung beans, lotus sprouts, alfalfa, green beans, peas, and peanuts. They are rich in protein (17–20% dry weight in peas and beans, 38–40% in soybeans and lupins) and dietary fiber (5–37% dry weight) (Kouris-Blazos and Belski 2016). The water absorption capability of soybean and pea fiber is lower than oat and wheat fiber (Kim and Paik 2012).

The functional properties of fruit and vegetable fibers are strongly influenced by the source, the botanical part used for extraction, physical status, and production technology (Balestra et al. 2019). Several authors have studied the recovery of dietary fibers from fruit and vegetable byproducts, including fruit peel and pomaces, such as citrus (Figuerola et al. 2005; Ubando-Rivera et al. 2005), Apple (Lim and Rabeta 2013), pear, peach (Grigelmo-Miguel and Martín-Belloso 1999), banana (Happi Emaga et al. 2007), mango (Ajila et al. 2008; Koubala et al. 2008; Martínez et al. 2012), guava (Jiménez-Escrig et al. 2001), grapes (Torres et al. 2002; Llobera and Cañellas 2007; Sánchez-Alonso et al. 2007), pineapple (Martínez et al. 2012), pomegranate, olive, artichoke (Colantuono et al. 2017), carrots (Chantaro et al. 2008), and coffee (Murthy and Naidu 2012). Fibers obtained from fruits and vegetable byproducts, which would otherwise be waste, can help manufactures meet consumer's demand for clean ingredient labels while reducing food waste. Thus, they can be used as clean label functional ingredients in low-fat food formulations (Jiménez-Colmenero and Delgado-Pando 2013).

Flax seeds are a potential source for functional food formulations. They contain approximately 28% fiber, of which a third is soluble. This ingredient can replace fat and being a natural product, allows clean labeling of the food, avoiding E-numbers (Eslava-Zomeño et al. 2016).

Many soluble fibers like pectins form gels and can contribute to the increased thickness or viscosity of products (Kim and Paik 2012). Their capacity to form a gel and the characteristics of that gel will depend on several factors, including concentration, temperature, presence of specific ions, and pH (Jiménez-Colmenero and Delgado-Pando 2013). Inulin can also structure low-fat foods, its property to act as a fat mimetic or fat replacer is based on its capacity to form microcrystals, which interact with each other, therefore forming small aggregates that occlude a great amount of water and ultimately may agglomerate into a gel network (Bayarri et al. 2011).

4.2.2.2.3 Gums

Gums and other hydrocolloids such as xanthan, guar and locust bean gums, gum arabic, and carrageenan, can be used as fat replacers either on their own, in a combination, or with other fat replacers. At a low concentration (0.1–0.5%), gums increase the viscosity of food systems while function as stabilizers and gelling agents. Gum-based fat mimetics tend not to influence the flavor profile of the product. Their applications include use in dairy products, salad dressings, soups, sauces, ice-cream, icings and glazes, and baked foods (O'Connor and O'Brien 2016). Although gums are extracted from natural products, some of them carry an E number, and hence, do not clearly fit in within the clean label category. Examples are carrageenan extracted from red seaweeds (E407) or xanthan gum obtained by fermentation (E415) (Alting and Van De Velde 2012).

4.2.2.2.4 Whole Foods and Purees

Recently, many products have used whole foods such as fruits and vegetables, legumes, or cereal-based ingredients as fat replacers. Entire foods are complete or partial food matrices that can be included in a food product, as natural clean label fat replacers. These foods are typically successful due to their highly creamy texture when mashed or processed. Foods such as avocado can achieve this due to its oil composition, banana for its high starch content, and legumes for their high starch and protein contents (Colla et al. 2018). Puree of plums, pears, and apples can also perform many of the fat functions due to their pectin, fiber, and sugar content. In particular, the complexes of fiber and pectin provide texture and body. The fruit sugars also provide additional solids and water-binding; in addition, health benefits may include antioxidant activity. These replacers may partially or entirely replace fat in cookies, muffins, cakes, and other bakery mixes (Shaltout and Youssef 2007).

4.2.2.3 Emulsions and Oleogels

In most foods, lipids are found in emulsions, such as in dairy products, chocolates, sauces, mayonnaise, soups, spreads, and dressings (Gallier and Singh 2012). To stabilize food emulsions, emulsifying and stabilizing agents are used. Emulsifiers are molecules with surface activity, which can rapidly adsorb at the oil/water interface during homogenization, to prevent the future aggregation of oil droplets. Stabilizing agents provide stability to the emulsion by increasing the viscosity of the continuous phase (Traynor et al. 2013; Li and Nie 2016).

However, manipulating the aggregation state of fat droplets in emulsions may provide a useful means of controlling their textural and mouthfeel properties while creating reduced-fat products. The fat droplets aggregation forms a three-dimensional network that can maintain high viscosity (Mao and Julian Mcclements 2012). This can be achieved by altering the droplet interface (surface hydrophobicity, thickness, or charge), solution conditions (pH and ionic strength), system composition (biopolymer type and concentration), and environmental conditions (temperature) (Damodaran 2005; Dickinson 2010). Wu and McClements (2015) used a negatively charged xanthan gum for its ability to induce aggregation of positively charged protein-coated fat droplets. This approach may be useful in developing cost-effective and "clean label" methods for formulating reduced-fat products with desirable quality attributes. Moreover, xanthan gum is already widely used in reduced-fat food products for its ability to increase the viscosity of aqueous solutions (Samavati et al. 2012; Lee et al. 2013).

Replacing fat with low oil content emulsions, like oil-in-water emulsions, can also be an attractive solution for designing low-calorie foods. In addition, using natural stabilizers and emulsifiers in the formulation of these emulsions allows the development of "clean label" foods. Thus, several authors have studied using "natural" polymers as emulsifiers and surfactants. Kasprzak et al. (2018) developed different oil-in-water emulsions and concluded that at least three non-chemically modified emulsifiers (waxy rice starch, non-waxy rice starch, and waxy maize starch) could stabilize oil-in-water emulsions. Zarai et al. (2016) showed that gelatin is another natural emulsifier that can be used when preparing emulsions and in the control of emulsion shelf life. In contrast, Chung et al. (2017) used a quillaja saponin as a natural emulsifier to form stable, clean label emulsions. Still, the research world continues to study using "natural" stabilizers and emulsifiers capable of enabling clean food labeling.

An alternative approach that may maintain consumer satisfaction is to attempt using strategies associated with controlling fat digestion (Espinal-Ruiz et al. 2014). An understanding of the fate of lipids within the human gastrointestinal tract is essential for designing functional foods to control the rate and extent of lipid digestion and absorption (Mun et al. 2017). Fat digestion is an interfacial process dependent on the adsorption of lipolytic enzymes to the surface of emulsified fat droplets (Golding and Wooster 2010). Foods containing lipids may either be ingested as oil-in-water emulsions, or they may be converted into oil-in-water emulsions in the mouth, stomach, and small intestine through the mechanical stresses they experience during digestion. This process is also influenced by various endogenous and dietary surface-active agents (Mun et al. 2017). In general, the human body has excess capacity for fat digestion, and the rate is controlled by the ability of lipase to bind to emulsion interfaces, which is controlled by the emulsion size and interfacial composition. Thus, over recent years particular research has focused on formulating foods, including emulsions with low-fat digestibility ingredients, on reducing or postpone fat digestion while maintaining sensory attributes (Norton et al. 2013). These emulsions can be prepared by manipulating the molecular characteristics of the lipids, the size of fat globules, the interfacial properties, and the structure and composition of the matrix (Borreani et al. 2020). Furthermore, Borreani et al. (2020) studied the in vitro intestinal digestion of oil-in-water emulsions stabilized with different molecules (tween, lecithin, xanthan gum, and methylcellulose). They showed the different lipolysis kinetics were not primarily related to differences in the droplet size/surface area available for the action of lipase, nor the viscosity of the emulsions. The differences found in the lipolysis kinetics were attributed to the nature and location of each molecule type in the oil-in-water emulsions. Various kinds of emulsifiers offer different resistance to bile salts displacement. Once bile salts are adsorbed at the oil-in-water interface, they aid lipase adsorption, thus enhancing lipid digestion. Nevertheless, interactions take place between lecithin or methylcellulose molecules and intestinal components (bile salts and lipase), preventing them from reaching the oil-in-water interface and/or inhibiting sufficient removal of lipolysis products; thus delays lipid digestion. Therefore, emulsions formulated with lecithin could be considered as a "clean label" fat replacer. Borreani et al. (2017a) also studied oil-in-water emulsions stabilized by cellulose ethers, showing that cellulose ethers delay lipid digestion compared to a conventional food emulsifier (calcium caseinate). Using these cellulose ethers made it possible to obtain oil-in-water emulsions with high consistency even during lipid digestion. Therefore, it was concluded that controlling the emulsion's structure and their digestion behavior could give emulsions that decrease lipid digestion, which could be used for weight management. Although cellulose ethers are obtained chemically, it would be interesting to study possible clean label cellulose derivatives. As it is a naturally occurring plant fiber and one of the most abundant materials in nature, cellulose is a renewable resource that has a long history of safe use. In fact, contrary to the message delivered by popular media, the whole broad class of hydrocolloids is ideal materials for clean label product development because of their natural and familiar origins. They are obtained from natural sources such as tree exudates, seeds, plant fragments, seaweed extracts, and microbial fermentation. Hydrocolloids are among the most used food ingredients, and most are used in their chemically unprocessed forms. Due to their multifunctional nature, hydrocolloids can contribute to the "clean label" image of reformulated products by simplifying ingredient lists (Patel et al. 2020).

Oleogels are structured materials prepared by gelation of liquid oil using oleogelators such as vegetable waxes, mono- and diglycerides, fatty acid alcohols or esters, phospholipids, and phytosterols. They can be used in the formulation of cakes, cookies, meat products, chocolate, and ice-cream to eliminate trans-fats, while also reducing the concentration of saturated fats and maintaining unsaturated oil concentration (Sedaghat Doost et al. 2020). Besides, direct oleogelation, in which the hydrophobic oleogelator is dispersed in the liquid oil by applying heat to their melting point followed by cooling, there is an indirect way for hydrophilic oleogelators. For example, the emulsion-template indirect method is a multi-step process where the oil droplets are packed into an interfacial coating of emulsifiers with a subsequent drying and mild shearing of the dried products to prepare the oleogel. Although using oleogels has more meaning for developing healthy products than in low-fat products, the oil foams can create healthier and low-fat food products. Edible oil foams are currently prepared by heating a solution containing a high-melting component in vegetable oil, cooling this solution to form an oleogel, and subsequently whipping it to obtain an air-in-oil system (Heymans et al. 2017). They can reduce the caloric content and the amounts of saturated and trans-fatty acids of food products. Incorporating air in a continuous fat phase will cause a product with reduced fat content on a volume basis. Aerated chocolate, for example, has a significantly reduced caloric content (per unit volume) without compromising the sensorial properties (Adred et al. 2008). Natural unsaturated oils like olive oil and corn oil were proposed by Klonoff (2007) as being one of the healthiest alternatives for replacing hydrogenated fats containing higher amounts of trans-fats. Anticipating the "clean label" trend, Binks and Marinopoulos (2017) demonstrated the ability to produce self-foaming oils produced by whipping edible vegetable oils at a specific temperature without additives. Notably, the solid fat content for optimal foaming (e.g., 20–30%) of these oils was higher than other edible oil foam systems (Mishima et al. 2016; Binks and Marinopoulos 2017).

Table 4.1 shows, as a summary, the clean label ingredients that can be used to replace fat in reduced-fat foods.

4.3 Clean Label Reduced-Fat Foods

In this section, the application of fat replacers in the preparation of different foods, as bakery, meat and dairy products, and sauces and dressings is discussed. Types of fat replacer and types of food are summarized in Table 4.2.

4.3.1 Fat Replacement in Bakery Products

4.3.1.1 Fat Replacement in Aerated Baked Products

Fiber is one of the leading clean label fat replacers, and many studies exist, related to using fiber to replace fat in bakery products. Replacing fat with fiber in aerated bakery products like muffins and cakes results in significant structural changes and, consequently, physical properties, such as reduced volume and increased crumb firmness (Rodríguez-García et al. 2014). Fat is multifunctional in baked products; it helps with the gas incorporation in cake batter during mixing, contributes to emulsification, and "shortens" the crumb, breaking gluten continuity to tenderize the crumb (Lai and Lin 2014). In reduced fat content products, the stability of air bubbles enclosed in the batter is lowered; they tend to coalesce and disappear, giving harder products (Quiles et al. 2018). However, fat provides important sensory

Table 4.1 Clean label fat replacers

Clean label fat replacers			
Fat substitutes	Fat mimetics		
Lipid-based molecules	**Protein-based mimetics**	**Carbohydrate-based mimetics**	**Emulsions and oleogels**
Vegetable oils	**Concentrate and isolate proteins**	**Starch-based**	**Low oil content emulsions** using natural stabilizers and emulsifiers.
(Olive, canola, avocado, sunflower …).	(Legume, cereal, oilseed, dairy, meat, egg proteins …)	Native and physically modified starches: corn, wheat, rice, oat, potato, tapioca (cassava), sago, …	**Emulsions with low-fat digestibility**
Improve the food nutritional profile and have a low hydrolysis rate.	*Provide nutritional value, physiological, and technological functionality.*	*Provide fat sensorial properties mouthfeel, creaminess, appearance, and thickness.*	**Oil foams**
Diacylglycerols		**Fibers**	*Decrease lipid digestion, reduce the caloric content.*
Formed by enzymatic hydrolysis of triacylglycerol present in animal tissues and seed oils.		*Soluble dietary fibers:*	
Reduce plasma triglyceride levels, body weight, waist circumference, and visceral and subcutaneous fat.		β-Glucan, *psyllium*, pectins, guar gum, arabinoxylans, inulin, …	
Waxes		*Insoluble dietary fibers:*	
Derived from animals (beeswax) or plants.		Cellulose, hemicellulose, chitosan, lignin, …	
They are poorly absorbed in the gastrointestinal tract.		*Not hydrolyzed by enzymes in the small intestine, showing beneficial physiological effects. Provide physicochemical properties like solubility, viscosity, water holding capacity, oil-binding ability, bulking, and fermentability.*	
Fish oil		**Gums**	
Improve the food nutritional profile and are low calorie fats.		(Xanthan, guar and locust bean gums, gum arabic, carrageenan, …)	
		Increase the viscosity of food systems. Function as stabilizers and gelling agents.	
		Whole foods and purees	
		(avocado, banana, legume, plum, pear, apple, …).	
		Provide texture, body, solids and water-binding, health benefits (antioxidant activity).	

properties to bakery products, such as color, taste, texture, and flavor, all contributing to consumer acceptance (Colla et al. 2018).

Fruit fibers, like peach fiber, have been used to replace fat in soft bakery products (Grigelmo-Miguel et al. 2001). The authors prepared fat-reduced muffins (up to 68% of replacement) using up to 10% of peach fiber and observed the muffins darkened and increased their hardness. However, only muffins with peach fiber up to 4% were acceptable by the consumers compared to the control. Dried pomace from fruit industry wastes can also replace fat in cakes and muffins. Berry pomace from blackcurrant and Aronia was used to replace 30% fat in sponge cakes, giving a better nutritional profile, which was acceptable to consumers (Quiles et al. 2018). Other studies extracted pectin from fruit pomace and used it as a fat replacer in bakery products. Lim et al. (2014) incorporated pectin from Yuja (Citrus Junos) pomace in cake formulations and showed that replacement with pectin up to 10% was effective at producing cakes as soft as the control cake, without a volume loss. Orange fiber, extracted from orange juice wastes, was used by Caggia et al. (2020) as a fat replacer in brioches. The 50% fat replacement gave brioches with better nutritional profiles and excellent textural characteristics up to 5 days of storage.

Another well-known fat replacer is inulin. Rodríguez-García et al. (2012) studied the effects of different fat replacement levels (0%, 35%, 50%, 70%, and 100%) of inulin in sponge cakes; they found that reductions up to 70% could be achieved. Thus, products could be labeled as "reduced in fat." Zahn et al. (2010) used up to 50% inulin as a fat replacer in muffins by with superior results.

Regarding cereal fiber, Díez-Sánchez et al. (2018) compared the effect of substituting 30% of fat with soluble or insoluble fiber on sponge cake quality, structure, and acceptability, and showed better results when soluble fiber was used. Brans and byproducts of grain processing have also been investigated for partial fat replacement. Inglett et al. (2004) used a combination of bran and barley flour and found it had similar rheological properties to coconut cream when preparing cakes and cookies.

Different hydrocolloids and gums from plants have been used to replace fat in sweet bakery products. Chia mucilage obtained from chia seed (Salvia hispanica L.) was used by Ferrari et al. (2015) and by Fernandes and Salas-Mellado (2017) to replace different percentages of fat in cakes; they found that replacement up to 25% and 50%, respectively, were technologically feasible without altering the cake's quality characteristics. Psyllium has also been an oil replacer because of its gelling and emulsifying properties (Belorio et al. 2019). Eslava-Zomeño et al. (2016) used a natural functional ingredient derived from flaxseed (OptiSolTM5300) to replace part of the fat in sponge cakes, obtaining reliable results up to 30% of fat replacement. Galactomannan extracted from the Cassia grandis seeds were used to prepare cakes and showed 75% of fat replacement with galactomannan at 1.0% concentration was achieved while maintaining acceptance of the product (Andrade et al. 2018). Other gums from microbial origin like xanthan gum have been used for fat replacement in cakes and muffins. Khouryieh et al. (2005) used xanthan gum together with maltodextrin and sucralose to prepare no low-fat sugar muffins. Zambrano et al. (2004) compared guar and xanthan gums for making low-fat

Table 4.2 Clean label fat-replacers in different food systems

Fat replacer	Applications	References
Fiber		
Oat fiber and carrageenan	Frankfurter	Cofrades et al. (2000)
Peach fiber	Muffin	Grigelmo-Miguel et al. (2001)
Corn fiber, maltodextrin and/or lupine extract	Biscuit	Forker et al. (2012)
Pectin		
Citrus	Cake	Lim et al. (2014)
Apple	Brioche	Caggia et al. (2020)
Pectin and banana powder	Cookie	Min et al. (2010)
	Cookie	Majeed et al. (2017)
Inulin		
	Cake	Rodríguez-García et al. (2012)
	Muffin	Rodríguez-García et al. (2014)
	Biscuit	Zahn et al. (2010)
	Sauce	Rodríguez-García et al. (2013)
	Dairy desserts	Błońska et al. (2014)
	Aerated yogurts	Krystyjan et al. (2015)
		Guardeño et al. (2012)
		Bayarri et al. (2011)
		Gomez-Betancur et al. (2020)
Gum		
Xanthan gum (XG)	Mayonnaise	Ma and Barbosa-Cánovas (1995)
XG and rice starch	Mayonnaise	Lee et al. (2013)
XG and maltodextrin	Muffin	Khouryieh et al. (2005)
XG or guar	Muffin	Zambrano et al. (2004)
XG and guar	Mayonnaise	Su et al. (2010)
XG and guar and citrus fiber	Mayonnaise	Thomareis and Chatziantoniou (2011)
Konjac glucomannan	Mayonnaise	Li et al. (2014)
Galactomannan	Cake	Andrade et al. (2018)
Chia mucilage	Cake	Ferrari et al. (2015)
Psyllium	Cake	Fernandes and Salas-Mellado (2017)
		Belorio et al. (2019)
Starch		
Starch and CMC	Muffin and cookie	Lee and Puligundla (2016)
Starch and apricot kernel flour	Yogurt	Román et al. (2019)
	Dairy dessert	Arancibia et al. (2015)
	Cookie	Özboy-Özbaş et al. (2010)
Oat		
Oat dextrin	Mayonnaise	Shen et al. (2011)

(continued)

Table 4.2 (continued)

Fat replacer	Applications	References
Oat Bran	Cookie	Lee et al. (2005)
Oatmeal extract	Salad dressing	Sumonsiri et al. (2020)
Protein		
Whey protein isolate	Mayonnaise	Liu et al. (2007)
Milk protein	Yogurt	Bernardino-Nicanor et al. (2015)
	Dairy dessert	Lesme et al. (2020)
	Fresh cheese	Borreani et al. (2017b)
		Lobato-Calleros et al. (2007)
Vegetable oil		
	Patties	Los et al. (2020)
	Paté	Vargas-Ramella et al. (2020)
Emulsion		
	Dairy dessert	Borreani et al. (2020)
	Meat product	Sanz et al. (2015)
		Urgu-Öztürk et al. (2020)
Oleogel		
	Meat batter	Alejandre et al. (2019)
	Muffin	Giacomozzi et al. (2018)
	Cookie	Oh and Lee (2018)
		Hwang et al. (2016)
		Patel et al. (2020)
Whole food		
Avocado	Cake	Nguyen et al. (2013)
Melon skin	Cookie	Wekwete and Navder (2008)
Apple puree	Cake	Al-Sayed and Ahmed (2013)
	Cookie	Colla et al. (2018)
Others		
β-glucans	Mayonnaise	Worrasinchai et al. (2006)
Chitosan	Cake	Rios et al. (2018)
FOS and probiotic strains	Sausages	Bis-Souza et al. (2020)
Berry pomace	Cake	Quiles et al. (2018)
Pea extract	Sausage	Conroy et al. (2018)
Flax-seed	Cake	Eslava-Zomeño et al. (2016)

muffins and obtained superior results with 50% of replacement, with xanthan gum being most liked by consumers. Animal origin hydrocolloids like chitosan from marine byproducts have been proposed as fat replacers (Rios et al. 2018); their results indicated that succinyl chitosan has excellent potential as a partial fat replacer in cakes, even reducing hardening during storage.

Moreover, native starches can be used at low concentrations to elaborate clean label low-fat bakery products. In this regard, native rice starch at 20% of fat replacement was suitable for preparing reduced-fat muffins and cookies (Lee and Puligundla 2016).

Oleogels are promising fat mimetics that can be used in bakery products. Giacomozzi et al. (2018) used oleogels from high oleic sunflower oil and monoglycerides as solid fat replacers in muffins; the fat replacement had a positive impact on the quality of the muffins and improved their nutritional profile (without trans-fat and low in saturated fat). Oh and Lee (2018) also found that fat could be replaced with up to 50% oleogels in muffins without negatively affecting the product's final qualities.

Finally, some whole foods, like avocados, have been used as fat replacers in cakes (Nguyen et al. 2013). Moreover, specific parts of some foods (melon skins) were used to replace 5–15% of fat in a cake formulation (Al-Sayed and Ahmed 2013); 10% fat replacement with the skins produced acceptable cakes, which were not significantly different to the control.

4.3.1.2 Fat Replacement in Cookies

Different clean label ingredients have been used to replace fat in biscuits and cookies; one example is pectin from different fruits. Pectin from apples was used to substitute 10%, 20%, or 30% of shortening in cookie formulations by Min et al. (2010). They observed that cookies elaborated with pectin were reduced in diameter and height, plus showed a lower fracturability. Furthermore, pectin and banana powder were used as carbohydrate-based fat replacers in cookies by Majeed et al. (2017).

Different vegetal parts or extracts have also been used, as in the study of Forker et al. (2012). They replaced 30% or 40% fat in short dough biscuits with a single fat replacer (corn fiber, maltodextrin, or lupine extract) or a binary combination of them. The flour obtained from fruit kernel was used by Özboy-Özbaş et al. (2010), who used apricot kernel flour and resistant starch (AKF/RS) for fat replacement in cookies. Spread ratios of cookies increased up to 20% replacement but then decreased when over 30% shortening was replaced by the AKF/RS mixture, and further cookie hardness decreased with the replacement of over 10%.

Among cookies and biscuits, short dough biscuits are formulated with high levels of shortening or fat relative to flour content, so they are good candidates for fat replacement. Rodríguez-García et al. (2013) evaluated the effects of inulin as a fat replacer in short dough biscuits and concluded that shortening might be partially replaced by up to 20%, with inulin giving similar biscuits than the full-fat counterpart. Short dough biscuits with partial replacement of fat by inulin were also successfully prepared by Błońska et al. (2014) and Krystyjan et al. (2015). Therefore, inulin is one of the most used clean label ingredients for fat replacement in biscuits.

Bran from different cereals has been used as fat replacers. For instance, various levels of shortening in cookies (10%, 20%, and 30% by weight) were replaced with 20% jet-cooked oat bran by Lee et al. (2005).

Hwang et al. (2016) studied the replacement of bakery fat with oleogels made with different natural waxes and vegetable oils. They found that cookie properties hardness, spread factor, and fracturability were not affected by the different waxes and oils. Furthermore, the replacement of shortening with wax oleogels increased the spread factor and reduced the snapping force, producing cookies with soft eating characteristics (Patel et al. 2020).

Regarding whole foods, using avocado puree as a fat replacer in oatmeal cookies was feasible at 50% fat replacement (Wekwete and Navder 2008). However, apple puree resulted in a reduction in sensory quality and consumer acceptance, even at a 10% replacement level (Colla et al. 2018).

4.3.2 Fat Replacement in Sauces and Dressings

The reduction of fat in sauces and dressings modifies their rheological and textural behavior and influences their stability during storage. In addition, the flavor perception of the reduced-fat dressing and mayonnaise products can also be significantly altered. Thus, food industries face considerable challenges (Ma and Boye 2013).

Mayonnaise is one of the most popular sauces worldwide; it contains a high-fat content in an oil-in-water emulsion, usually containing over 70% vegetable oil. Different compounds have been studied for the reduction of fat content in mayonnaise. One of the most studied ingredients is gums. Xanthan gum was used in the 1990s by Ma and Barbosa-Cánovas (1995), who studied the rheological properties at different concentrations (0.1–1.5%). Lee et al. (2013) also used xanthan gum at 0.2% together with gelatinized rice starch, obtaining good emulsion stability at up to 30% of fat replacement. Mixtures of polysaccharide gums, including xanthan gum and citrus fiber, and variable concentrations of guar gum were used to formulate up to 50% low-fat mayonnaise; the mixture of xanthan gum – guar gum (XG-GG) was more acceptable to consumers (Su et al. 2010). This XG-GG mixture also revealed synergistic action in terms of viscosity (Thomareis and Chatziantoniou 2011). Different concentrations of konjac glucomannan have been used to substitute fat up to 60% (Li et al. 2014). The effects of the substitution ratio were determined by rheology, color measurements, and structural observations, which indicated that over 30% of replacement was not acceptable.

β-glucans and oat dextrins have also been studied as fat replacers, giving satisfactory results at replacement levels of 50% and 30%, respectively (Worrasinchai et al. 2006; Shen et al. 2011). Regarding proteins, whey protein isolate (WPI) was used with low-methoxy pectin as a fat mimetic in mayonnaise by Liu et al. (2007). WPI was also used combined with Opuntia robusta mucilage by Bernardino-Nicanor et al. (2015) to develop a low-fat mayonnaise with functional properties like those of commercial mayonnaise.

Another sauce studied in terms of fat replacement is white sauce. White sauce is used in a variety of cooked and precooked dishes and is the basis for other sauces. The ingredients of a white sauce are basically flour, milk, and fat. Fat was replaced by inulin in white sauces by Guardeño et al. (2012), and their conducted sensory tests suggested fat could be totally substituted by inulin, producing no differences in texture and flavor acceptability.

Salad dressings are oil-in-water emulsions made of vegetable oil (around 30%), vinegar, sugar, egg, modified starch, water, and salt and are usually served with a salad for flavor enhancement. Some efforts have been made to reduce fat content in dressings. Sumonsiri et al. (2020) used oatmeal extract, produced by boiling ground oats, to replace the fat (3–22%) in salad dressings; their results showed that products with lower fat content and higher protein content were accepted by consumers.

4.3.3 Fat Replacement in Meat Products

Fat mimetics have been used in meat products for many reasons including, (1) for reduction of animal fats and cholesterol, (2) improvement of the nutritional profile (increasing the proportion of unsaturated fatty acids), and (3) lowering of the total calories (Jiménez-Colmenero et al. 2015). Products that have undergone fat replacement include meat suspensions (sauces, batters, and pastes) and processed meat products (sausages, patties, and frankfurters) (Kouzounis et al. 2017; Wolfer et al. 2018).

Improving the lipid component (reduce the fat content and provide a better fatty acid profile) by reformulating traditional foods is technologically challenging and faces difficulty achieving consumer acceptance. Given that meat and meat products are some of the most important sources of dietary fat, new approaches to stabilize and structure edible oils for use as fat animal alternatives, to develop healthier lipid meat products, are especially interesting (Jiménez-Colmenero et al. 2015).

Various types of fibers have been studied singly or in combination with other ingredients to formulate reduced-fat meat products, coarse ground and restructured products, and meat emulsions (Bodner and Sieg 2009; Kim and Paik 2012; Mehta et al. 2013).

Dietary fiber can be effectively incorporated in processed meat products, as a clean label functional ingredient, with relevant effects on meat technological, nutritional, and sensory traits. Fibers are mainly used in meat products to improve the water holding capacity to retain added water, thus acting as an extender (like in meat batters). Fibers improve the binding and forming of comminuted meat products like burgers and breaded patties, whereas they improve the emulsion stability and fat retention of emulsified meats (like frankfurters and bologna). They are a powerful tool to modulate the texture and sensory properties of the finished product toward the desired profile (to add bite or increase tenderness and juiciness). Some soluble fractions can be useful for the formulation of low-fat meat products because of their "fat mimetic" behavior (Balestra and Petracci 2019). Cofrades et al. (2000) studied the effect of carrageenan and oat fiber on the textural properties of frankfurters and found that oat fiber was more effective in improving the objective and sensorial texture, while carrageenan mainly developed sensorial textural attributes.

The combination of dietary fiber and probiotic strains to reduce the fat content of fermented sausages has been used for the development of innovative and healthier meat products. Pork back fat was partially replaced by fructooligosaccharides (FOS), and the probiotic strains Lactobacillus paracasei and Lactobacillus rhamnosus, and had positive effects on fermented sausages as they also contribute to the stability and safety of low-fat Spanish Salchichón (Bis-Souza et al. 2020).

Pea extract may be used in sausage reformulations to increase protein content and reduce fat content (Conroy et al. 2018). Reducing the levels of fat by adding pea extract caused varied sensory acceptance depending on the age of the consumers, especially regarding product flavor and texture.

Regarding vegetable oils substituting animal fat, the utilization of pre-emulsified hazelnut oil (HO) plus hazelnut powder (HP) presents a good opportunity for introducing healthier oils and protecting the overall quality of emulsified meat products (Urgu-Öztürk et al. 2020). To improve the fat composition of beef patties formulations, Los et al. (2020) replaced 50% and 33% of animal fat with pomace olive oil (POO) and canola oil (CO), respectively, obtaining good sensory results.

Two promising structured oil gel delivery systems (organogels and hydrogel emulsions) prepared with different technological strategies, showed adequate properties to be used as total fat replacements in comminute meat batters. Using liquid CO by itself (i.e., unstructured) resulted in undesirable attributes (e.g., texture and color) that could be improved by structuring the oils organogels or hydro-gelled emulsions. These gel systems also improved the fatty acid profile of meat batters and reduced lipid oxidation (Alejandre et al. 2019).

The influence of pork fat substituted with microencapsulated healthier oils (tigernut, chia, and linseed oils) on the physicochemical and nutritional properties of deer pâté was evaluated by Vargas-Ramella et al. (2020). The fatty acid composition of the pâtés was modified, decreasing the total amount of saturated fatty acid (SFA), and increasing polyunsaturated fatty acid (PUFA) (chia and linseed pâtés) or monounsaturated fatty acid (MUFA) contents (tigernut pâtés). Therefore, the substitution of pork back fat by vegetable oils is a good strategy to obtain healthier pâtés, which would meet the dietary recommendations of international organizations and the requirements of consumers regarding foods with positive health benefits. Besides reducing fat content, this substitution allowed to enhance the nutritional value of this meat product. Therefore, lower cholesterol and a balanced fatty acid composition (high unsaturated fatty acids, low saturated fats, and ratios n-6/n-3) were obtained. The increase of PUFA content could explain the softer textures, the susceptibility of modified pâtés to oxidation, and the lower acceptability values.

4.3.4 Fat Replacement in Milk Products

Starches have been used to substitute fats, improve the textural properties, or increase satiety in yogurts and acidified milk gels. The substitution of milk with up to 33% flour increased the firmness and consistency of the set gel, but these changes were reversed at greater substitution levels, due to the lack of a connected protein structure. Besides, greater syneresis was observed in the gels as the milk content was reduced, an effect that was reversed by increasing the percentage of solids in the mixture (Román et al. 2019). The addition of native starches from different origins to reduced-fat yogurts contributed to the formation of more stable dispersed acidified milk gelled systems (Lobato-Calleros et al. 2014).

As stated before, although being chemical-sounding ingredients, cellulose ethers derive from cellulose, a naturally occurring plant fiber. Different cellulose derivatives have been used to prepare reduced-fat dairy products. Arancibia et al. (2015) developed different dairy matrices varying in thickener type and concentration: carboxymethyl cellulose (1.1% and 1.3%) and starch (3.5% and 4.0%), and changing milk type (whole or skimmed). The results obtained in fat-free matrices containing thickeners highly depended on the type of thickener used. Furthermore, the effect of emulsions when using different hydrocolloids (xanthan gum, hydroxypropyl methylcellulose, and methylcellulose) on the structural, textural and sensory properties, and lipid digestibility of fat-reduced panna cotta was investigated by Borreani et al. (2020). Their results may contribute to the manufacture of reduced lipid digestion dairy foods, which could be used for weight management. In addition, using different cellulose types, with and without incorporating glycerol, gives stable emulsions with different rheological properties, microstructure, and thermal behavior, opening a wide variety of food applications, including dairy products, where the emulsions can be used as replacers of conventional fat (Sanz et al. 2015).

Another fat replacer used in dairy products is inulin. Bayarri et al. (2011) made fat-reduced custards using inulin with different chain lengths and distinct types of milk. The creaminess, roughness, and mouth coating of the low-fat custards did not differ in the intensity of these attributes for whole-milk custard, although their visual consistency and thickness were perceived as significantly lower than a whole-milk sample. The addition of long-chain inulin gave rise to a low-fat product with the same thickness intensity as the full fat sample, though less creamy, rougher, and with more mouth-coating. Gomez-Betancur et al. (2020) used long-chain inulin and microencapsulated calcium carbonate as an alternative in the production of low-fat and aerated yogurts due to their structuring effect and positive impact over the rheological and sensory texture attributes. Together, this could facilitate the development of functional and novel products attractive to consumers, especially those with special nutritional requirements.

Milk proteins are of particular importance to replace fat or enhance the structure of fat-reduced yogurts thanks to their texturing properties. When proteins are used to replace fat in yogurts, besides their gelation properties, they impact the organoleptic quality of the product. In fact, milk proteins bind flavor compounds to different extents, leading to the modification of flavor release during storage and possibly to an unbalanced and unpleasant flavor compared to the original flavor of the product (Lesme et al. 2020). Borreani et al. (2017b) made panna cottas, designed to aid weight management, with the combination of reducing fat content (to obtain a reduced energy density product) and adding whey protein (to increase satiety capacity). The addition of extra dairy proteins and the reduction of cream content in panna cottas led to a denser and more aggregated matrix than the control sample and therefore increased the firmness values of the panna cottas, which could be related with increased satiety.

The total or partial substitution of the milk fat with emulsified CO and/or whey protein concentrate produced cheeses with different structures than a full-milk fat cheese (Lobato-Calleros et al. 2007). Predominate whey protein concentrate caused a denser, compact, and continuous protein matrix. In contrast, when emulsified CO predominated, a looser, more disrupted protein matrix was formed. Thus, these different cheese microstructures exhibited differing textural characteristics.

4.4 Conclusions

Lipids overconsumption can lead to serious medical consequences. Replacing fats in food formulation with non-lipid ingredients (fats mimetics) or lipids with reduced digestibility (fats substitutes) is an important route of action to formulate healthier food products. However, fats are responsible for many high-quality attributes in the food, and its replacement is very complicated. The challenge for the food industry lies in designing healthier food products by using clean label fat replacers, without compromising the organoleptic properties of the food. Many natural fat replacers (e.g., starch, fibers, proteins, plant waxes, …) have been used as functional food ingredients to produce different kind of products (bakery, meat, or dairy products, …) well accepted by the consumers. Nevertheless, research is still ongoing in this field, being essential to remember that not all additives are harmful, and "natural" is not always right.

References

Adred D, Crilly J, Homan J (2008) Aerated fat-continuous products. Patent Application No 12/287.957

Agyei-Amponsah J, Macakova L, DeKock HL, Emmambux MN (2019) Sensory, tribological, and rheological profiling of "Clean Label" starch–lipid complexes as fat replacers. Starch/Staerke 71:1–9. https://doi.org/10.1002/star.201800340

Ajila CM, Leelavathi K, Prasada Rao UJS (2008) Improvement of dietary fiber content and antioxidant properties in soft dough biscuits with the incorporation of mango peel powder. J Cereal Sci 48:319–326. https://doi.org/10.1016/j.jcs.2007.10.001

Akoh CC (2007) Fatty acids in foods and their health implications. In: Fat-based fat substitutes, 3rd edn. Institute of Food Technologists, Chicago, IL, pp 461–471

Alejandre M, Astiasarán I, Ansorena D, Barbut S (2019) Using canola oil hydrogels and organogels to reduce saturated animal fat in meat batters. Food Res Int 122:129–136. https://doi.org/10.1016/j.foodres.2019.03.056

Al-Sayed HMA, Ahmed AR (2013) Utilization of watermelon rinds and sharlyn melon peels as a natural source of dietary fiber and antioxidants in cake. Ann Agric Sci 58:83–95. https://doi.org/10.1016/j.aoas.2013.01.012

Alting A, Van De Velde F (2012) Proteins as clean label ingredients in foods and beverages. In: Natural food additives, ingredients and flavourings. Woodhead Publishing, Cambridge, pp 197–211. https://doi.org/10.1016/B978-1-84569-811-9.50009-8

Álvarez D, Delles RM, Xiong YL et al (2011) Influence of canola-olive oils, rice bran and walnut on functionality and emulsion stability of frankfurters. LWT Food Sci Technol 44:1435–1442. https://doi.org/10.1016/j.lwt.2011.01.006

Andersson AAM, Armö E, Grangeon E et al (2004) Molecular weight and structure units of (1→3, 1→4)-β-glucans in dough and bread made from hull-less barley milling fractions. J Cereal Sci 40:195–204. https://doi.org/10.1016/j.jcs.2004.07.001

Andrade FJET, De Albuquerque PBS, De Seixas JRPC et al (2018) Influence of Cassia grandis galactomannan on the properties of sponge cakes: a substitute for fat. Food Funct 9:2456–2468. https://doi.org/10.1039/c7fo01864a

Arancibia C, Castro C, Jublot L et al (2015) Colour, rheology, flavour release and sensory perception of dairy desserts. Influence of thickener and fat content. LWT Food Sci Technol 62:408–416. https://doi.org/10.1016/j.lwt.2014.08.024

Asioli D, Aschemann-Witzel J, Caputo V et al (2017) Making sense of the "clean label" trends: a review of consumer food choice behavior and discussion of industry implications. Food Res Int 99:58–71. https://doi.org/10.1016/j.foodres.2017.07.022

Apostolos S. Thomareisa, Soumela Chatziantoniou Evaluation of the consistency of low-fat mayonnaise bysqueezing flow viscometry. Procedia Food Science 1 (2011) 1997 – 200 EFSA Panel on Dietetic Products, Nutrition and Allergies (NDA). Scientific Opinion on the substantiation of health claims related to EPA, DHA, DPA and maintenance of normal blood pressure (ID 502), maintenance of normal HDL-cholesterol concentrations (ID 515), maintenance of normal (fasting) blood concentrations of triglycerides (ID 517), maintenance of normal LDL-cholesterol concentrations (ID 528, 698) and maintenance of joints (ID 503, 505, 507, 511, 518, 524, 526, 535, 537) pursuant to Article 13(1) of Regulation (EC) No 1924/2006.

Balestra F, Petracci M (2019) Technofunctional ingredients for meat products. In: Sustainable meat production and processing. Academic Press, London, pp 45–68

Balestra F, Bianchi M, Petracci M (2019) Applications in meat products. In: Dietary fiber: properties, recovery, and applications. Academic Press, London, pp 313–344

Bayarri S, González-Tomás L, Hernando I et al (2011) Texture perceived on inulin-enriched low-fat semisolid dairy desserts. Rheological and structural basis. J Texture Stud 42:174–184. https://doi.org/10.1111/j.1745-4603.2010.00280.x

Belorio M, Sahagún M, Gómez M (2019) Psyllium as a fat replacer in layer cakes: batter characteristics and cake quality. Food Bioprocess Technol 12:2085–2092. https://doi.org/10.1007/s11947-019-02362-3

Bernardino-Nicanor A, Hinojosa-Hernández EN, Juárez-Goiz JMS et al (2015) Quality of Opuntia robusta and its use in development of mayonnaise-like product. J Food Sci Technol 52:343–350. https://doi.org/10.1007/s13197-013-0989-8

Binks BP, Marinopoulos I (2017) Ultra-stable self-foaming oils. Food Res Int 95:28–37. https://doi.org/10.1016/j.foodres.2017.02.020

Bis-Souza CV, Pateiro M, Domínguez R et al (2020) Impact of fructooligosaccharides and probiotic strains on the quality parameters of low-fat Spanish Salchichón. Meat Sci 159:107936. https://doi.org/10.1016/j.meatsci.2019.107936

Błońska A, Marzec A, Błaszczyk A (2014) Instrumental evaluation of acoustic and mechanical texture properties of short-dough biscuits with different content of fat and inulin. J Texture Stud 45:226–234. https://doi.org/10.1111/jtxs.12068

Bodner J, Sieg J (2009) Fiber. In: Ingredients in meat products: properties, functionality and applications. Springer, New York, NY, pp 83–109

Borreani J, Espert M, Salvador A et al (2017a) Oil-in-water emulsions stabilised by cellulose ethers: stability, structure and in vitro digestion. Food Funct 8:1547–1557. https://doi.org/10.1039/c7fo00159b

Borreani J, Hernando I, Salvador A, Quiles A (2017b) New hydrocolloid-based emulsions for replacing fat in panna cottas: a structural and sensory study. J Sci Food Agric 97:4961–4968 https://doi.org/10.1002/jsfa.8373

Borreani J, Hernando I, Quiles A (2020) Cream replacement by hydrocolloid-stabilized emulsions to reduce fat digestion in panna cottas. LWT-Food Sci Technol 119:108896. https://doi org/10.1016/j.lwt.2019.108896

Caggia C, Palmeri R, Russo N et al (2020) Employ of citrus byproduct as fat replacer ingredient for bakery confectionery products. Front Nutr 7:1–9. https://doi.org/10.3389/fnut.2020.00046

Câmara AKFI, Pollonio R (2015) Reducing animal fat in Bologna sausage using pre-emulsified linseed oil: technological and sensory properties. J Food Qual 38:201–212

Chantaro P, Devahastin S, Chiewchan N (2008) Production of antioxidant high dietary fiber powder from carrot peels. LWT Food Sci Technol 41:1987–1994. https://doi.org/10.1016/j.lwt.2007.11.013

Cheung TTL, Junghans AF, Dijksterhuis GB et al (2016) Consumers' choice-blindness to ingredient information. Appetite 106:2–12. https://doi.org/10.1016/j.appet.2015.09.022

Chung C, Sher A, Rousset P, McClements DJ (2017) Use of natural emulsifiers in model coffee creamers: physical properties of quillaja saponin-stabilized emulsions. Food Hydrocoll 67:111–119. https://doi.org/10.1016/j.foodhyd.2017.01.008

Cofrades S, Hughes E, Troy DJ (2000) Effects of oat fibre and carrageenan on the texture of frankfurters formulated with low and high fat. Eur Food Res Technol 211:19–26. https://doi.org/10.1007/s002170050583

Colantuono A, Vitaglione P, Ferracane R et al (2017) Development and functional characterization of new antioxidant dietary fibers from pomegranate, olive and artichoke byproducts. Food Res Int 101:155–164. https://doi.org/10.1016/j.foodres.2017.09.001

Colla K, Costanzo A, Gamlath S (2018) Fat replacers in baked food products. Foods 7:192. https://doi.org/10.3390/foods7120192

Conroy PM, O'Sullivan MG, Hamill RM, Kerry JP (2018) Impact on the physical and sensory properties of salt-and fat-reduced traditional Irish breakfast sausages on various age cohorts acceptance. Meat Sci 143:190–198. https://doi.org/10.1016/j.meatsci.2018.04.037

Cruz-Requena M, Escobedo-García S, Salas-Tovar JA et al (2019) Definitions and regulatory perspectives of dietary fibers. In: Dietary fiber: properties, recovery, and applications Academic Press, London, pp 1–25. https://doi.org/10.1016/B978-0-12-816495-2.00001-0

Cui SW, Wu Y, Ding H (2013) The range of dietary fibre ingredients and a comparison of their technical functionality. In: Fibre-rich and wholegrain foods: improving quality. Woodhead Publishing Limited, Cambridge, pp 96–119. https://doi.org/10.1533/9780857095787.1.96

Damodaran S (2005) R: concise reviews/hypotheses in food science protein stabilization of emulsions and foams. Food Sci 70:54–66

Delgado-Pando G, Cofrades S, Rodríguez-Salas L, Jiménez-Colmenero F (2011) A healthier oil combination and konjac gel as functional ingredients in low-fat pork liver pâté. Meat Sci 88:241–248. https://doi.org/10.1016/j.meatsci.2010.12.028

Dickinson E (2010) Flocculation of protein-stabilized oil-in-water emulsions. Colloids Surf B: Biointerfaces 81:130–140. https://doi.org/10.1016/j.colsurfb.2010.06.033

Díez-Sánchez E, Llorca E, Quiles A, Hernando I (2018) Using different fibers to replace fat in sponge cakes: in vitro starch digestion and physico-structural studies. Food Sci Technol Int 24:533–543

Doan CD, To CM, De Vrieze M et al (2017) Chemical profiling of the major components in natural waxes to elucidate their role in liquid oil structuring. Food Chem 214 717–725. https://doi.org/10.1016/j.foodchem.2016.07.123

Domínguez R, Agregán R, Gonçalves A, Lorenzo JM (2016) Effect of fat replacement by olive oil on the physico-chemical properties, fatty acids, cholesterol and tocopherol content of pâté. Grasas Aceites 67:e133. https://doi.org/10.3989/gya.0629152

Domínguez R, Pateiro M, Sichetti Munekata PE et al (2017) Influence of partial pork backfat replacement by fish oil on nutritional and technological properties of liver pâté. Eur J Lipid Sci Technol 119:1–7. https://doi.org/10.1002/ejlt.201600178

Eslava-Zomeño C, Quiles A, Hernando I (2016) Designing a clean label sponge cake with reduced fat content. J Food Sci 81:C2352–C2359. https://doi.org/10.1111/1750-3841.13446

Espinal-Ruiz M, Parada-Alfonso F, Restrepo-Sánchez LP et al (2014) Impact of dietary fibers [methyl cellulose, chitosan, and pectin] on digestion of lipids under simulated gastrointestinal conditions. Food Funct 5:3083–3095. https://doi.org/10.1039/c4fo00615a

Euromonitor International (2016) Report extract: lifestyle 2016: new survey insights and system refresher. Euromonitor International, London

EFSA Journal 2009;7(9):1263.

Fellows PJ (2017) Properties of food and principles of processing. In: Food processing technology. Woodhead Publishing, Cambridge

Fernandes SS, Salas-Mellado MM (2017) Addition of chia seed mucilage for reduction of fat content in bread and cakes. Food Chem 227:237–244. https://doi.org/10.1016/j.foodchem.2017.01.075

Figuerola F, Hurtado ML, Estévez AM et al (2005) Fibre concentrates from apple pomace and citrus peel as potential fibre sources for food enrichment. Food Chem 91:395–401. https://doi.org/10.1016/j.foodchem.2004.04.036

Forker A, Zahn S, Rohm H (2012) A combination of fat replacers enables the production of fat-reduced shortdough biscuits with high-sensory quality. Food Bioprocess Technol 5:2497–2505. https://doi.org/10.1007/s11947-011-0536-4

Gallier S, Singh H (2012) The physical and chemical structure of lipids in relation to digestion and absorption. Lipid Technol 24:271–273. https://doi.org/10.1002/lite.201200240

Giacomozzi AS, Carrín ME, Palla CA (2018) Muffins elaborated with optimized monoglycerides oleogels: from solid fat replacer obtention to product quality evaluation. J Food Sci 83:1505–1515. https://doi.org/10.1111/1750-3841.14174

Gomez-Betancur AM, Carmona-Tamayo R, Martínez-Álvarez OL et al (2020) Effect of fat substitution using long-chain inulin and fortification with microencapsulated calcium in the rheological and sensory properties of yogurt mousse. J Food Process Eng:1–12. https://doi.org/10.1111/jfpe.13433

Grigelmo-Miguel N, Martín-Belloso O (1999) Comparison of dietary fibre from byproducts of processing fruits and greens and from cereals. LWT Food Sci Technol 32:503–508. https://doi.org/10.1006/fstl.1999.0587

Grigelmo-Miguel N, Carreras-Boladeras E, Martín-Belloso O (2001) Influence of the addition of peach dietary fiber in composition, physical properties and acceptability of reduced-fat muffins. Food Sci Technol Int 7:425–431

Guardeño LM, Hernando I, Llorca E et al (2012) Microstructural, physical, and sensory impact of starch, inulin, and soy protein in low-fat gluten and lactose free white sauces. J Food Sci 77:859–865. https://doi.org/10.1111/j.1750-3841.2012.02798.x

Golding, Matt; Wooster Tim. (2010). The influence of emulsion structure and stability on lipid digestion. Current Opinion in Colloid & Interface Science 15(1):90–10.

Hall RH (2012) Applications of natural ingredients in savoury food products. In: Natural food additives, ingredients and flavourings. Woodhead Publishing, Cambridge, pp 281–317. https://doi.org/10.1016/B978-1-84569-811-9.50013-X

Happi Emaga T, Andrianaivo RH, Wathelet B et al (2007) Effects of the stage of maturation and varieties on the chemical composition of banana and plantain peels. Food Chem 103:590–600. https://doi.org/10.1016/j.foodchem.2006.09.006

Hargrove J, Greespan P, Hartle D (2004) Nutritional significance and the metabolism of very long chain fatty alcohols and acids from dietary waxes. Exp Biol Med 229:215–226

Herbert S, Bouchet B, Riaublanc A et al (1999) Multiple fluorescence labelling of proteins, lipids and whey in dairy products using confocal microscopy. Lait 79:567–575. https://doi.org/10.1051/lait:1999646

Heymans R, Tavernier I, Dewettinck K, Van der Meeren P (2017) Crystal stabilization of edible oil foams. Trends Food Sci Technol 69:13–24. https://doi.org/10.1016/j.tifs.2017.08.015

Hwang HS, Singh M, Lee S (2016) Properties of cookies made with natural wax-vegetable oil organogels. J Food Sci 81:C1045–C1054. https://doi.org/10.1111/1750-3841.13279

Inglett GE, Carriere CJ, Maneepun S, Tungtrakul P (2004) A soluble fibre gel produced from rice bran and barley flour as a fat replacer in Asian foods. Int J Food Sci Technol 39:1–10. https://doi.org/10.1111/j.1365-2621.2004.00739.x

Jiménez-Colmenero F, Delgado-Pando G (2013) Fibre-enriched meat products. In: Fibre-rich and wholegrain foods. Woodhead Publishing, Cambridge, pp 329–347

Jiménez-Colmenero F, Salcedo-Sandoval L, Bou R et al (2015) Novel applications of oil-structuring methods as a strategy to improve the fat content of meat products. Trends Food Sci Technol 44:177–188. https://doi.org/10.1016/j.tifs.2015.04.011

Jiménez-Escrig A, Rincón M, Pulido R, Saura-Calixto F (2001) Guava fruit (Psidium guajava L.) as a new source of antioxidant dietary fiber. J Agric Food Chem 49:5489–5493. https://doi.org/10.1021/jf010147p

Joly G, Anderstain B (2009) Ingredients in meat products: properties, functionality and applications. Springer, New York, NY. pp 25–55.

Kanjilal S, Prasad RBN, Kaimal TNB et al (1999) Synthesis and estimation of calorific value of a structured lipid- potential reduced calorie fat. Lipids 34:1045–1055. https://doi.org/10.1007/s11745-999-0456-7

Kasprzak MM, Macnaughtan W, Harding S et al (2018) Stabilisation of oil-in-water emulsions with non-chemical modified gelatinised starch. Food Hydrocoll 81:409–418. https://doi.org/10.1016/j.foodhyd.2018.03.002

Khouryieh HA, Aramouni FM, Herald TJ (2005) Physical and sensory characteristics of no-sugar-added/low-fat muffin. J Food Qual 28:439–451. https://doi.org/10.1111/j.1745-4557.2005.00047.x

Kilcast D (2004) Texture in food. In: Solid foods. Woodhead Publishing Limited, Duxford

Kim HJ, Paik H-D (2012) Functionality and application of dietary fiber in meat products. Korean J Food Sci Anim Resour 32:695–705. https://doi.org/10.5851/kosfa.2012.32.6.695

Klonoff DC (2007) Replacements for trans fats - will there be an oil shortage? J Diabetes Sci Technol 1:415–422. https://doi.org/10.1177/193229680700100316

Koubala BB, Kansci G, Mbome LI et al (2008) Effect of extraction conditions on some physicochemical characteristics of pectins from "Améliorée" and "Mango" mango peels. Food Hydrocoll 22:1345–1351. https://doi.org/10.1016/j.foodhyd.2007.07.005

Kouris-Blazos A, Belski R (2016) Health benefits of legumes and pulses with a focus on Australian sweet lupins. Asia Pac J Clin Nutr 25:1–17. https://doi.org/10.6133/apjcn.2016.25.1.23

Kouzounis D, Lazaridou A, Katsanidis E (2017) Partial replacement of animal fat by oleogels structured with monoglycerides and phytosterols in frankfurter sausages. Meat Sci 130:38–46. https://doi.org/10.1016/j.meatsci.2017.04.004

Kris-Etherton PM, Lefevre M, Mensink RP et al (2012) Trans fatty acid intakes and food sources in the U.S. population: NHANES 1999-2002. Lipids 47:931–940. https://doi.org/10.1007/s11745-012-3704-z

Krystyjan M, Gumul D, Ziobro R, Sikora M (2015) The effect of inulin as a fat replacement on dough and biscuit properties. J Food Qual 38:305–315. https://doi.org/10.1111/jfq.12148

Lai H, Lin T (2014) Bakery products science and technology, 2nd edn. Wiley, New York, NY, pp 1–761

Lazaridou A, Biliaderis CG (2004) Cryogelation of cereal β-glucans: structure and molecular size effects. Food Hydrocoll 18:933–947. https://doi.org/10.1016/j.foodhyd.2004.03.003

Lee H (2017) Change of hydrolysis rate on hydrogenated palm kernel oil and shea butter blendings using in vitro digestion system. J Korean Soc Food Sci Nutr 46:1205–1215

Lee YT, Puligundla P (2016) Characteristics of reduced-fat muffins and cookies with native and modified rice starches. Emirates J Food Agric 28:311–316. https://doi.org/10.9755/ejfa.2015-05-227

Lee S, Kim S, Inglett GE (2005) Effect of shortening replacement with oatrim on the physical and rheological properties of cakes. Cereal Chem 82:120–124. https://doi.org/10.1094/CC-82-0120

Lee I, Lee S, Lee N, Ko S (2013) Reduced-fat mayonnaise formulated with gelatinized rice starch and xanthan gum. Cereal Chem 90:29–34. https://doi.org/10.1094/CCHEM-03-12-0027-R

Lee WJ, Zhang Z, Lai OM et al (2020) Diacylglycerol in food industry: synthesis methods, functionalities, health benefits, potential risks and drawbacks. Trends Food Sci Technol 97:114–125. https://doi.org/10.1016/j.tifs.2019.12.032

Lesme H, Rannou C, Famelart MH et al (2020) Yogurts enriched with milk proteins: texture properties, aroma release and sensory perception. Trends Food Sci Technol 98:140–149. https://doi.org/10.1016/j.tifs.2020.02.006

Lett AM, Norton JE, Yeomans MR (2016) Emulsion oil droplet size significantly affects satiety: a pre-ingestive approach. Appetite 96:18–24. https://doi.org/10.1016/j.appet.2015.08.043

Li JM, Nie SP (2016) The functional and nutritional aspects of hydrocolloids in foods. Food Hydrocoll 53:46–61. https://doi.org/10.1016/j.foodhyd.2015.01.035

Li J, Wang Y, Jin W et al (2014) Application of micronized konjac gel for fat analogue in mayonnaise. Food Hydrocoll 35:375–382. https://doi.org/10.1016/j.foodhyd.2013.06.010

Lim ASL, Rabeta MS (2013) Proximate analysis, mineral content and antioxidant capacity of milk apple, malay apple and water apple. Int Food Res J 20:673–679

Lim J, Ko S, Lee S (2014) Use of Yuja (Citrus junos) pectin as a fat replacer in baked foods. Food Sci Biotechnol 23:1837–1841. https://doi.org/10.1007/s10068-014-0251-9

Liu H, Xu XM, Guo SD (2007) Rheological, texture and sensory properties of low-fat mayonnaise with different fat mimetics. LWT Food Sci Technol 40:946–954. https://doi.org/10.1016/j.lwt.2006.11.007

Llobera A, Cañellas J (2007) Dietary fibre content and antioxidant activity of Manto Negro red grape (Vitis vinifera): pomace and stem. Food Chem 101:659–666. https://doi.org/10.1016/j.foodchem.2006.02.025

Lobato-Calleros C, Reyes-Hernández J, Beristain CI et al (2007) Microstructure and texture of white fresh cheese made with canola oil and whey protein concentrate in partial or total replacement of milk fat. Food Res Int 40:529–537. https://doi.org/10.1016/j.foodres.2006.10.011

Lobato-Calleros C, Ramírez-Santiago C, Vernon-Carter EJ, Alvarez-Ramirez J (2014) Impact of native and chemically modified starches addition as fat replacers in the viscoelasticity of reduced-fat stirred yogurt. J Food Eng 131:110–115. https://doi.org/10.1016/j.jfoodeng.2014.01.019

Los PR, Marson GV, Dutcosky SD et al (2020) Optimization of beef patties produced with vegetable oils: a mixture design approach and sensory evaluation. Food Sci Technol 40:12–20. https://doi.org/10.1590/fst.22518

Ma L, Barbosa-Cánovas GV (1995) Rheological characterization of mayonnaise. Part I: Slippage at different oil and xanthan gum concentrations. J Food Eng 25:397–408. https://doi.org/10.1016/0260-8774(94)00011-W

Ma Z, Boye JI (2013) Advances in the design and production of reduced-fat and reduced-cholesterol salad dressing and mayonnaise: a review. Food Bioprocess Technol 6:648–670. https://doi.org/10.1007/s11947-012-1000-9

Majeed M, Anwar S, Khan MU et al (2017) Study of the combined effect of pectin and banana powder as carbohydrate based fat replacers to develop low fat cookies. Foods Raw Mater 5:62–69. https://doi.org/10.21179/2308-4057-2017-2-62-69

Mao Y, Julian Mcclements D (2012) Fabrication of reduced fat products by controlled heteroaggregation of oppositely charged lipid droplets. J Food Sci 77:144–152. https://doi.org/10.1111/j.1750-3841.2012.02680.x

Martínez R, Torres P, Meneses MA et al (2012) Chemical, technological and in vitro antioxidant properties of mango, guava, pineapple and passion fruit dietary fibre concentrate. Food Chem 135:1520–1526. https://doi.org/10.1016/j.foodchem.2012.05.057

Mcdonagh P (2012) Native, modified and clean label starches in foods and beverages. In: Natural food additives, ingredients and flavourings. Elsevier, Amsterdam

Mehta N, Ahlawat SS, Sharma DP, Dabur RS (2013) Novel trends in development of dietary fiber rich meat products—a critical review. J Food Sci Technol 52:633–647. https://doi.org/10.1007/s13197-013-1010-2

Min B, Bae IY, Lee HG et al (2010) Utilization of pectin-enriched materials from apple pomace as a fat replacer in a model food system. Bioresour Technol 101:5414–5418. https://doi.org/10.1016/j.biortech.2010.02.022

Mishima S, Suzuki A, Sato K, Ueno S (2016) Formation and microstructures of whipped oils composed of vegetable oils and high-melting fat crystals. J Am Oil Chem Soc 93:1453–1466 https://doi.org/10.1007/s11746-016-2888-4

Mudgil D, Barak S (2019) Classification, technological properties, and sustainable sources. In Dietary fiber: properties, recovery, and applications. Elsevier Inc., Philadelphia, PA, pp 27–58

Mun S, Kim J, McClements DJ et al (2017) Fluorescence imaging of spatial location of lipids and proteins during digestion of protein-stabilized oil-in-water emulsions: a simulated gastrointestinal tract study. Food Chem 219:297–303. https://doi.org/10.1016/j.foodchem.2016.09.158

Murthy PS, Naidu MM (2012) Recovery of phenolic antioxidants and functional compounds from coffee industry byproducts. Food Bioprocess Technol 5:897–903. https://doi.org/10.1007/s11947-010-0363-z

Maria Herminia Ferrari Felisberto, Adriana Lucia Wahanik , Cristiane Rodrigues Gomes-Ruffi, Maria Teresa Pedrosa Silva Clerici, Yoon Kil Chang, Caroline Joy Steel l. (2015) Use of chia (Salvia hispanica L.) mucilage gel to reduce fat in pound cakes. LWT-Food Science and Technology, 63:1049–1055.

Nguyen D, Carotenuto M, Khan S et al (2013) Effect of avocado fruit puree as fat replacer on the physical, textural and sensory properties of shortened cakes. J Acad Nutr Diet 113:A59. https://doi.org/10.1016/j.jand.2013.06.206

Norton JE, Fryer PJ, Norton IT (2013) Introduction to food formulation engineering. In: Formulation engineering of foods. Wiley, New York, NY, pp 1–8

O'Connor TP, O'Brien NM (2016) Cheese: low-fat and reduced-fat cheese ☆. In: Fat replacers. Elsevier, Amsterdam

O'Sullivan MG (2017) Nutritionally optimised low fat foods. In: A handbook for sensory and consumer-driven new product development. Elsevier, Amsterdam, pp 177–196

Ognean CF, Darie N, Ognean M (2006) Fat replacers – review. J Agroaliment Process Technol XII:433–442

Oh IK, Lee S (2018) Utilization of foam structured hydroxypropyl methylcellulose for oleogels and their application as a solid fat replacer in muffins. Food Hydrocoll 77:796–802. https://doi.org/10.1016/j.foodhyd.2017.11.022

Osborn S (2015) Labelling relating to natural ingredients and additives. In: Advances in food and beverage labelling: information and regulations. Elsevier Ltd, Cambridge, pp 207–221. https://doi.org/10.1533/9781782420934.3.207

Özboy Özbas O, Ardiç M (2016) Dietary fibers as functional ingredients. Actual Direct Sci Res XXI Cent Theory Pract 2:465–468. https://doi.org/10.12737/6203

Özboy-Özbaş Ö, Seker IT, Gökbulut I (2010) Effects of resistant starch, apricot kernel flour, and fiber-rich fruit powders on low-fat cookie quality. Food Sci Biotechnol 19:979–986. https://doi.org/10.1007/s10068-010-0137-4

Patel AR, Nicholson RA, Marangoni AG (2020) Applications of fat mimetics for the replacement of saturated and hydrogenated fat in food products. Curr Opin Food Sci 33:61–68. https://doi.org/10.1016/j.cofs.2019.12.008

Peng X, Yao Y (2017) Carbohydrates as fat replacers. Annu Rev Food Sci Technol 8:331–351. https://doi.org/10.1146/annurev-food-030216-030034

Pourashouri P, Shabanpour B, Razavi SH et al (2014) Impact of wall materials on physicochemical properties of microencapsulated fish oil by spray drying. Food Bioprocess Technol 7:2354–2365. https://doi.org/10.1007/s11947-013-1241-2

Prabhavathi Devi BLA, Gangadhar KN, Prasad RBN et al (2018) Nutritionally enriched 1,3-diacylglycerol-rich oil: low calorie fat with hypolipidemic effects in rats. Food Chem 248:210–216. https://doi.org/10.1016/j.foodchem.2017.12.066

Quiles A, Campbell GM, Struck S et al (2018) Fiber from fruit pomace: a review of applications in cereal-based products. Food Rev Int 34:162–181. https://doi.org/10.1080/87559129.2016.1261299

Rahmati NF, Mazaheri Tehrani M, Daneshvar K, Koocheki A (2014) Influence of selected gums and pregelatinized corn starch on reduced fat mayonnaise: modeling of properties by central composite design. Food Biophys 10:39–50. https://doi.org/10.1007/s11483-014-9356-1

Rios RV, Garzón R, Lannes SCS, Rosell CM (2018) Use of succinyl chitosan as fat replacer on cake formulations. LWT-Food Sci Technol 96:260–265. https://doi.org/10.1016/j.lwt.2018.05.041

Rodríguez-Carpena JG, Morcuende D, Estévez M (2012) Avocado, sunflower and olive oils as replacers of pork back-fat in burger patties: effect on lipid composition, oxidative stability and quality traits. Meat Sci 90:106–115. https://doi.org/10.1016/j.meatsci.2011.06.007

Rodríguez-García J, Puig A, Salvador A, Hernando I (2012) Optimization of a sponge cake formulation with inulin as fat replacer: structure, physicochemical, and sensory properties. J Food Sci 77:C189–C197. https://doi.org/10.1111/j.1750-3841.2011.02546.x

Rodríguez-García J, Laguna L, Puig A et al (2013) Effect of fat replacement by inulin on textural and structural properties of short dough biscuits. Food Bioprocess Technol 6:2739–2750. https://doi.org/10.1007/s11947-012-0919-1

Rodríguez-García J, Salvador A, Hernando I (2014) Replacing fat and sugar with inulin in cakes: bubble size distribution, physical and sensory properties. Food Bioprocess Technol 7:964–974. https://doi.org/10.1007/s11947-013-1066-z

Rogers M (2019) Fat replacers. In: Encyclopedia of food chemistry. Elsevier, Amsterdam, pp 96–100

Román L, Matia-Merino L, Reguilón MP, Gómez M (2019) Extruded maize flour as texturizing agent in acid-unheated skim milk gels. Food Bioprocess Technol 12:990–999. https://doi.org/10.1007/s11947-019-02252-8

Samavati V, Emam-Djomeh Z, Mohammadifar MA et al (2012) Stability and rheology of dispersions containing polysaccharide, oleic acid and whey protein isolate. J Texture Stud 43:63–76. https://doi.org/10.1111/j.1745-4603.2011.00317.x

Sánchez-Alonso I, Jiménez-Escrig A, Saura-Calixto F, Borderías AJ (2007) Effect of grape antioxidant dietary fibre on the prevention of lipid oxidation in minced fish: evaluation by different methodologies. Food Chem 101:372–378. https://doi.org/10.1016/j.foodchem.2005.12.058

Sanz T, Falomir M, Salvador A (2015) Reversible thermal behaviour of vegetable oil cellulose ether emulsions as fat replacers. Influence of glycerol. Food Hydrocoll 46:19–27. https://doi.org/10.1016/j.foodhyd.2014.11.030

Sedaghat Doost A, Nikbakht Nasrabadi M, Kassozi V et al (2020) Recent advances in food colloidal delivery systems for essential oils and their main components. Trends Food Sci Technol 99:474–486. https://doi.org/10.1016/j.tifs.2020.03.037

Shaltout OE, Youssef M (2007) Fat replacers and their applications in food products: a review. Alex J Food Sci Technol 4:29–44

Shen R, Luo S, Dong J (2011) Application of oat dextrine for fat substitute in mayonnaise. Food Chem 126:65–71. https://doi.org/10.1016/j.foodchem.2010.10.072

Su HP, Lien CP, Lee TA, Ho JH (2010) Development of low-fat mayonnaise containing polysaccharide gums as functional ingredients. J Sci Food Agric 90:806–812. https://doi.org/10.1002/jsfa.3888

Sumonsiri N, Panjun B, Naksuk S et al (2020) Effect of oatmeal as a fat replacer on physical properties and sensory acceptance of creamy salad dressing. E3S Web Conf 141:1–5. https://doi.org/10.1051/e3sconf/202014102006

Surampudi P, Enkhmaa B, Anuurad E, Berglund L (2016) Lipid lowering with soluble dietary fiber. Curr Atheroscler Rep 18:75. https://doi.org/10.1007/s11883-016-0624-z

Talukder S (2015) Effect of dietary fiber on properties and acceptance of meat products: a review. Crit Rev Food Sci Nutr 55:1005–1011. https://doi.org/10.1080/10408398.2012.682230

Torres JL, Varela B, García MT et al (2002) Valorization of grape (Vitis vinifera) byproducts. Antioxidant and biological properties of polyphenolic fractions differing in procyanidin composition and flavonol content. J Agric Food Chem 50:7548–7555. https://doi.org/10.1021/jf025868i

Traynor MP, Burke R, Frías JM et al (2013) Formation and stability of an oil in water emulsion containing lecithin, xanthan gum and sunflower oil. Int Food Res J 20:2173–2181. https://doi.org/10.21427/D7D189

Ubando-Rivera J, Navarro-Ocaña A, Valdivia-López MA (2005) Mexican lime peel: comparative study on contents of dietary fibre and associated antioxidant activity. Food Chem 89:57–61. https://doi.org/10.1016/j.foodchem.2004.01.076

Urgu-Öztürk M, Öztürk-Kerimoğlu B, Serdaroğlu M (2020) Design of healthier beef sausage formulations by hazelnut-based pre-emulsion systems as fat substitutes. Meat Sci 167:108162. https://doi.org/10.1016/j.meatsci.2020.108162

Varela P, Fiszman SM (2013) Exploring consumers' knowledge and perceptions of hydrocolloids used as food additives and ingredients. Food Hydrocoll 30:477–484. https://doi.org/10.1016/j.foodhyd.2012.07.001

Vargas-Ramella M, Pateiro M, Barba FJ et al (2020) Microencapsulation of healthier oils to enhance the physicochemical and nutritional properties of deer pâté. LWT-Food Sci Technol 125:109223. https://doi.org/10.1016/j.lwt.2020.109223

Wekwete B, Navder KP (2008) Effects of avocado fruit puree and Oatrim as fat replacers on the physical, textural and sensory properties of oatmeal cookies. J Food Qual 31:131–141. https://doi.org/10.1111/j.1745-4557.2008.00191.x

Winkler-Moser JK, Anderson J, Byars JA et al (2019) Evaluation of beeswax, candelilla wax, rice bran wax, and sunflower wax as alternative stabilizers for peanut butter. J Am Oil Chem Soc 96:1235–1248. https://doi.org/10.1002/aocs.12276

Wolfer TL, Acevedo NC, Prusa KJ et al (2018) Replacement of pork fat in frankfurter-type sausages by soybean oil oleogels structured with rice bran wax. Meat Sci 145:352–362. https://doi.org/10.1016/j.meatsci.2018.07.012

Worrasinchai S, Suphantharika M, Pinjai S, Jamnong P (2006) β-Glucan prepared from spent brewer's yeast as a fat replacer in mayonnaise. Food Hydrocoll 20:68–78. https://doi.org/10.1016/j.foodhyd.2005.03.005

Wu BC, McClements DJ (2015) Design of reduced-fat food emulsions: manipulating microstructure and rheology through controlled aggregation of colloidal particles and biopolymers. Food Res Int 76:777–786. https://doi.org/10.1016/j.foodres.2015.06.034

Wylie-Rosett J (2002) Fat substitutes and health: an advisory from the nutrition committee of the American Heart Association. Circulation 105:2800–2804. https://doi.org/10.1161/01.CIR.0000019402.35632.EB

Yanai H, Tomono Y, Ito K et al (2007) Diacylglycerol oil for the metabolic syndrome. Nutr J 6:1–6. https://doi.org/10.1186/1475-2891-6-43

Zahn S, Pepke F, Rohm H (2010) Effect of inulin as a fat replacer on texture and sensory properties of muffins. Int J Food Sci Technol 45:2531–2537. https://doi.org/10.1111/j.1365-2621.2010.02444.x

Zambrano F, Despinoy P, Ormenese RCSC, Faria EV (2004) The use of guar and xanthan gums in the production of "light" low fat cakes. Int J Food Sci Technol 39:959–966. https://doi.org/10.1111/j.1365-2621.2004.00864.x

Zarai Z, Balti R, Sila A et al (2016) Helix aspersa gelatin as an emulsifier and emulsion stabilizer: functional properties and effects on pancreatic lipolysis. Food Funct 7:326–336. https://doi.org/10.1039/c5fo00963d

Chapter 5
Recent Advances in Oleofoam Stability and Its Application

Areeba Rana, Waqar Ahmed, Saima Naz, Muhammad Inam-Ur-Raheem, Muhammad Kashif Iqbal Khan, Muhammad Abid, Muhammad Asim Shabbir, and Rana Muhammad Aadil

5.1 Introduction

Solid fat plays a vital role in transforming and enhancing the taste, tenderness, satiety, and texture of many foodstuffs and is considered one of the obligatory constituents of processed foods (Pehlivanoğlu et al. 2018). Trans and saturated fatty acids are present in these solid fats. The risk of various severe diseases (heart diseases, obesity, and diabetes) is expected to occur (Luo et al. 2019). Therefore, the production of healthier products is promoted because of consumer awareness and health consciousness. One of the most suitable methods to accomplish this is to reformulate the fat composition of food products. The animal fat present in different products is replaced with other lipids to lessen the saturated fat and cholesterol contents suitable for human consumption (Gómez-Estaca et al. 2019a).

Oleogel is an organized fat medium with distinct structural attributes like cohesiveness, shape, ductility, and firmness, and exhibits fewer saturated fats. Oleogel is considered as the essential "shaping oil" for bakery products since it contains a reduced amount of fat. Moreover, with advanced studies, multifactor systems, and functional components, it is regarded as a carrier medium. Oleogelators are used to accomplish the oleogelation of liquid oils such as mono and diglycerides, fatty

A. Rana (Deceased) · W. Ahmed (Deceased) · M. Inam-Ur-Raheem · M. A. Shabbir (✉)
R. M. Aadil (✉)
National Institute of Food Science and Technology, University of Agriculture, Faisalabad, Pakistan

S. Naz
Department of Clinical Nutrition, Nur International University, Lahore, Punjab, Pakistan

M. K. I. Khan
Department of Food Engineering, University of Agriculture, Faisalabad, Pakistan

M. Abid
Institute of Food and Nutritional Sciences, Pir Mehr Ali Shah Arid Agriculture University, Rawalpindi, Pakistan

C. M. Galanakis (ed.), *The Age of Clean Label Foods*,
https://doi.org/10.1007/978-3-030-96698-0_5

alcohols (several long-chain alcohols analogous to fatty acids), and waxes. Currently, due to the low cost, gelling ability, and availability of the plant wax esters, their use as oleogelators is consistently increasing (Papadaki et al. 2019).

Depending upon the aqueous volume fraction and the existence of the crystalline constituents that physically arrest the contact besides the droplet movement, these emulsions (water in oil) can be stabilized contrary to droplet coalescence and phase separation (Wijarnprecha et al. 2019). Furthermore, not only the gelator molecule and nature of oil but also the mechanical and the thermal conditions (Palla et al. 2017), time and temperature (Heymans et al. 2018), and contact angle (Murakami et al. 2018) affect the structure of the oil.

5.2 Oleofoam

Due to the unique chemical and physical attributes, oleogel development became an emerging technique in the recent era. Because of the three-dimensional linkage of oleogel, it constitutes the positive platform for replacing fat and carriers for hydrophobic bioactive molecules. Mixing the liquid oil with a structuring agent followed by further reformation results in a very stable three-dimensional linkage the process of oleogel formation. The flexibility and strength of the gel are mainly due to that linkage (Rosen-Kligvasser and Davidovich-Pinhas 2021). However, during the storage period, oleogels are not stable, so there is a need to improve their stability; for this purpose, the most important thing is to control the crystalline linkage. So, to modify the oleogel configuration, ultrasonic standing wave fields are considered a good alternative. Crystals are developed during crystallization, and these emerging crystals go towards the ultrasonic standing wave nodal planes. Factors like cooling rate, concentration, and oleogelator category do not affect the formation of uniform and condensed microcrystals bands. The longer the wavelength of these ultrasonic standing waves, the more will be the thickness of these bands. In decreasing the migration kinetics of fat-soluble colorant, these novel arrangements act as physical obstacles as compared to statically crystallized oleogels (Valoppi et al. 2020).

5.3 Oleogel Optimization

The goal of oleogel optimization aims to create edible oleogels with the highest oil binding capacity (OBC) and textural and rheological qualities comparable to those of industrial shortening (Abdolmaleki et al. 2020). Different processing trials for producing oleogels from high oleic sunflower oil and Myverol were investigated to generate a fat with rheological and textural properties similar to margarine. The oleogels made under multi-objective optimization settings are known to have the same hardness value as margarine and rheological model characteristics that were also similar (Palla et al. 2017).

A research study was conducted to optimize the oleogel formation based on xanthan gum (XG, 0–1 g/100 g), sodium caseinate (CN, 04 g/100 g), and guar gum (GG, 0–1 g/100 g) using the dehydrating technique (oven dehydration, freeze dehydration) to acquire the desired rheological and textural characteristics using response surface methodology. The results shown by regression modeling indicated that recovery values of oven dehydration technique led to less structure retrieval than those of freeze dehydrating technique with the most feasible construction consisting of 4 g/100 g CN, 0.98 g/100 g GG, and 0.43 g/100-g XG. Outcomes indicated that a similar sunflower oil and industrial shortening feature is reasonable at 94.5 g/100 g to fabricate oleogels (Abdolmaleki et al. 2020).

A study was conducted by Park et al. (2018) to evaluate the effect of oleogel processing on the oxidative stability of an oleogel cream cheese. The oleogelation system was developed using rice bran wax and high oleic soybean for manufacturing cheese products. Two commercial cream-cheese products and a cream-cheese sample without gel were considered controls. High-performance liquid chromatography data analysis showed no considerable changes in total tocopherol constituents of oleogel cream-cheese and non-gel cheese cream products. During 2 weeks of storage, the amount of all three-tocopherol isomers stayed constant. In oleogelation samples, the oxidation analysis showed more unstable compounds than high oleic sunflower oil; there was only a slight dissimilarity in the volatile compounds between the oleogel cream-cheese and the controls. The results indicated that little plant-based oleogelation might support their potential application in dairy foodstuffs due to storage and thermal treatment.

In another study, the impact of monoglycerides concentration, high-intensity ultrasound application (HIU), and cooling rate on physical attributes of oleogels from monoglycerides and HOSO (high oleic sunflower oil) was studied. After 24 h and at 25 °C, different attributes (melting profile, microstructure, solid fat content characteristics, elasticity, and oil binding capacities) were assessed after preparing the sample. Reduction in the elasticity was observed at a slow cooling rate after 24 h as compared to using a cooling rate of 10 °C/min. Linkage in OBC increased by enhancing the cooling rate and monoglyceride concentration and applying high-intensity ultrasound. Melting enthalpy of oleogel was enriched with monoglyceride concentration after storage. However, no increase in solid fat content took place. According to rheology attributes and texture, the complex structure was usually improved with the rise in the cooling rate the intricate design was usually improved for both sonicated and non-sonicated conditions. To increase the OBC and rigidity of the network at 0.1 °C/min, greater monoglyceride concentration and HIU were most effective (Giacomozzi et al. 2019). HIU did not affect solid fat content and thermal behavior, and in all oleogels, the anticipated β′ polymorphism was achieved. Moreover, using HIU, the physical characteristics of monoglyceride oleogels could be considerably enhanced to get appropriate lipids with low amounts of saturated fats for different food applications (Giacomozzi et al. 2020).

5.4 Oleofoam Preparation

Oleogels are oil-continuous gels in which liquid oil is immobilized by a low amount of a non-triglyceride gelator within a three-dimensional structured linkage. If oleogelators in different processed food products can gel oil in the presence of other constituents, for example dispersed water droplets or sugar, the purpose of oleogelator is considered to be completed (Wijarnprecha et al. 2019). Oleogelation is an innovative knowledge to organize oils. The oleogelation broadly substitutes trans-fats and saturated fats. According to the nutritional point of view, extra virgin olive oil is widely accepted as it is used to produce a novel oleogel that showed firmness of 3.8 N (Papadaki et al. 2020). Mostly the food manufacturing industries use solid stock fats (palm fat, fully and partially hydrogenated vegetable oils, and animal fat) to achieve claim for functional lipids that are oxidatively stable and provide optimal textural and functional characteristics like firmness, plasticity, and mouth feels but these are hazardous to health (Abdolmaleki et al. 2020).

In a related research study, different stearic acid (SA) ratios were used to prepare innovative soy lecithin (SL) based oleogel and emulsion. Oleogels were produced using two various gelator concentrations (20 wt% and 30 wt%) and 1 wt% of H_2O with SL:SA ratios (0:10, 3:7, 5:5, 7:3, 10:0). The rigidity and thermal characteristics of the samples were examined by using a textural analyzer and a differential scanning calorimeter respectively. To stabilize the three-dimensional complex, soy lecithin oleogel emulsions were structured primarily by SA crystals that connected synergistically with SL reverse micelles linkage. More SA content added to the rigidity of the SL based oleogel and emulsion. However, oleogel emulsion based on soy lecithin was firmer than soy lecithin based oleogel in samples consisting of both SL and SA, and represented the synergist impact of oleogelators. When the primary purpose is to obtain the semisolid characteristics, then a more stable alternative to oleogels can be innovative hybrid SL oleogel and SL based oleogel emulsion articulated with SA (Gaudino et al. 2019).

Gao and Wu (2020) carried research to establish an innovative oleogel system. Candelilla wax, oil, distilled water, and potato starch were used to create the oleogel, and its properties were examined. The resultant oleogel product was a low-cost, low-firmness, and low-fat product. In the oleogel system, the constituent of wax and starch was 5 wt%. Type II starch-lipid oleogel system was formed by a potato starch/candelilla wax ratio from 1.22 to 5.67, and the pure water content was 45 times the content of starch. The result obtained was that potato starch/candelilla wax-established oleogels are a typical lipid structure. Oil and wax content can be reduced by starch/wax-based oleogels in oleogel systems and exhibited great potential to manufacture oleogel products that constitute low-fat and fewer hardness values.

In a similar study, xanthan and gelatin were used to produce a foam-templated oleogel. Therefore, oleogel was manufactured at different concentrations of biomacromolecules, and characteristics of the solution, cryogel, and correlated oleogels were determined. The results indicated that the addition of xanthan increases the foam stability and viscosity of the solution. Increased cryogel linkage density and inflexibility occur due to increased bio-macromolecular concentration, but moisture sorption remains unaffected. All oleogels have >92% oil binding capability. At higher foam stability (96.87 ± 4.42), reduced foam density (0.016 ± 0.00), and subsequently appropriate oil sorption (46.10 ± 4.40), the oleogel consisting of 2% xanthan and 3% gelatin was designated as the most appropriate sample. Corresponding measurements showed that the oleogel with 60% structural improvement and thixotropic performance can bind the oil at a temperature less than 100 °C. During 2 months of storage, oleogel linkage was able to save the fat from oxidative degradation (Abdollahi et al. 2019).

5.5 Foam Stability

The base of foam stability is linked with three chief mechanisms: (1) coalescence (2) drainage and (3) diffusive coarsening (Muin et al. 2021). Foam drainage describes the liquid stream in interstitial spaces between bubbles driven by and counteracted by gravity and capillary pressure (Plateau border suction). Due to drainage, the circular exteriors of foams are designed so that dehydrated polyhedral shape cells appear (Heymans et al. 2017). It is broadly recognized that by increasing the bulk viscosity of the fluid, drainage can be slowed down (Muin et al. 2021). After preparing foam, coalescence is grasped as a process that has a higher possibility to happen below a precarious liquefied section (or film thickness) happening when the bubbles are dehydrating or maturing (Heymans et al. 2017). By adding particles and surfactants, coalescence can be minimized (Wang and Brito-Parada 2020), occupying the gas-liquid boundaries. Coalescence is gas rearrangement among foams that subsequently rise in normal bubble size. This can either be obsessed by Laplace pressure difference (disproportionation or Ostwald ripening) or the rupture of the thin liquid film where two contiguous foams merge may be used to obsess the coalescence (Heymans et al. 2017). Finally, diffusive coarsening can be inhibited with foam formation with a gas having less permeability or greater surface compression modulus (Muin et al. 2021).

In a research study, to develop a deeper understanding of the oleofoam stability, different monoglyceride concentrations (3, 5, 7, 9, and 11 wt%) were employed to structure oleofoams through whipping oleogels. Increasing monoglyceride concentrations credited more firmly organized bubbles, more substantial interfacial elasticity, and more concentrated bubble size distributions in oleofoams. Furthermore, thermal values indicated that higher heat was needed to destroy a more robust foam structure (Du et al. 2021). A brief summary of formation of stable foam oils is shown in Table 5.1.

Table 5.1 Summary of formation of stable oil foams

Oils	Crystal particles	Sources	References
Salad oil (mixture of soybean and rapeseed oil)	High melting fat crystal	Fully hydrogenated rapeseed oil rich in behenic acid (FHR-B)	Mishima et al. (2016)
HOSO + other vegetable oil	Long-chain carboxylic acid (myristic acid C14 carboxylic acid)	Sunflower oil containing high oleic acid and other vegetable oil	Binks et al. (2016)
Sunflower oil	High fraction of fully soluble mono-glycerides	Sunflower oil having high oleic acid content (18:1) Cis-9	Gunes et al. (2017)
AMF (anhydrous milk fat)	Di glyceride fatty acid ester	Mixture of triglycerides. Melting temperature −40 °C to +40 °C	Goibier et al. (2019)
MCT oil (medium chain triglycerides) C10:C8=30:70 BASF, Myritol318	Crystalline glycerides mono-palmitin and tri-palmitin crystals	Medium chain triglycerides (C10:C8=30:70 BASF, Myritol318)	Mishra et al. (2020a)

5.5.1 Tools for Describing the Foam Stability

Pore network modeling (PNM) and microcomputed tomography (I-CT) are the most important tools to describe foam stability. Unlike the outdated methodologies, these methods permit the sequential and three-dimensional measurement of Plateau interface thickness and the volume of nodes, which proves to be very helpful to study foam drainage. The synchronization value of the nodes indicated deviance from Plateau's law as the foam became mistier. Moreover, the PNM outcomes have revealed power law between liquid fraction and permeability with an exponent of 2.6, which line up with a channel-dominated flow pattern. The planned techniques give an innovative way to observe the relevant factors that affect foam stability (Muin et al. 2021).

5.6 Pickering Stabilization by Crystalline Particle

By developing either physical obstacle through surface-active colloidal particles or developing steric interfacial films through water dissolvable hydrocolloids and proteins, or dropping interfacial tensions by small molecular weight surfactants, emulsion droplets' stabilization can be understood. The first method of emulsions stabilization was discussed as Pickering emulsions, entitled by the scientist's name who examined the procedure in the journal, even though Walter Ramsden in 1903 first documented this conclusion. In both the water and oil phases, colloidal constituents' parts should stay stable to stabilize Pickering emulsion, however, three-phase contact angle θ revealed appropriate fractional wettability (Xiao et al. 2016).

Interfacially stabilized on aqueous fat-centered foams, called oleofoams, are hardly met compared to the vast amount of aqueous bubbles stabilized by particulate or molecular emulsifiers. No such case in the literature thoroughly described the substantial classification of the interfacial influence on oleofoam constancy. Here to satisfy this gap some techniques are described. Fatty crystals produced by a significant section of completely solvable monoglyceride in oil do not merely form a congested, narrowly packed crystals layer around the foam bubble created through beating (Pickering effect) and adsorb at the oil–air border (Gunes et al. 2017). With a high air portion (up to 75%), soft textures, fine foams, or stable ones for shaving foams could be gained, unprecedented. Non-circular forms of a bubble can be maintained in absence of bulk effects by thin, congested crystal layers on the bubble for times much lengthier than the characteristic–capillary relaxation time for bare bubbles, which is the accurate indication for Pickering-type interfacial stabilization. Beating is essential for bubble covering with a coat of crystals, by associating to foams attained by depressurization. The crystallization performance of monoglyceride in oil can be deliberated as the model system comparable to that in many lipids. In the case of fat-based foaming, approaches are very broad-spectrum mainly in foodstuffs and were used in patent applications (Gunes et al. 2017).

Crystalline particles stabilize oleofoams based on oleogels at the air bubble exterior and in the bulk. The vital parameter in oleofoams stabilization is the size of crystalline particles. The formation procedure (shear rate and cooling rate) can adjust its properties. To conclude, the consequence of changing the weight ratio between fatty alcohols and fatty acids (long-chain) on oleofoam, two ideal weight ratios R = 8:2 and R = 7:3, created the best foams in terms of foam stability, overrun, and foam firmness. Weight ratios influenced the crystal size and the number of crystalline particles existing in the oleogel. Mixed crystals help to create and stabilize the foams. The oleogel hardness and stability is interrelated with their subsequent oleofoam characteristics (Callau et al. 2020).

The difference in foam stability could be well reflected by distributions of bubble size for oleofoams prepared under various monoglycerides concentrations. Firstly, numerous optical micrographs under 20 × 10 magnification were taken immediately during the foaming process. After that, the imaging software (Image-Pro Plus 6.0, Media Cybernetics, Inc., America) was used to calculate bubble sizes. Determination of bubble size by the software revealed that the measurement of traced bubbles after the color adjustment of the picture for decreasing the spatial calibration, noisy point, was executed using the maximum radius (maximum distance between objects centroid and outline) and gauge on the image, was designated. In each sample, diameters of nearly 2500 bubbles were calculated, by plotting the relative frequency histogram bubble size distributions were represented. It is important to note that to get homogeneity all samples were acquired under the same whipping time and beating rate (Du et al. 2021).

After the practical consequences of different techniques, plant-based oils (salad oil) and high melting lipid crystals (wholly hydrogenated rapeseed oil containing a high amount of behenic acid) were used to develop the whipped oil. This whipped oil was created without any additional emulsifier. To detect fine fat crystal particles of the most stable polymorph of β (β-fat crystal), fully hydrogenated rapeseed oil containing a high amount of behenic acid and their adsorption at the air-oil superficial before, during, and after the development of beaten oil, Microprobe Fourier-transform infrared spectroscopy, synchrotron rays microbeam, differential scanning calorimetry, polarized optical, and microscopy X-ray diffraction (SR-μ-XRD) were employed. The results revealed the following points: (1) for developing whipped oil, an organogel preparation consists of tiny fibrous β-fat crystals and salad oil using a distinctive tempering method was required. (2) During the developmental procedure of whipping oil at the air-oil exterior, the β-fat crystals were adsorbed to encapsulate the air bubbles. (3) After an aeration time of half-hour at 20 °C, the overrun values of whipped oil reached greater than 200% (4) β-fat crystal's lamellar planes adjacent to the oil-air exterior were organized almost parallel to the oil-air superficial plane (Mishima et al. 2016).

5.7 Factors Affecting Oleofoam Stability

5.7.1 Time and Temperature

By controlling the temperature, crystals of high melting triglycerides in a continuous phase of low melting triglycerides became easy to obtain. This system was then aerated to produce highly stable oil foams. They proved an optimum solid fat content of around 20–30% for foaming (Goibier et al. 2019). During the manufacturing and storage, the properties of crystal-oil mixtures were affected by time and temperature. Therefore, to formulate oleogels different tempering protocols were set up, and by using differential scanning calorimetry, rheology, and X-ray diffraction, their resulting microstructure was examined. After the manufacturing process and 28 days of storage, the attributes of both systems in terms of foamability and foam stability were reviewed. That study showed that time-temperature relation had an enormous impact on foam characteristics. The weaker structure was observed by the partially crystallized mixtures, because of their less viscosity they capture high air, and at the interface, crystallization would preferentially occur. These bubbles were more significant in size and less firm and unstable. It was suggested that by interfacial contributions, their rheological characteristics were primarily shown. More vital gel linkage was demonstrated by the stored mono-glyceride-oil blends and completely crystallized oil blends that contained less air, more stable during the storage, and smaller air bubbles (Heymans et al. 2018). Smaller lipid crystals were formed when the cooling rate became faster and caused a higher degree of supercooling. Lower complex modulus values revealed that this process led to a less rigid structure. As a result, at the oil-air interfaces adsorption of lipidic crystals was facilitated. According to the Pickering mechanism, air bubbles are readily stabilized by the tiny lipidic crystals. Consequently, a significant reduction in the whipping time (up to 80%) was observed (Truong et al. 2019).

Table 5.2 shows brief summary of various alteration that may affect foam stability as a result of variations in time and temperature. Temperature responsive foams are the important when there is a need for foam destruction. Different oleofoam appearances discussed at different temperature ranges from 25 to 60 °C. Accompanied by foam collapse all oil foams showed liquid drainage at the temperature of 45 °C. In the continuous phase and at the interface, melting of crystals occur. At temperature

Table 5.2 Alteration in the foam stability at different time and temperature

Oil	Oil's specification	Condition	Oil foam	References
Coconut oil (RBD)	Refined breached and deodorized oil having free FA < 0.02%	23 °C	Excellent foam stability although less viscous	Binks and Marinopoulos (2017)
High oleic sunflower oil	Monounsaturated fatty acid (18:1) M. P = −17 °C	Heat to 90 °C for 10 min and then cool to room temperature	Gel formed having elastic modulus in range of 10–10^{8} for C < 20%	Gunes et al. (2017)
n-heptane +methyl salicylate	n-heptane at 25 °c having surface tension 19.66 mNm^{-1} and methyl salicylate 39.2 mNm^{-1}	Contact angle 55°	Maximum oil foaming efficiency	Murakami et al. (2018)
DAG and FHPO	Palm oil with acid value = 0.2 mg M. P = 55 ± 1.1 °C	Temp = 45 °C	Oil foam show liquid drainage accompanied by foam collapse	Lei et al. (2020)
Anhydrous milk fat (AMF)	Mixture of triglyceride melting temp = 43 ± 0.03 °C	Temp = 20 °C for 5 min at rotation rate of 1000 rpm	A coarse foam containing large air bubbles	Gcibier et al. (2019)
Refined soybean oil (total of 10 physically blended oil)	RSBO without antioxidants	Temp = 170 ± 1 °C frying duration = 3 min	Maximum foam height (MFH)	Li et al. (2020)
Mixture of triglycerides oil with monoglycerides (MG) and diglycerides	Middle chain triglyceride oil(C_{10}–C_8) 30:70 BASF, Myritol 318	ΔT = −9 °C Time = after 48 h	Complete bulk crystallization	Mishra et al. (2020a)
Mono-glyceride-sun flower oil mixture	MG extracted from hydrogenated palm stearin contain 96.5% MG M.P = 63.69 °C	T = 80 °C Time = 30 min Fully crystalized MG-oil mixture store for 5 h	More robust gel network which include less air, smaller air bubble stable during storage	Heymans et al. (2018)

45 °C, larger bubble size indicated by Diacylglycerol-based foam than fully hydrogenated palm oil (FHPO) foams because of low melting points of diacylglycerols (DAG) molecules. All foams were changed completely into transparent solution when the temperature was raised to 60 °C. At the temperature ranges from 25 and 35 °C, the microscope images of bubbles revealed that the FHPO-DAG foams had the smallest size. DAG can accelerate the crystal nucleation rate the possible reason for that was DAG's surface activity and hence smaller crystals formed. Nonetheless, the FHPO-DAG foams showed minor bubbles at 50 °C possibly because of the oleogel lower melting temperature (Lei et al. 2020).

When temperature changes from −9 to 0 °C a macroscopic layer of tri-palimitin crystals form. At $\Delta T = 3$ °C, between 72 and 96 h, a slight decrease in meniscus bending depth and a meniscus whitening between is visible with no macroscopic layer. At $\Delta T = -9$ °C after 48 h, the M.P.-MCT (medium chain triglycerides) oil system shows complete bulk crystallization. Low bulk crystallization occurs at 0 °C medium and $\Delta T = -3$ and an impervious MCT oil surface is indicated (Mishra et al. 2020a).

5.7.2 *Gel and Crystal Formation of Myristic Acid in HOSO*

Using visual interpretation, the influence of temperature on solubility of myristic acid in HOSO was determined. HOSO remains liquid down to at least −20 °C and neat MA liquefies at 53.5 °C. At about 80 °C solid myristic acid at different proportions was softened in HOSO and a clear uniform liquid appeared after the 10 min. This myristic acid molecule is described as a one-phase solution. At a rate of 0.1 °C min^{-1} the solution was chilled, until at precipitation temperature first crystal seemed and giving rise to an unclear dispersion. Dispersion of myristic acid crystals in equilibrium with a molecular solution at its solubility is described as a two-phase assortment. Upon further cooling for 5 °C and 5 min, at the dissolution temperature until the onset of clearing the dispersion was heated at 0.1 °C min^{-1}. In measurements of the solubility boundary temperature, a hysteresis of a few degrees was observed by cooling and heating. Crystals start to develop under the process of cooling. By heating, no energy obstacle is expected in crystal dissolution. Hence, the equilibrium temperature and the dissolution temperature of the solubility interface should be closer to each other. As a result of this using an equation, we describe the temperature solubility relationship:

$$\ln x = \frac{\Delta fosH}{R}\left(\frac{1}{Tf} - \frac{1}{T}\right)$$

In this equation x represents the mole fraction of solute in a saturated solution at absolute temperature (T), Tf represents its temperature of fusion for MA (326.6 k). R refers to gas constant, and Δ fosH describes the enthalpy of fusion for myristic acid (45 kJ mol^{-1}). The relative molar mass of HOSO to be 912 g mol^{-1}, to compute the difference of MA solubility with temperature Eq. (1) was used (Binks et al. 2016).

5.7.3 *Yield Stress*

The crystal network formations are yield stress shear-thinning fluids (viscosity decreases under shear strain). Crystal linkage is accountable for the appearance of yield stress in crystal-melt suspensions. Gas bubble stabilization was studied mainly in palm kernel oil crystal-melt suspension and the impact of morphology and crystal concentration on the yield stress of crystal-melt suspensions was examined. Particularly shear influenced the crystallization in the laminar flow; thus, to manufacture the palm kernel oil crystal-melt suspensions diverse shear rates were applied. Higher yield stress was found with the increase in the shear rate at constant crystal content. This behavior is expected to form at higher shear rates from the increased aspect ratio and crystal agglomerates or smaller sizes of crystals. The correlated impact on gas bubble stabilization by crystal network formation was explored. Within lipid melts, gas bubbles become stable through crystal linkage formation as a result stable lipid microfoams obtained with an adjustable structure–rheology association through tailored crystallization (Mishra et al. 2020b). Weaker structures manufactured by partially crystallized blends that hold more air due to their less viscosity and border crystallization would favorably occur. It is proposed that interfacial assistances dominate their rheological attributes (Heymans et al. 2018). All the flow curves are analyzed using the Herschel-Buckley Eq. (2) model as follows:

$$\tau = \tau 0 + \mathrm{k}\dot{\mathrm{Y}}n$$

In this equation τ represents the shear stress (Pa), k describes the consistency coefficient (Pa sn), τo refers to the yield stress (Pa), $\dot{\mathrm{Y}}$ is the shear rate (s^{-1}) and n shows the flow behavior index (dimensionless) (Paximada et al. 2016).

5.7.4 *Contact Angle*

Contact angle (three-phase contact angle) is defined as the angle at the three-phase boundary of the continuous phase, dispersed phase, and solid particle. If the three-phase contact angle θ is less than 90° then O/W emulsions will come into being (Yang et al. 2017). In a Pickering emulsion, particles interact with their neighboring particles and the particle's surface on the value of the contact angle θ made at the oil-water interface these factors affect the precise location of each particle at the interface (Dickinson 2017).

The impact of contact angle is understood by considering the example of oligomer of tetrafluoroethylene (OTFE) that is used to stabilize the oleofoam. Before exposure to air, particles of OTFE were dispersed in oil blends, to eliminate the oil-repellency nature of constituents and to assess accurately the influence of contact angle on the bubbling behavior. By increasing methyl salicylate mole fraction, from 42° (for pure n-heptane) to 89° (for pure methyl salicylate), θY value became higher. When θY = 55° after aeration, the air volume assimilated in scatterings in oil mixtures indicates a maximum. When θY is higher, unusual behavior of foaming detected due to flocculation of an OTFE particles in bulk oils. There is a connection between the concentration of oligomerized lychee fruit extract and the degree of flocculation. In bulk oil, the degree of the flocculation lowers the effective concentration of OTFE, leading to partial stabilization of bubble particles. This outcome proposes that using particles as a stabilizer effective oil foaming was attained by optimizing both degrees of flocculation and the particle contact angle in oils (Murakami et al. 2018).

In oil, for relatively less tension, a dispersion of particles is obtained in the case of particles of tetrafluoroethylene which wet the constituent parts (liquid/air <30 mN m^{-1} such as benzene, squalane, etc.). With intermediary tension (45mN m^{-1} g liquid/air >30 mN m^{-1}) oleofoams are obtained for liquids such as eugenol or sunflower oil. The interface's curvature is inverted for liquids like glycerol and the water that are polar with higher tension. A powder-like material is formed containing glycerol or water droplets coated by particles. This powder-like material is similar to dry water or dry oil, rupture of the droplets, and fluid release occurs by applying shear, but it does not flow. Oil foams are obtained for cos (θ) below 0.8 (i.e. θ above 40 °C). Thus, the surface tension of the oil significantly affects the wetting behavior of OTFE and particles of tetrafluoroethylene. To regulate such particles' surface activity, suitable oil and oil blends require to be selected at the liquid-air surface to get oleofoam (Fameau and Saint-Jalmes 2017). pH of the dispersed phase affected the contact angle, the charge of stabilized particles, and the wettability (Xiao et al. 2020).

5.8 Food Application

Most food products are complex colloidal systems that comprise multiple phases and interfaces created by close contact of dispersed and continuous phases. Foams are one such arrangement where disseminated air bubbles are assimilated and stabilized in a constant phase. Because of the enormous potential of edible foams in reformulating food products with reduced calories, many research has been taken in that field. Besides, due to their distinctive texture and mouth sensation, foams have also expanded fame in the discipline of gastronomy. However, investigations in the zone of food foams have not been acknowledged at the same level of consideration as edible emulsions or gels. The primary aim of this effort is to examine air-oil boundaries stability to generate non-aqueous foams using edible plant-based oil and food-permitted emulsifiers. The challenge of dispersing a large volume of air bubbles in a hydrophobic continuous phase (liquid oil) is explored to create foam oleogels (Karp 2016). In most of the aerated foods, the bulk contribution to the stabilization dominates for a more extended period (i.e., hours to months) (Gunes et al. 2017).

Various applications of oleogels in industrial processing of food products are shown in Table 5.3. Vegetable oil and fat are the essential components of many foodstuffs that are aerated in nature like numerous bakery products. Driven by the requirement for replacing trans and saturated lipid approaches, the field of lipid mimetics has made remarkable improvements over the last decade. Polymeric linkages of ethyl cellulose, emulsion-templated linkages of proteins and polysaccharides, colloidal and self-accumulated fibrillar complexes of polar fat crystals, and solid o/w emulsions of oil imprisoned within crystallized lamellar mesophases, structure the lipid mimetics. To make several food stuffs with lipid mimetics wide-ranging research has been performed. These comprise bakery foodstuffs (laminated pastries and cakes), chocolate, spreads, margarine, and meat products. To produce lipid mimetics, using the clean-label nature of the constituents and their demonstrated functionality in food products is recommended (Patel et al. 2020).

5.8.1 Baking

Products such as liquid oils, butter, and margarine are well known in which lipid is the continuous phase. In baking, they are generally used as lipid constituents. They can also be used as a frying medium or even on bread or pancakes used directly like the frosting. In these products, lipid phase is usually a combination of fat (solid at room temperature) and liquid oil (fat that is liquid at room temperature). The fat phase is structured by solid fat (we can also be called it hard stock fat) and, if the water is present as droplets, then in the continuous phase, forming a fat crystal network aids in stabilizing the emulsion. It is good if the solid lipid dissolves around at body temperature and melts down if this does not happen; then there may be a glistening mouth feel. Besides the constant requirement to decrease manufacturer cost, these days reduced fat-containing compositions are in great demand by health-conscious consumers. To reduce fat one method is to substitute part of the lipid with an edible gas. This bubbling food configuration includes products like ice cream, mousses, and beaten cream, but these also contain many other ingredients and are like complex colloidal construction (Binks and Marinopoulos 2017).

Typically, gas is in the form of disseminated bubbles, but the stability is frequently less regarding fizzy oils/fats. Consequently, coalescence and coarsening of bubbles and liquid oil drainage can occur during storage that decreases the aesthetic worth of the product. That's why there is a need to develop highly stable, aerated fat-continuous products. One more problem is that some consumers do not like to consume a product containing synthetic constituents such as sugars, fatty acids, and synthetic E-number emulsifiers. A novel method is introduced to stabilize the aerated oils without any additive. The primary aim is to stabilize the oil foams with fat crystals and manipulate the temperature of a fat/oil so additives are not needed. For

Table 5.3 Applications of oleogels in various food products

Non-aqueous foam	Products	Key role	References
Palm oil (palm stearin and palm mid fraction were blended with refined Sal fat in various ratios viz., 90:10, 85:15, 80:20, 75:25, 70:30, 65:35, 60:40 and 50:50)	Cocoa butter equivalent	• Region specificity • Lower energy consumption • Better product control	Hithamani (2020)
Shortening (edible solid fat product)	Breads, cakes, cookies, pastries	• Key role to convert gluten and starch granules to prevent their adherence	Demirkesen and Mert (2020)
Emulsified rapeseed oil (solidification occur at 18.23 °C)	Biscuits	• Suitable replacements for palm fat in terms of processability as well as analytical and sensory assessment	Schmid et al. (2020)
Ternary fat mixture of CB, Sat, SatO and OSatO	Dark chocolate	• Fat blending ratio of CB/OSatO/SatSatO at 50/30–20/20–30 is the most attractive candidate in terms of hardness and fat bloom stability	Watanabe et al. (2021)
Oleogel with small amount of solid phase (HOSO and structuring agents)	Biscuits	• Able to hold liquid oil phase in the biscuit matrix better than (CLX 3% and 2%)	Onacik-Gür and Żbikowska (2020)
Fat mimetics (polymer network of ethyl cellulose, emulsion template network of protein and polysaccharide	Bakery products (cakes laminated pastries) laminated pastries, margarines, spreads	• Significant reduction of saturated fat and elimination of hydrogenated fat can be achieved	Patel et al. (2020)
Amylose lipid complex (corn starch and SA)	White pan bread	• Potentially useful in bakery industry for preparing low caloric and low-fat products	Lee et al. (2020)
Oleogel (beeswax at 10% (m/m) under sunflower seed oil	Gluten-free cake	• Potential to reduce the saturated fatty acids (SFA) content of gluten-free baked products	Demirkesen and Mert (2019)
Chia mucilage gel	Meat	• Improves meat emulsion stability • Potential fat substitute • Resistant of thermal treatment	Câmara et al. (2020)

(continued)

Table 5.3 (continued)

Non-aqueous foam	Products	Key role	References
Mango kernel fats(1,3distearoyl-2-oleogyl-glycerol) including physical blends non-interesterified blend	Dark chocolate	• Higher heat stability • Denser structure	Jin et al. (2019)
Emulsion gel and PBF 14% and 10% respectively	Bologna sausages	• Reduction in saturated fat up to 14% • More homogenous batter and compact structure • Greater chewiness	de Souza Paglarini et al. (2019)
Oleogel (canola oil structured with hydroxyl methyl cellulose (HPMC) into solid like oleogel)	Meat patties	• Cooking loss reduced • Texture become softer	Oh et al. (2019)
Carnauba wax (CB) based oleogel with adipic acid	Cake, beef burger	• Improve thermal behavior and crystallinity of oleogel • Products with lower saturated and trans-fatty acid	Aliasl khiabani et al. (2020)
Soybean oil (a pure liquid form or a pre-emulsified foam stabilized by NC)	Meat sausages	• Higher emulsion stability • Stronger textural properties • Higher water and fat binding capacities	Qi et al. (2020)
Hybrid hydrogel (sodium alginate pectin-citric acid mixture) use as fat mimetic	Chocolates	• Low-fat • Heat resistant chocolate	Francis and Chidambaram (2019)
16% of pork back fat replaced with inulin gelled emulsion of corn and rapeseed oil	Fermented sausages	• Significant decrease in saturated fatty acid • Increase in PUFA n-6, n-3 type	Glisic et al. (2019)
Whey and lipid extract of chlorella microalgae	Fermented poultry meat products	• High nutritional value	Bazarnova et al. (2019)
Modified fat (immobilized Sn1,3 specific pyLip lipase	Trans-free formation of base stocks of spreads, margarines and confectionery fat	• Improved oxidative stability • Steaper M.P • Change in crystalline structure	Kavadia et al. (2019)

this purpose, average chain length triglycerides fats/oil (mainly C3–C12) was chosen, consisting of primarily saturated fatty acid chains. Their melting points are around or higher than room temperature. Meanwhile, such oils contain triglycerides having the diverse composition and different chain length, so making use of discriminatory crystallization of those triglycerides that are of higher melting points within liquid oil of those triglycerides that are of lesser melting point is optimal, such as, an oil dispersal of crystals (Binks and Marinopoulos 2017).

Due to the comparatively less cost and the constructive influence of palm oil on shelf life and textural properties of baked products, it is frequently used in the bakery. Due to the significant concerns about societal and biological sustainability in the recent era, there is a great need to use alternatives. It is tough to replace palm fat for the baking industry because palm fat retains distinctive characteristics. By using innovative physical technologies, unhydrogenated rapeseed oil was treated with Pickering emulsion, wax crystallization and stable bubbling to simulate the characteristics of palm oil. The study revealed that the fat replacement products worked like palm fat in bakery experiments. In contrast, the lipid substitute foodstuffs' initial viscosity was less than palm lipid. When emulsified oil was used to bake the biscuits using rapeseed oil accompanied with wax crystals, the processability and sensory and systematic evaluation were considered appropriate substitutes for palm lipid (Schmid et al. 2020).

Most of the nutrients in food fat have a higher caloric density, so one of the crucial trends in foodstuffs modernization is to reduce the fat and cholesterol content. Numerous studies have been studied to replace margarine with several substitutes. The most significant plant source of alpha-linolenic acid was integrated to replace the margarine flaxseed oil. At intermediary levels, different researchers re-counted that by exchanging shortening with hydroxyl propyl methylcellulose and methylcellulose-structured canola oil, cookie emulsions with required practical actions could be prepared (Lee et al. 2020).

In sponge cake batter, fat and protein from eggs and wheat flour at the air-liquid boundary were studied through a foaming protocol. Using sugar, flour, eggs, and agents responsible for increasing the loaf volume to produce a sponge cake only or add from these constituents and exogenous fats (ELs, i.e., mono-arrangement DAG and polyglycerol esters of fatty acids). The bubbling protocol comprised of beating diluted sponge cake mixture into the foam. For the air-liquid interface (ALI) in sponge cake mixtures, the foam was observed to be optimal for the ALI in sponge cake mixtures. After whipping and chemically analyzed foam fractions were isolated at multiple time points (Pycarelle et al. 2020).

By enzymatic interesterification of fat with palm stearin fraction and palm, mid fraction combinations in numerous proportions different bakery and confectionery lipids were synthesized. After that their free fatty acids, slip melting point, fatty acid configuration, complex lipid constituents, and triglyceride configuration were determined. There was no trans-fats in the mixtures and were high in oleic (25.2–39.9%), palmitic (13.9–58.5%), and stearic (7.7–36.7%) shown by fatty acid configuration. Between 20 and 25 °C combinations of sal: palm mid fraction (50:50 and 25:75), sal: palm stearin fraction (50:50) indicated high content of solid lipid with less melting point range. Plasticity of foodstuffs increased after interesterification, relative to bakery lipids (commercial). Interesterification created a substantial alteration in the triacylglycerol configuration of the combination's studies. The resulting interesterified foodstuffs and mixtures prepared indicated favorable features and could be used to replace marketable confectionery and bakery lipids with no trans-fats (Hithamani 2020).

To enhance the nutritional quality of the product, use high oleic rapeseed in the form of oleogels organized by waxes (yellow and white beeswax, candelilla, and rice bran), ethyl cellulose, and monoacylglycerols. In this arrangement, oil has solid uniformity and keeps dietetic value. In terms of viscosity and consistency, the resultant oleogels were considerably different from palm oil. However, similar attributes like control sample (with palm oil) use oleogel with 5% of monoacylglycerol like margarine to obtain biscuits. To get the foodstuff, the addition of 2% of rice bran wax, 5% of monoacylglycerol, 2 and 3% of candelilla wax, and 3% of ethyl cellulose, allowed to obtain foodstuffs from which fat fraction migration was smaller or similar then the control sample with palm oil (Onacik-Gür and Żbikowska 2020).

Within the laboratory scale crystallization unit oleogels produced with beeswax was co-crystallized with marketable cake margarine. Then, in gluten-free cake preparations, the resulting amalgams were assessed. The batter samples' overrun values were reduced; when the beeswax oleogel was only used, less air-holding capacity was shown. When margarine was replaced entirely with beeswax, foodstuffs permeability and substantial volume of the samples were also reduced. However, the baked products with better properties were obtained compared to those of control products if used the 45%, 30%, and 15% replacement of margarine with beeswax oleogel (Demirkesen and Mert 2019).

To associate the characteristics of cookies formed with oleogel based cookies and commercial cookies the rice bran wax at different concentrations (3, 5, 7, and 9 wt%) oleogels (expeller-pressed)corn germ oil was manufactured. Their structural attributes, including, rheological characteristics, color, thermal behavior, hardness, and crystal structure were assessed. To inspect the performance of oleogels for different potential food applications it was incorporated into the cookies instead of using the marketable margarine. Overall, at concentration ≥3 wt% oleogels in both refined and expeller rapeseed corn germ oil could be manufactured by rice bran wax. Refined corn oil can produce a firmer gel than crude corn oil. There was no significant difference in the characteristics of commercially prepared cookies and oleogel based cookies but only the color difference was observed. Results show that the marketable shortening can be replaced in the bakery industry with the oleogels (Zhao et al. 2020).

5.8.2 Meat Emulsion and Sausages

The importance of pork back fat was already discussed for the bakery industry. Similarly, in the preparation of dry fermented sausages, it is also traditionally used and contributes to the characteristics of the final foodstuffs. During seasoning and dehydrating processes, in addition to its significant technological purpose, this lipid considerably influences the formation, texture, and appearance of the characteristic flavor and scent of dry fermented sausages, thus in the meat industry, its replacement in these foodstuffs is the main challenge. Sixteen percent of pork back fat was substituted with corn or rapeseed oil-based inulin gelled emulsions to yield fermented sausages having less fat with better fatty acid configuration. A substantial reduction in saturated fatty acids caused by the addition of emulsions and an increase in polyunsaturated fatty acids (PUFA), omega 6, and omega 3 types ($P < 0.05$). Only in sausages reformed with inulin rapeseed oil (5.87) an enhanced omega-6/omega -3 ratio was showed. Thiobarbituric acid did not show any mark of fat oxidation. In the reformed sausages reactive substance constituents were noticed. However, after 1—the month of storage and ripening the higher susceptibility of these sausages to lipolysis and oxidation in contrast to the control sausages shown by the considerably greater peroxide value and total acid number in modified sausages ($P < 0.05$) (Glisic et al. 2019).

Emulsions stabilized by whey with the addition of microalgae fat are reasonably appropriate for developing food preparations. Broiler chicken meat comprises high-ranking, highly palatable that can be easily digested animal protein, but essential fats are less in this meat. It would be valuable to develop recipes that will be enriched with essential fatty acids for seasoning chicken meat semi-finished products and other food products. Now consumers get more interested in world cuisine and are increasingly inclined to experiment with food. For modernization, packaging experimentations, and flavors the seasoning sauces classification is the most promising niche (Bazarnova et al. 2019).

Soybean oil emulsion gels (EG) were examined using stabilizers, functional ingredients, and gelling agents (sodium tripolyphosphate, chia, flour and soy protein isolate, inulin, carrageenan, and sodium caseinate) and in Bologna sausage used to exchange 50 and 100% of pork back fat. As control sample take two formulations with only pork backfat (20 and 10%). Microstructural, technological and nutritional features of Bologna sausages were assessed. The same lightness was shown by the emulsion gel articulated with chia flour as pork black fat. A more significant amount of polyunsaturated fatty acids, improved omega 3 constituent and lessens the saturated lipids up to 41% are shown by the sausages comprising EG(14%) or a mixture of EG and PBF (14% and 10%, respectively). When EG were added to the control foodstuffs more uniform batter and a condensed configuration were detected. The Bologna sausages comprising partial replacement of PBF exhibited greater firmness, chewiness, and shear force ($p < 0.05$), concerning texture characteristics. Products analogous to the control formulation were achieved when emulsion gel was used in amalgamation with PBF, particularly in rheological properties and texture, built an improved fatty acid profile (de Souza Paglarini et al. 2019).

Chia mucilage gels in emulsified meat model system use as a functional constituent. To replace 50% of pork back fat in the meat 3 model systems, three different concentrations of chia mucilage gel (15%, 20%, and 25%), were applied in two levels (2.5% and 5.0%). Verification of the two control treatments (FC1 and FC2 comprising 20 and 10% fat, respectively) was also done. The frequency range studied values of tan δ fluctuated between 0.37 and 0.40; represent a weak elastic gel-like behavior. After the thermal treatment mechanical characteristics of chia mucilage gels were well-preserved. With the addition of this gel, the meat emulsion firmness was enhanced. By considerably improved solidity, reduced cohesiveness values ($P < 0.05$), and elasticity compared to control system 1 all formulations with 5% chia mucilage were categorized. In emulsified meat foodstuffs, chia mucilage demonstrates its ability to replace saturated fat with enhanced technological features and additional healthier entitlements (Câmara et al. 2020).

To examine the consequence of nanocellulose (NCs) on microstructural and other properties of the sausages, use soybean oil in unadulterated pre-emulsified form or liquid form stabilized with NCs to substitute pork back fat in some meat sausages partially and examined the influence of nanocellulose types (cotton cellulose nanofiber, sisal cellulose nanofiber, and cotton cellulose nanocrystal) on the physicochemical and microstructural characteristics of the sausages. The physicochemical attributes, comprising rheological behavior, water holding capacity, cooking loss and textural features were assessed. The outcomes indicate that the sausages with pre-emulsified oil presented much-enhanced water and lipid binding capabilities, with considerably increased stiffness, chewiness, and springiness. it is summarized that NC-based emulsions are a feasible fat exchanger for meat sausages having similar stability and textural properties (Qi et al. 2020).

There is another way to reduce the saturated fatty acid in meat products. In pork patties for partial substitution of the solid lipid, fraction oleogel was used (H-75 for 75% replacement and H-25 for 25% replacement). Alteration in fatty acid profile and a substantial reduction of n-6/n-3 ratios occurs due to the integration of oleogel. For both degrees of lipid replacement, results demonstrated that both/ the patties are of the same quality produced with the control and oleogel incorporation, concerning the textural properties such as cohesiveness, firmness, and chewiness (Martins et al. 2019).

Using foam structured hydroxypropyl methylcellulose (HPMC) as oleogelator, canola oil was effectively changed to solid oleogels. By increasing the level of HPMC, a distinct increase in the textural and rheological characteristics of the HPMC oleogels and improvement in their oxidative stability were observed during storage. Using HPMC oleogels instead of beef fat contributed much progress by reducing the cooking loss and soft texture in terms of characteristics of meat patties. Furthermore, in meat patties, the contents of saturated fats holding beef tallow sample is about 42% and in HPMC oleogels containing meat patties were considerably condensed to 15%. In modern eras, oleogelation is used to convert the liquid plant source lubricant (oil) to compact-like gel without altering its chemical configuration. There are still many studies required to simplify the use of oleogel that are hydrocolloid-based (Oh et al. 2019).

For fresh meat foodstuffs development, the appropriateness of beeswax oleogels and ethyl cellulose manufactured with a healthy fat blend (fish oils, linseed, and olive) as lipid exchangers was studied. These oleogel remained stable for at least 1 month of refrigerated storage (3 ± 1 °C). However, during chilled storage, particularly in the ethyl cellulose case, there was a chance of some fat oxidation. Burgers (low-fat pork) manufactured by the developed oleogels with total replacement of PBF were observed to be softer than control and without significant variations in optical properties. 3.6 times increase of the polyunsaturated fatty acid/saturated fatty acid proportion and a 23-fold reduction of the omega-6/omega-3 proportion was observed in the fatty acid profile of the reformulated burger as compared to the control. The results of acceptability test indicate low scored values for the burgers produced with ethyl cellulose, with ratings below the neutral point in contrast to the ones manufactured with beeswax oleogel. From the above discussion, the potential of constituents developed to manufacture healthier fresh meat foodstuffs with an enhanced fatty acid profile. However, to enhance oxidative stability and sensory characteristics there is still a need to work on these strategies (Gómez-Estaca et al. 2019b).

To validate the applicability of ethyl cellulose oleogels, the thermo-oxidative behavior is crucial to characterize. Concerning textural properties, and dynamic chemical variations that occurred during the manufacturing process, holding of the gels, and heating should be taken into consideration for the composition of the gel network. Oxidative stability of ethyl cellulose oleogel indicated that at 6.0% ethyl cellulose 100 concentrations in oleogels, the movement of liquid oil confined in gel complex was hindered by its stable gel linkage and high viscosity. Processing temperature ≤120 °C for 32 less than 2 h to minimize the oxidation was recommended when integrated into food systems (Fu et al. 2020).

5.8.3 Chocolate and Confectionery Products

Fat-based bubbles are included in chocolate to change mouthfeel and textural characteristics that are important for end-users. Such foams are of current interest in foods because of reduced fat/oil content and these foams are lighter in weight than oil alone (Binks and Marinopoulos 2017). It was found that the most stable β-form of the double chain length structure (β-2) was formed by the ternary fat mixtures of CB/SatSatO/OSatO with a ratio of CB/ (SatSatO+ OSatO) of 1:1), in which CB refers to the cocoa butter, O and Sat refers to the oleic acid moieties and saturated fatty acids (stearic:S + palmitic:P), respectively, which revealed sharp melting profiles and sufficient hardness around body temperature without the tempering processes. Dark chocolates during 1 year for experimental examination at a temperature between 15 °C and 30 °C show no lipid bloom with cocoa butter alternative fats at proportions of CB/SatSatO/OSatO 50/20/30–50/0/50. Overall, it was observed that without tempering procedures, fat combinations made of OSatO/CB/SatSatO, which are high in oleic acid moieties can be used equally in place of the cocoa butter lipid (Watanabe et al. 2021).

Mango kernel fat-based chocolate fat, interesterified blends, physical blend, and its control non-interesterified blends prepared with mango kernel fat-third stearin contained high-StOSt (1,3-stearoyl-2-oleoyl-glycerol) levels. At 10–30 °C physical blend and interesterified blends exhibited good compatibilities with optimal triglycerides compositions (i.e., appropriate ratios of POP/StOSt/POSt). The dark chocolate formed by using these fats in contrast to traditional cocoa butter chocolate was shown by significantly higher heat stability. The non-interesterified blends crystals showed spherulitic growth that was the same as cocoa butter, while the physical blends and interesterified blends demonstrated a plate-like increase. This plate-like growth contributed to the denser structure, at hot temperature preventing bloom from the melted fat migration, and at room temperature further polymorphic transformation and recrystallization (Jin et al. 2019).

The hybrid hydrogel was prepared in chocolate by combining chemical and physical cross-linking approaches and using sodium alginate-pectin-citric acid mixture as fat mimetic. From the rheological, structural, crystal polymorph, and thermal points of view, different concentrations (0, 25, 50, 75, and 100% (v/v)) of the hybrid hydrogel dispersed chocolate formulations were characterized. It was observed that in chocolate, up to 50% (v/v) hydrogel dispersion exhibited the highest melting resistance (80 °C), minor surface roughness required from V (β polymorph) cocoa butter polymorphism, a glossy appearance, and non-Newtonian behavior well described by Casson model. Consequently, by using this approach, consumer preference can be met towards low-fat chocolates by producing the chocolate that is heat stable and enhancing the chocolate confections with the particles of hybrid hydrogel (Francis and Chidambaram 2019).

The research was done about the effects of candelilla wax oleogel on the quality of noodles compared to the control without oleogel, in terms of microstructural, topographical, and rheological attributes. The tensile and viscoelastic measurements indicated higher viscous characteristics of noodles with increasing levels of oleogels doughs. The more excellent thermal conductivity was presented by the oleogel noodles associated with a softer texture, in contrast to control. Moreover, these oleogel noodles possessed higher porous microstructural attributes. Furthermore, by microcomputed tomography, the 3-dimensional volumetric examination indicated that oleogel during cooking allowed cooked noodles to have higher pore size distribution and less compact clearly showing gains to facilitate the rehydration. Considerable reduction in cooking loss and water absorption occurs in noodles with oleogels (Oh and Lee 2020).

5.9 Conclusion

This review provides a glance at the oleofoam, its stability, and factors affecting its strength. As in the previous studies we observe that drainage, coalescence, and diffusive coarsening are the problems that affect the oleofoam stability. In this review, we discuss the solution to minimize these problems. Increasing the fluid's bulk viscosity, adding particles and surfactants, and in foam, formation using gas having less permeability or more significant surface compression, these problems can be reduced respectively. Moreover, the effect of time, temperature, cooling rate, and shear rate on the foam stability was observed in this review. At greater monoglyceride concentration, increasing the OBC and rigidity of the network at a 0.1 °C/min HIU application was more effective. To reduce the health hazards (heart problems, obesity, cancer, diabetes)created by the fat (saturated fat), primarily the palm fat is used widely in the baking industry and the pork back fat that is primarily used in the meat industry, we discussed different methods to replace these fats with less saturated fat (polyunsaturated fat) possessing the same quality as by these two mentioned fats.

References

Abdollahi M, Amir S, Goli H, Soltanizadeh N (2019) Physicochemical properties of foam-templated oleogel based on gelatin and xanthan gum. Eur J Lipid Sci Technol 122:1900196. https://doi.org/10.1002/ejlt.201900196

Abdolmaleki K, Alizadeh L, Nayebzadeh K, Hosseini SM, Shahin R (2020) Oleogel production based on binary and ternary mixtures of sodium caseinate, xanthan gum, and guar gum: optimization of hydrocolloids concentration and drying method. J Texture Stud 51(2):290–299. https://doi.org/10.1111/jtxs.12469

Aliasl Khiabani A, Tabibiazar M, Roufegarinejad L, Hamishehkar H, Alizadeh A (2020) Preparation and characterization of carnauba wax/adipic acid oleogel: a new reinforced oleogel for application in cake and beef burger. Food Chem 333:127446. https://doi.org/10.1016/j.foodchem.2020.127446

Bazarnova J, Kuznetsova A, Toumi A (2019) Use of lipids of chlorella microalgae in poultry meat marinades and sauces recipes. Agron Res 17:1287–1298. https://doi.org/10.15159/AR.19.097

Binks BP, Marinopoulos I (2017) Ultra-stable self-foaming oils. Food Res Int 95:28–37. https://doi.org/10.1016/j.foodres.2017.02.020

Binks BP, Garvey EJ, Vieira J (2016) Whipped oil stabilised by surfactant crystals. Chem Sci 7(4):2621–2632. https://doi.org/10.1039/c6sc00046k

Callau M, Sow-Kébé K, Jenkins N, Fameau AL (2020) Effect of the ratio between fatty alcohol and fatty acid on foaming properties of whipped oleogels. Food Chem 333:127403. https://doi.org/10.1016/j.foodchem.2020.127403

Câmara AKFI, Okuro PK, da Cunha RL, Herrero AM, Ruiz-Capillas C, Pollonio MAR (2020) Chia (Salvia hispanica L.) mucilage as a new fat substitute in emulsified meat products: technological, physicochemical, and rheological characterization. LWT Food Sci Technol 125:109193. https://doi.org/10.1016/j.lwt.2020.109193

Demirkesen I, Mert B (2019) Utilization of beeswax oleogel-shortening mixtures in gluten-free bakery products. J Am Oil Chem Soc 96(5):545–554. https://doi.org/10.1002/aocs.12195

Demirkesen I, Mert B (2020) Recent developments of oleogel utilizations in bakery products. Crit Rev Food Sci Nutr 60(14):2460–2479. https://doi.org/10.1080/10408398.2019.1649243

Dickinson E (2017) Biopolymer-based particles as stabilizing agents for emulsions and foams. Food Hydrocoll 68:219–231. https://doi.org/10.1016/j.foodhyd.2016.06.024

Du L, Jiang Q, Li S, Zhou Q, Tan Y, Meng Z (2021) Microstructure evolution and partial coalescence in the whipping process of oleofoams stabilized by monoglycerides. Food Hydrocoll 112:106245. https://doi.org/10.1016/j.foodhyd.2020.106245

Fameau AL, Saint-Jalmes A (2017) Non-aqueous foams: current understanding on the formation and stability mechanisms. Adv Colloid Interf Sci 247:454–464. https://doi.org/10.1016/j.cis.2017.02.007

Francis FP, Chidambaram R (2019) Hybrid hydrogel dispersed low fat and heat resistant chocolate. J Food Eng 256:9–17. https://doi.org/10.1016/j.jfoodeng.2019.03.012

Fu H, Lo YM, Yan M, Li P, Cao Y (2020) Characterization of thermo-oxidative behavior of ethylcellulose oleogels. Food Chem 305:125470. https://doi.org/10.1016/j.foodchem.2019.125470

Gao Y, Wu S (2020) Development and evaluation of a novel oleogel system based on starch-water-wax-oil. Food Funct 11(9):7727–7735. https://doi.org/10.1039/d0fo01785j

Gaudino N, Ghazani SM, Clark S, Marangoni AG, Acevedo NC (2019) Development of lecithin and stearic acid based oleogels and oleogel emulsions for edible semisolid applications. Food Res Int 116:79–89. https://doi.org/10.1016/j.foodres.2018.12.021

Giacomozzi AS, Palla CA, Carrín ME, Martini S (2019) Physical properties of monoglycerides oleogels modified by concentration, cooling rate, and high-intensity ultrasound. J Food Sci 84(9):2549–2561. https://doi.org/10.1111/1750-3841.14762

Giacomozzi A, Palla C, Elena M, Martini S (2020) Tailoring physical properties of monoglycerides oleogels using high-intensity ultrasound. Food Res Int 134:109231. https://doi.org/10.1016/j.foodres.2020.109231

Glisic M, Glisic M, Boskovic M, Baltic MZ, Trbovic D, Suvajdzic B, Vasilev D (2019) Fat replacement and PUFA enrichment challenges in fermented sausage production. IOP Conference Series: Earth and Environmental Science 333(1). https://doi.org/10.1088/1755-1315/333/1/012061

Goibier L, Pillement C, Monteil J, Faure C, Leal-Calderon F (2019) Emulsification of non-aqueous foams stabilized by fat crystals: towards novel air-in-oil-in-water food colloids. Food Chem 293:49–56. https://doi.org/10.1016/j.foodchem.2019.04.080

Gómez-Estaca J, Herrero AM, Herranz B, Álvarez MD, Jiménez-Colmenero F, Cofrades S (2019a) Characterization of ethyl cellulose and beeswax oleogels and their suitability as fat replacers in healthier lipid pâtés development. Food Hydrocoll 87:960–969. https://doi.org/10.1016/j.foodhyd.2018.09.029

Gómez-Estaca J, Pintado T, Jiménez-Colmenero F, Cofrades S (2019b) Assessment of a healthy oil combination structured in ethyl cellulose and beeswax oleogels as animal fat replacers in low-fat, PUFA-enriched pork burgers. Food Bioprocess Technol 12(6):1068–1081. https://doi.org/10.1007/s11947-019-02281-3

Gunes DZ, Murith M, Godefroid J, Pelloux C, Deyber H, Schafer O, Breton O (2017) Oleofoams: properties of crystal-coated bubbles from whipped oleogels-evidence for pickering stabilization. Langmuir 33(6):1563–1575. https://doi.org/10.1021/acs.langmuir.6b04141

Heymans R, Tavernier I, Dewettinck K, Van der Meeren P (2017) Crystal stabilization of edible oil foams. Trends Food Sci Technol 69:13–24. https://doi.org/10.1016/j.tifs.2017.08.015

Heymans R, Tavernier I, Danthine S, Rimaux T, Van Meeren P, Dewettinck K (2018) Food-grade monoglyceride oil foams: the effect of tempering on foamability, foam stability and rheological properties. Food Funct 9(6):3143–3154. https://doi.org/10.1039/c8fo00536b

Hithamani G, Reddy YS, Khatoon S (2020) Lipase-mediated interesterification of oils and fats for the preparation of bakery and confectionery fats. Available at Research Square. https://doi.org/10.21203/rs.3.rs-67079/v1

Jin J, Jin Q, Wang X, Akoh CC (2019) Improving heat and fat bloom stabilities of "dark chocolates" by addition of mango kernel fat-based chocolate fats. J Food Eng 246:33–41. https://doi.org/10.1016/j.jfoodeng.2018.10.027

Karp K (2016) Fabrication and characterization of arrested non-aqueous foam. Lodz University of Technology.

Kavadia MR, Yadav MG, Vadgama RN, Odaneth AA, Lali AM (2019) Production of trans-free interesterified fat using indigenously immobilized lipase. Prep Biochem Biotechnol 49(5):444–452. https://doi.org/10.1080/10826068.2019.1566142

Lee HS, Kim KH, Park SH, Hur SW, Auh JH (2020) Amylose lipid complex as a fat replacement in the preparation of low-fat white pan bread. Foods 9(2):1–12. https://doi.org/10.3390/foods9020194

Lei M, Zhang N, Lee WJ, Tan CP, Lai OM, Wang Y, Qiu C (2020) Non-aqueous foams formed by whipping diacylglycerol stabilized oleogel. Food Chem 312:126047. https://doi.org/10.1016/j.foodchem.2019.126047

Li X, Cheng X, Wu G, Huang J, Zhang H, Jin Q, Wang X (2020) Individual and combined effects of frying load and deteriorated polar compounds on the foaming of edible oil. Food Res Int 134:109206. https://doi.org/10.1016/j.foodres.2020.109206

Luo SZ, Hu XF, Jia YJ, Pan LH, Zheng Z, Zhao YY, Mu DD, Zhong XY, Jiang ST (2019) Camellia oil-based oleogels structuring with tea polyphenol-palmitate particles and citrus pectin by emulsion-templated method: preparation, characterization and potential application. Food Hydrocoll 95:76–87. https://doi.org/10.1016/j.foodhyd.2019.04.016

Martins AJ, Lorenzo JM, Franco D, Vicente AA, Cunha RL, Pastrana LM, Quiñones J, Cerqueira MA (2019) Omega-3 and polyunsaturated fatty acids-enriched Hamburgers using sterol-based Oleogels. Eur J Lipid Sci Technol 121(11):1–8. https://doi.org/10.1002/ejlt.201900111

Mishima S, Suzuki A, Sato K, Ueno S (2016) Formation and microstructures of whipped oils composed of vegetable oils and high-melting fat crystals. J Am Oil Chem Soc 93(11):1453–1466. https://doi.org/10.1007/s11746-016-2888-4

Mishra K, Bergfreund J, Bertsch P, Fischer P, Windhab EJ (2020a) Crystallization-induced network formation of tri- and monopalmitin at the middle-chain triglyceride oil/air Interface. Langmuir 36(26):7566–7572. https://doi.org/10.1021/acs.langmuir.0c01195

Mishra K, Dufour D, Windhab EJ (2020b) Yield stress dependent foaming of edible crystal-melt suspensions [research-article]. Cryst Growth Des 20(2):1292–1301. https://doi.org/10.1021/acs.cgd.9b01558

Muin SR, Rabbani A, Bournival G, Ata S, Armstrong RT (2021) Utilization of microcomputed tomography and pore network modeling to characterize foam dynamics. Chem Eng Sci 230:116174. https://doi.org/10.1016/j.ces.2020.116174

Murakami R, Kobayashi S, Okazaki M, Bismarck A, Yamamoto M (2018) Effects of contact angle and flocculation of particles of oligomer of tetrafluoroethylene on oil foaming. Front Chem 6:435. https://doi.org/10.3389/fchem.2018.00435

Oh I, Lee S (2020) Rheological, microstructural, and tomographical studies on the rehydration improvement of hot air-dried noodles with oleogel. J Food Eng 268:109750. https://doi.org/10.1016/j.jfoodeng.2019.109750

Oh I, Lee JH, Lee HG, Lee S (2019) Feasibility of hydroxypropyl methylcellulose oleogel as an animal fat replacer for meat patties. Food Res Int 122:566–572. https://doi.org/10.1016/j.foodres.2019.01.012

Onacik-Gür S, Żbikowska A (2020) Effect of high-oleic rapeseed oil oleogels on the quality of short-dough biscuits and fat migration. J Food Sci Technol 57(5):1609–1618. https://doi.org/10.1007/s13197-019-04193-8

Palla C, Giacomozzi A, Genovese DB, Carrín ME (2017) Multi–objective optimization of high oleic sunflower oil and monoglycerides oleogels: searching for rheological and textural properties similar to margarine. Food Struct 12:1–14. https://doi.org/10.1016/j.foostr.2017.02.005

Papadaki A, Cipolatti EP, Aguieiras ECG, Pinto MCC, Kopsahelis N, Freire DMG, Mandala I, Koutinas AA (2019) Development of microbial oil wax-based oleogel with potential application in food formulations. Food Bioprocess Technol 12(6):899–909. https://doi.org/10.1007/s11947-019-02257-3

Papadaki A, Kopsahelis N, Freire DMG, Mandala I, Koutinas AA (2020) Olive oil oleogel formulation using wax esters derived from soybean fatty acid distillate. Biomol Ther 10(1):1–9. https://doi.org/10.3390/biom10010106

Park C, Bemer HL, Maleky F (2018) Oxidative stability of rice bran wax Oleogels and an Oleogel cream cheese product. J Am Oil Chem Soc 95(10):1267–1275. https://doi.org/10.1002/aocs.12095

Patel AR, Nicholson RA, Marangoni AG (2020) Applications of fat mimetics for the replacement of saturated and hydrogenated fat in food products. Curr Opin Food Sci 33:61–68. https://doi.org/10.1016/j.cofs.2019.12.008

Paximada P, Koutinas AA, Scholten E, Mandala IG (2016) Effect of bacterial cellulose addition on physical properties of WPI emulsions. Comparison with common thickeners. Food Hydrocoll 54:245–254

Pehlivanoğlu H, Demirci M, Toker OS, Konar N, Karasu S, Sagdic O (2018) Oleogels, a promising structured oil for decreasing saturated fatty acid concentrations: production and food-based applications. Crit Rev Food Sci Nutr 58(8):1330–1341. https://doi.org/10.1080/10408398.2016.1256866

Pycarelle SC, Bosmans GM, Nys H, Brijs K, Delcour JA (2020) Stabilization of the air-liquid interface in sponge cake batter by surface-active proteins and lipids: a foaming protocol based approach. Food Hydrocoll 101. https://doi.org/10.1016/j.foodhyd.2019.105548

Qi W, Wu J, Shu Y, Wang H, Rao W, Xu HN, Zhang Z (2020) Microstructure and physiochemical properties of meat sausages based on nanocellulose-stabilized emulsions. Int J Biol Macromol 152:567–575. https://doi.org/10.1016/j.ijbiomac.2020.02.285

Rosen-Kligvasser J, Davidovich-Pinhas M (2021) The role of hydrogen bonds in TAG derivative-based oleogel structure and properties. Food Chem 334:127585. https://doi.org/10.1016/j.foodchem.2020.127585

Schmid T, Baumer B, Rüegg R, Näf P, Kinner M, Müller N (2020) Evaluation of innovative technological approaches to replace palm fat with physically modified Swiss rapeseed oil in bakery products. Int J Food Sci Technol 55(8):2990–2999. https://doi.org/10.1111/ijfs.14564

de Souza Paglarini C, de Figueiredo Furtado G, Honório AR, Mokarzel L, da Silva Vidal VA, Ribeiro APB, Cunha RL, Pollonio MAR (2019) Functional emulsion gels as pork back fat replacers in Bologna sausage. Food Struct 20:100105. https://doi.org/10.1016/j.foostr.2019.100105

Truong T, Prakash S, Bhandari B (2019) Effects of crystallisation of native phytosterols and monoacylglycerols on foaming properties of whipped oleogels. Food Chem 285:86–93. https://doi.org/10.1016/j.foodchem.2019.01.134

Valoppi F, Salmi A, Ratilainen M, Barba L, Puranen T, Tommiska O, Helander P, Heikkilä J, Haeggström E (2020) Controlling oleogel crystallization using ultrasonic standing waves. Sci Rep 10(1):1–13. https://doi.org/10.1038/s41598-020-71177-6

Wang H, Brito-Parada PR (2020) Coalescence dynamics of particle-laden bubbles. Langmuir 36(19):5394–5399. https://doi.org/10.1021/acs.langmuir.0c00938

Watanabe S, Yoshikawa S, Sato K (2021) Formation and properties of dark chocolate prepared using fat mixtures of cocoa butter and symmetric/asymmetric stearic-oleic mixed-acid triacylglycerols: impact of molecular compound crystals. Food Chem 339:127808. https://doi.org/10.1016/j.foodchem.2020.127808

Wijarnprecha K, de Vries A, Santiwattana P, Sonwai S, Rousseau D (2019) Microstructure and rheology of oleogel-stabilized water-in-oil emulsions containing crystal-stabilized droplets as active fillers. LWT Food Sci Technol 115:108058. https://doi.org/10.1016/j.lwt.2019.04.059

Xiao J, Wang X, Perez Gonzalez AJ, Huang Q (2016) Kafirin nanoparticles-stabilized Pickering emulsions: microstructure and rheological behavior. Food Hydrocoll 54:30–39. https://doi.org/10.1016/j.foodhyd.2015.09.008

Xiao J, Zhang M, Wang W, Li S, Wang Y, Du G, Zhang K, Li Y (2020) Using Flammulina velutipes derived chitin-glucan nanofibrils to stabilize palm oil emulsion: a novel food grade Pickering emulsifier. Int J Biol Macromol 164:4628–4637. https://doi.org/10.1016/j.ijbiomac.2020.09.073

Yang Y, Fang Z, Chen X, Zhang W, Xie Y, Chen Y, Liu Z, Yuan W (2017) An overview of Pickering emulsions: solid-particle materials, classification, morphology, and applications. Front Pharmacol 8:1–20. https://doi.org/10.3389/fphar.2017.00287

Zhao M, Lan Y, Cui L, Monono E, Rao J, Chen B (2020) Formation, characterization, and potential food application of rice bran wax oleogels: expeller-pressed corn germ oil versus refined corn oil. Food Chem 309:125704. https://doi.org/10.1016/j.foodchem.2019.125704

Chapter 6
Clean Label Interventions in Active and Intelligent Food Packaging

Abdul Waheed Khan, Ume Roobab, Kainat Shehzadi, Muhammad Inam-Ur-Raheem, and Rana Muhammad Aadil

6.1 Introduction

Packaging protects food from surrounding environmental factors such as oxygen, water vapor, volatile compounds, and spoilage microorganisms, avoid nutritional losses and desirable quality factors such as flavor volatile compounds, and extend the shelf life of food (Gaikwad et al. 2018). At the same time, active and intelligent packaging technologies provide these functionalities all in one package and other new ways of maintaining the nutritional and sensorial quality of packaged food (Ahmed et al. 2017). Traditional synthetic plastic polymers/petrochemicals are low cost, readily available, and have excellent optical, mechanical, and barrier properties; however, they are hardly biodegradable (Di Filippo et al. 2021), and increases the severe environmental issues due to plastic waste production (Wang and Wang 2017). Besides, synthetic additives incorporated during packaging formulations prevent oxidation and improve the shelf life of food but are undesirable to consumers who prefer clean label products. For this purpose, biobased eco-friendly packaging has been developed, which reduces environmental waste by focusing on recyclability, reusability, and biodegradability (Herbes et al. 2018).

Abdul Waheed Khan and Ume Roobab contributed equally with all other contributors.

A. W. Khan · M. Inam-Ur-Raheem · R. M. Aadil (✉)
National Institute of Food Science and Technology, University of Agriculture, Faisalabad, Pakistan

U. Roobab
School of Food Science and Engineering, South China University of Technology, Guangzhou, China

K. Shehzadi
Institute of Food Science and Nutrition, University of Sargodha, Sargodha, Pakistan

C. M. Galanakis (ed.), *The Age of Clean Label Foods*,
https://doi.org/10.1007/978-3-030-96698-0_6

Bio-based packaging materials (used for coatings, multi-layers, and films) can be classified based on their origin/source, degree of polymerization, chemical structure, type/shape of molecular linkage, extraction techniques, etc. (Table 6.1 summarizes the range of biobased materials used for clean label food packaging). Hence these various properties determine the final attributes/physicochemical aspects of overall packaging (Reichert et al. 2020). Bio-based clean label packaging (includes cellulose, starch, proteins, lipids, and waxes) is biodegradable and can be used safely as a carrier of nutrients, antimicrobials, antioxidants, flavorings, etc., in direct contact with the food matrix. Advanced forms of non-synthetically produced polymers include bio-composites, coatings, multilayers, surface modifications, additives incorporation to improve gas/moisture barrier quality, and biodegradability (Müller et al. 2020). In contrast, bio-based materials have a low barrier and mechanical properties, which reduced the shelf life of food as compared to traditional food packaging materials, thus limited its use. The problem can be addressed by incorporating different clean label nanomaterial or active agents in bio-based packaging such as plant extracts or oils to improve the antimicrobial activity, prevent oxidation, and creating barriers to microbial spoilage (Al-Tayyar et al. 2020).

This led researchers to investigate more clean label techniques for bio-based food packaging such as utilizing cellulose and chitosan (structural material of crustaceans, insects, and fungi) as biodegradable flexible food packaging material. Various strategies have been applied to use chitosan in edible food coatings with clean label fillers such as natural extracts and essential oils as compared to non-biodegradable synthetic plastic-based food packaging materials (Priyadarshi and Rhim 2020). The chitosan and gelatin-based edible films have high gas/aroma barrier capability to retain aromatic components such as 1-hexanal, 2-hexen-1-ol, 1-hexanol, 3-hexanone, and phenol. To prevent aroma loss, polymeric materials can be selected for food packaging based on their aroma barrier capabilities (Benbettaïeb et al. 2019).

Despite the considerable effort in the past, the carbon-rich agricultural wastes (having fruits and vegetable pomace, peels, husk, and straws) have been used as sustainable raw materials in the production of bio-based blends in comparison to synthetic plastics such as polypropylene (PP) and polyethylene (PET) (Maraveas 2020). Bio-based plastics such as PLA have been progressively looking for a clean label substitute for the traditional plastic but PLA film shows a low O_2 barrier and poor antifogging drawbacks. However, Zhang et al. (2019b) modified PLA films via layer-by-layer assembly of poly(vinyl alcohol) (PVA) and tannic acid to develop all-organic coatings having high transparency, high O_2 barrier, and antifogging advantages. Utilization of unpurified agar-based extracts for clean label food packaging films is a cost-effective and sustainable way to grabbed attention because of no added plasticizer in contrast to the purified agar films. Whereas the impurities (polysaccharides and proteins) provided greater light blocking capacity, which prevents β-carotene degradation by UV light (Martínez-Sanz et al. 2019). Furthermore, the traditional composite films consist of nano-silver as an antibacterial component that is potentially hazardous due to leakage of silver into the human body (Gu et al. 2020).

Table 6.1 Advanced bio-based materials used as clean label alternatives

Packaging characteristics	Advantage	References
Pinhao/cassava/polybutylene adipate co-terephthalate blends with the natural extracts (rosemary and green tea) are an excellent alternative matrix (substitute to cassava starch)	Have good mechanical and barrier properties.	Müller et al. (2020)
Gelatin-chitosan bio-based composite film	Have antioxidants, mechanical, and physicochemical characteristics but UV irradiation for 10 min reduced mechanical strength and barrier properties of the film against water vapor. While long-term radiation caused some cracks in the film's network.	Rezaee et al. (2020)
Rice starch (extracted from broken grains) and fish protein (recovered from *Withemouth Croacker* (*Micropogonias furnieri*) with the addition of oregano essential oil	Starch/protein (50/50) was the most suitable for packaging materials development, which had homogeneous and cohesive texture, lower solubility (8.0%), water vapor permeability, and intermediate mechanical properties (tensile strength, elongation) with inhibition of peroxidase suggesting its promising application as anti-browning packaging in fruits and vegetables.	Romani et al. (2017)
Cellulose acetate-based films with the addition of polysaccharides such as sodium alginate and carrageenan	Increased tensile strength, thermal stability, and antibacterial activity inhibit the growth of *E. coli*, *Staphylococcus aureus*, and *Pseudomonas syringae*.	Rajeswari et al. (2020)
Cellulose/lignin composite films	Higher UV-shielding property (block the ultraviolet lights at only 5 wt.% lignin content).	Guo et al. (2019)
Unpurified agar-based extracts obtained from *Gelidium sesquipedale* seaweed-based films	Sustainable, has excellent water barrier performance, which is better than purified agar. Showed resistance to high humidity conditions and retained film structure.	Martínez-Sanz et al. (2019)
Red cabbage extract and sweet whey into starch films	Presented high antioxidant capacity (due to anthocyanins), good mechanical properties; however, low solubility and water vapor permeability. Were suitable for ground beef packaging.	Ribeiro Sanches et al. (2021)
Yerba mate extract (10 wt.% or 20 wt.%) into starch film	Showed color changes in different pH, indicating their promising role as active and smart packaging for food, following the new trends for biodegradable and functional packaging.	Ceballos et al. (2020)

(continued)

Table 6.1 (continued)

Packaging characteristics	Advantage	References
Sunflower hulls phenolic compounds and cellulose fibers into starch film	The phenolic extract provided antioxidant properties to the films while the cellulose fibers reinforced the starch films with a threefold increase in Young's modulus and improved film integrity.	Menzel (2020)
Betalains-rich red pitaya (*Hylocereus polyrhizus*) peel extract (0.25–1.00 wt%) onto starch/PVA films	The incorporation of the extract enhanced the water vapor barrier and UV-Vis barrier ability, mechanical, antioxidant and antimicrobial potential of the films. The film containing 1.00 wt% of the extract was more sensitive to ammonia than other films.	Qin et al. (2020)
Curcumin and green tea extract into gelatin-based bioactive coatings	Showed an excellent release behavior in contact with fatty food simulants, zein-based coatings would be more adequate for the packaging of high-water content food products.	Alehosseini et al. (2019)
Soy protein isolates, micro fibrillated cellulose, and clove essential oil as active nanocomposite films	Showed antioxidant and antimicrobial activity against pathogens. Micro fibrillated cellulose reinforced the protein matrix, increasing the mechanical strength and Young's modulus of the films and improving their barrier properties to water vapor and oxygen. Simultaneously, clove essential oil caused some plastification effect of protein and nanocomposite matrix (mainly observed in the mechanical properties, solubility, and water content).	Ortiz et al. (2018)
Corn starch-chitosan composite films with 3% incorporation of copolymer pluronic F127	Increased hydrophobicity and decreased water vapor permeability (42 to 3%).	Fonseca-García et al. (2021)
Green tea and basil extracts (5 wt.%) into cassava starch	These films degraded in soil under 2 weeks and were thermally stable up to 240 °C.	Medina et al. (2017)
Peanut shell and skin extracts (rich rutin and 4-*O*-caffeoulquinic acid) into starch-chitosan films	The tensile strength of film increased with the addition of peanut skin extracts, it decreases viscosity, water vapor permeability, and swelling. While expanding the color L* and opacity of film.	Meng et al. (2020)
Cellulose films based on durian rind	The regenerated cellulose film exhibited a fair appearance with a smooth surface and excellent transparency with high rigidity and tensile strength. That film could be 100% broken down in the soil in 4 weeks, higher biodegradation percentage than that of cellophane, suggesting outstanding biodegradability.	Zhao et al. (2019)

(continued)

Table 6.1 (continued)

Packaging characteristics	Advantage	References
Cellulose acetate films with the addition of carotenoids (norbixin, zeaxanthin, or lycopene) at the concentration of 0.03%, 0.05%, and 0.1%	The addition of 0.1% lycopene or zeaxanthin showed higher tensile strength and elongation at break. Films with 0.1% of incorporation of norbixin or lycopene showed a better protective effect on sunflower oil stability than zeaxanthin. More significant barrier to the transmission of UV–Vis, has an antioxidant effect and pigment stability.	Assis et al. (2020)
Carboxymethyl cellulose films	Biodegradable materials for food packaging, contain hydrophobicity, water barrier, and mechanical properties	Shahbaz et al. (2016)
Okra mucilage/corn starch films	Compact and uniform structure, low water vapor permeability, low solubility (15%), good thermal and mechanical properties, swelling capacity (95%), and no toxicological responses compared to films made with corn starch alone.	Araújo et al. (2018)
Valonea tannin (0–15% w/w) into soy protein isolate films	Showed improved UV light barrier properties, and excellent antioxidant properties via soya by-product.	Wang and Wang (2017)
Alginate and chitosan-based coatings on paperboard	Achieved wettability, water absorptiveness, and grease resistance.	
Alginates-edible films having sunflower oil, glycerol, or virgin coconut oil	A clean label alternative to plastic packaging.	Dewi and Husni (2020), Marismandani and Husni (2020)
Cellulose nanofibril coatings	Pouches made of these multilayer films retained their atmosphere and shape and protected ground dry foods from further oxidation during storage. Even the water vapor/oxygen barrier property was unaffected by irradiation (used to sterilize that multilayer films packed foods).	Vähä-Nissi et al. (2017)

On the other hand, intelligent packaging systems provide information on packed food and its environment in a convenient, and integrated way. Freshness indicators are linked directly with the quality aspects of packed food. Any alteration of packed food characteristics indicates microbial growth and/or oxidation of nutrients. Several metabolites (glucose, organic acids, carbon dioxide, biogenic amines, and sulfuric compounds) produced during microbial activities react with the sensor which indicates quality alteration by colorimetric indicators (visible color change), spoilage gases, or electrical signals as in the case of radio-frequency identification (RFID) packaging or tags (Dalmoro et al. 2017). This chapter will create awareness and interest in green and clean label active and intelligent aspects of food packaging.

6.2 Clean Label Active Packaging

Since the last decade, active or interactive packaging films made with bio-based materials or include clean label materials or freshness additives. Along with that, active packaging is designed to interact directly with the food contents and packaging environment (Remya et al. 2017). Traditionally, food packaging in direct contact could be a potential risk of toxicity by the migration of chemicals or synthetic compounds from film to the food especially in the case of edible films and coatings (Iordanskii et al. 2018). Many new technologies have been developed today, which are expected to provide clean label active food packaging solutions as mentioned below.

6.2.1 Antimicrobial Food Packaging

Food safety concerns with the increasing number of microbial cross-contamination events and the need for extending shelf-life of packaged food in addition to the overall quality necessitates the development of novel antimicrobial or preservative strategies (Rokbani et al. 2019). Subsequently, biofilms or active bioplastic food packaging with the addition of natural antimicrobial agents simultaneously prevent the spoilage microbes and extend the food shelf life. Antimicrobial active packaging is an imperial type of food packaging, whose effectiveness depends on incorporated active antimicrobial agent or peptides and antimicrobial polymers that could inhibit pathogenic or spoilage microorganisms. An overall summary of advanced antimicrobial agents incorporated onto bio-based films is given in Table 6.2.

Chitosan has the advantage to be used in food packaging as a biodegradable marine polymer and antimicrobial agent with potential for antioxidant aspects and biocompatibility of the package. Chitosan surface coatings are nontoxic, eco-friendly polymer, provide barrier properties against corrosion with excellent film quality such as tensile strength, mechanical characteristics, etc. Further, the addition of other natural or clean label agents such as essential oil extracts into chitosan films can enhance its antimicrobial ability (Remya et al. 2017).

According to El-Fattah et al. (2016), 15% extracted chitosan as natural organic filler with epoxy coating was useful against Gram-positive bacteria. The synergistic effect of chitosan-epoxy coatings (2–20%) not only provides antimicrobial activity but also improved the viscoelastic attributes, mechanical and chemical resistance in the composite. Similarly, the chitosan-starch edible composite film (1.2% v/v total weight, 0.4% chitosan, and 1.5% glycerol as a plasticizer) with the addition of turmeric extract (0.375%) showed antimicrobial activity against *Escherichia coli* with improved physical properties of film such as tensile strength (18.1 kg/mm^2), and elongation (30.24%) (Diana et al. 2019).

Table 6.2 Bio-based antimicrobial food packaging materials

Packaging	Characteristics	Antimicrobial function	Other advantages	References
PLA films	Plasticizers: 10 wt.% triethyl citrate and glycerol triacetate. Fillers: halloysite nanotubes (3 wt.%) and chitosan (1 wt.%)	Have antibacterial efficacy against *E. coli* (80%) and *S. aureus* (70%).	Acted as an excellent reinforcing filler and improved the tensile strength and tensile modulus (~12 times higher elongation at break compared to pure PLA), good barrier properties against moisture and UV rays.	Singh et al. (2020)
	Active agents (thymol, kesum, and curry at 10% wt.) 2 g of PLA resins (5% wt.) was mixed with 40 mL of dichloromethane	Showed positive results against Gram-positive bacteria (S. aureus); however, shows adverse effects against Gram-negative bacteria (*E. coli* and *Aspergillus Brasiliensis* due to embedded active agents within PLA matrix.	Used in food packaging to enhance the shelf life of meats, fruits, and vegetable products. It preserved chicken meat until 15 days by retaining its odorless and firm textural properties.	Mohamad et al. (2020)
	PLA blended with poly (3-hydroxybutyrate)	Antimicrobial activities against *S. aureus* and *E. coli.*	Good barrier to water vapor and increased antioxidant activities.	Burgos et al. (2017)
PVA films	Cellulose nanocrystals and chitosan nanoparticle onto PVA films	The antifungal activity was investigated against *Colletotrichum gloeosporioides*, *Lasiodiplodia theobromae*, and *C. gloeosporioides.*	Exhibited high tensile strength, thermostability biodegradability, and excellent antifungal properties against post-harvest pathogens.	Dey et al. (2020)

(continued)

Table 6.2 (continued)

Packaging	Characteristics	Antimicrobial function	Other advantages	References
Chitosan-based active packaging	Combination of chitosan with propolis glycolic extract, carboxymethyl or microfibrillated cellulose	*L. innocua* (1 log units reduction in 48 h at 4 °C).	Increased polyphenols retention ten times. Paper sheets produced with the highest molecular weight chitosan, at pH 7 showed the best-wet strength (7.4 ± 0.5 Nm/g) and wet resistance (13.3 ± 1.2%).	Rollini et al. (2017)
	Combination of chitosan and quinoa-protein printed with chitosan-tripolyphosphate nanoparticles and thymol loaded nanoparticles	Exhibited higher antimicrobial activity against Gram-positive bacteria (*L. innocua*, *S. aureus*), and harmful bacteria (*Salmonella typhimurium*, *Enterobacter aerogenes*, *Pseudomonas aeruginosa*, and *E. coli*).	Chitosan printed films were more effective than hybrid films and exhibited decreased elongation at break and increased tensile strength. In contrast, the printed hybrid film showed increases in both extension and tensile strength compared with the control films. It enables the development of new packaging materials to extend the shelf life of fresh fruits.	Caro et al. (2016), Yoksan and Chirachanchai (2010)
	Combination of chitosan and kojic acid	Showed significant antibacterial effectiveness against *S. aureus* and *E. coli*.	Chitosan with higher molecular weights presented higher viscosity, improved mechanical properties, and lower water vapor permeability kojic acid addition decreased apparent density and water content.	Liu et al. (2020)

(continued)

Table 6.2 (continued)

Packaging	Characteristics	Antimicrobial function	Other advantages	References
	Combination of chitosan, polypyrrole nanoparticles, and microfibrillated cellulose	Bactericidal activity against *E. coli* (5.43 log units reduction). The positive charges Ions may attach to the negatively charged bacterial cell wall, thus destroying the cell wall, and eventually killing the bacteria.	Enhances antioxidants, good thermal stability, oxygen barrier properties, and inhibited 2,2-diphenyl-1-picrylhydrazyl assay (DPPH) activity increases to 54%. This property depends on the increased activity of the hydrogen atoms.	Gao et al. (2020)
	Zataria multiflora Boiss essential oil (1–4% v/v) and nanoliposome (1–4%) onto chitosan/whey protein isolate edible films	Good antimicrobial effect against *L. monocytogenes*, *Bacillus cereus*, *S. aureus*, *E. coli* O157:H7, and *vibrio parahaemolyticus*. The most significant antimicrobial effect was achieved at a 4% concentration of essential oil against *L. monocytogenes* in both films.	Zataria multiflora Boiss. essential oil is a rich source of different antimicrobial compounds especially thymol and carvacrol.	Shoja Gharehbagh et al. (2017)
Cellulose-based active packaging	Commercial cellulose derivatives (hydroxypropyl methylcellulose or carboxymethyl cellulose) mixed with snail mucus extracted from (*Helix Aspersa Muller*)	Antimicrobial activity against both Gram-positive and Gram-negative bacteria.	Snail mucus addition decreased film solubility in water. Natural organosolv lignin acts as a reinforcing and bioactive agent, proved an excellent antioxidant activity, and useful to prevent undesirable oxidation reactions in packaged food.	Di Filippo et al. (2021)

(continued)

Table 6.2 (continued)

Packaging	Characteristics	Antimicrobial function	Other advantages	References
	Sugarcane bagasse nanocellulose (0.8% w/w) and nisin (640, 1280, 1920, 2560, or 3200 mg/L) nanohybrid films	The hybrid film with a nisin content of 1920 mg/L or above can effectively inhibit the growth of *L. monocytogenes* to the tolerable limit of contamination (100 CFU/g, Regulation (EC) 2073/2005) within 7 days. It completely inhibited *L. monocytogenes* during 7 days of storage at 4 °C.	The concentration of nisin has a remarkable influence on the mechanical, light transmission, gas barrier, and antimicrobial properties of these films. The films with 1920 mg/L nisin exhibited sound light transmission, relatively high tensile strength, low oxygen permeability, and low water vapor transmission rates.	Yang et al. (2020)
	Curcumin onto cellulose nanofibril-oil composite films	inhibiting the growth of foodborne bacteria such as *E. coli.*	Biodegradable structure with increased thermal stability (300 °C) and the embedded oil influenced the mechanical and wetting properties. Curcumin encapsulation as an antioxidant increased radical scavenging activity by 30%.	Valencia et al. (2019)
	Epsilon-poly-L-lysine (ε-PL) (0.5 and 1% w/v) onto cellulose paper-chitosan double-layer films	1% ε-PL exhibited 1.5 log units reductions in *L. monocytogenes* for 12 days meat stored at 4 °C as compared to 0.5% ε-PL.	Films impregnated with ε PL, to control foodborne pathogens in meat.	Razavi et al. (2020)
	Oxidized regenerated cellulose microparticles as a filler onto poly(ε-caprolactone) films	Bactericidal activity against *E. coli*, *S. aureus*, and *L. monocytogenes*. 50% of the total colony-forming units on salami was dead after 14 days.	4% microparticles exhibited decreased oxygen (93%) and water (70%) permeability.	Aydemir Sezer et al. (2016)

(continued)

Table 6.2 (continued)

Packaging	Characteristics	Antimicrobial function	Other advantages	References
Alginate bioactive films	Sodium alginate films with addition of castor oil (1–3 mL/100 g of alginate)	Antibacterial activity towards Gram-positive bacteria (*S. aureus* and *Bacillus subtilis*) no inhibitory effect was observed for Gram-negative bacteria (*S. typhi* and *E. coli*).	The water vapors permeability was significantly reduced ($p < 0.05$) after castor oil incorporation and better mechanical properties and thermal stability.	Abdel Aziz et al. (2018)
	Hydrolyzed cottonseed proteins onto alginate films	Antibacterial effect against *S. aureus*, *Colletotrichum gloeosporioides*, and *Rhizopus oligosporus* but not against *E. coli*	The increase in the PH concentration increased the barrier properties to visible light, and the film color became darker, reddish, and yellowish. Active films released more than 60% of their peptides in 30 min showed promise as active packaging to preserve fatty foods.	de Oliveira Filho et al. (2019)
	Natamycin onto alginate/pectin films	The excellent compatibility of natamycin with alginate, the low diffusion coefficient, and the excellent physical properties indicate that this film has tremendous potential to be used in antimicrobial packaging, such as in cheese wrappings to inhibit mold spoilage.	The addition of natamycin promoted an increase in soluble matter in water, in the water vapor permeability and the opacity and a decreased in tensile strength when compared to films without the added anti-microbial agent.	Bierhalz et al. (2012)

(continued)

Table 6.2 (continued)

Packaging	Characteristics	Antimicrobial function	Other advantages	References
Gelatin-based packaging film	A protein concentrate of microalga spirulina (*Spirulina platensis*) onto gelatin films	Antimicrobial effect against aerobes and psychrotrophs.	The preservative effect of films proves that they are a potential source of high added-value and preservative bio compounds. It enhanced the quality of refrigerated Atlantic mackerel (*Scomber scombrus*).	Stejskal et al. (2020)
Pectin-based active films	Papaya puree and cinnamaldehyde emulsions onto methylester pectin edible films	Showed antimicrobial properties against *E. coli*, *Salmonella enterica*, *L. monocytogenes*, and *S. aureus*.	Papaya puree reduced films' resistance and rigidity while increased extensibility and water vapor permeability. Cinnamaldehyde nano-emulsions balanced the plasticizing effect of papaya puree by increasing rigidity and decreasing extensibility and permeability to pectin films' water vapor.	Otoni et al. (2014)
	Acca sellowiana waste by-product (feijoa peel flour) onto pinhão starch (2% w/w), citric pectin (3% w/w) films	Antimicrobial activity against *E. coli*, *Salmonella typhimurium*, and *Pseudomonas aeruginosa*.	Contains bioactive antioxidants The packaging maintained the quality of apples during storage, with constant weight after 5 days of storage.	Sganzerla et al. (2020)

(continued)

Table 6.2 (continued)

Packaging	Characteristics	Antimicrobial function	Other advantages	References
Starch-based active films	Epigallocatechin-3-gallate, blueberry ash fruit, and macadamia skin extracts onto pea starch-guar gum edible films	Showed the highest activity against *S. epidermidis* and *B. subtilis* at 93.8 g/mL and *Salmonella typhimurium* and *Klebsiella pneumoniae* at 750 g/mL.	Micrographs of films presented a continuous and smooth structure with no development of cracks in the film. The addition of active agents caused a significant increase in tensile strength and Young's modulus and a reduction in elongation percentage at break of films.	Saberi et al. (2017)
	Eugenol onto Chinese yam starch films	The film displayed good eugenol dose-dependent antibacterial activity against *E. coli* than *L. monocytogenes* and *S. aureus*.	3% eugenol extended the shelf-life of meat beyond 50% and has excellent active packaging potential for extending the shelf-life of foodstuffs.	Cheng et al. (2019)
	Grapefruit seed extract (0–1.5% v/v) corn starch/chitosan nano-clay	showed high antifungal activity against *Aspergillus niger* of stored bread sample at 25 °C.	1.5% showed better physical, mechanical, and thermal properties than other bio-nanocomposite.	Jha (2020)

(continued)

Table 6.2 (continued)

Packaging	Characteristics	Antimicrobial function	Other advantages	References
	Rambutan peel extract and cinnamon oil onto cassava and acetylated starch/ whey protein isolate films	The extract contained large amounts of geraniin with mixed polyphenolic compounds, whereas oil composed of cinnamaldehyde which exhibited antioxidant and antibacterial activities. It delayed microbial growth in salami.	Reduced water vapor permeability (depending on starch types) and showed redness stabilization of salami during storage. Antioxidant capacity was controlled by release mechanisms, whereas antibacterial ability in foods was dependent on food matrix components.	Chollakup et al. (2020a, b)
	Sappan and cinnamon herbal extracts onto thermoplastic cassava starch films	Release of herbal phenolic compounds to food surfaces inhibited microbial growth, giving lower TVC values (0.2–0.5 log units) on day 3. The TVC values increased, reaching 8 log units on day 6 of storage, concurrent with darkening of beef samples.	Preserved redness. The non-homogeneous microstructures caused the dispersion of clumps giving decreased mechanical properties and increased water vapor permeability but decreased oxygen permeability due to increased hydrophilicity of the film matrices.	Khumkomgool et al. (2020)

Sharma et al. (2020) applied ultrasonicated-polyphenolic extracts (200 W/20 min) of *Moringa oleifera* seed meal (3.12 mg/g) and incorporate (3%) them in chitosan coating. *Moringa oleifera* seed meal extracts (1–5%) reduced 1.4 log units *B. cereus* and *E. coli* on day-3 and 1.78 log units on day-6 in raw chicken sausages during storage. Antimicrobial thin film coatings can provide an effective approach to enhance food quality and safety. On the other hand, Ashrafi et al. (2018) prepared the innovative chitosan and kombucha tea (1–3% w/w) active film, which reduced *E. coli* and *Staph. aureus* from 5.36 to 2.11 log units in 4 days storage of minced beef, as well as extending the shelf life up to 3 days.

Alkan Tas et al. (2019) designed antimicrobial thin (225 nm) film coatings by the incorporation of ten bilayers chitosan and carvacrol (an active component of essential thyme oil) loaded halloysite nanotubes (tubular clay nanoparticles) onto the polyethylene surface. The antimicrobial component of coated polyethylene films (carvacrol) encapsulated, and sustainability was released into the package, which induced an 85% reduction of pathogenic bacteria (*Aeromonas hydrophila*) and 48% lower counts of aerobic bacteria on chicken meat surfaces. Kim et al. (2018) showed the impact of grapefruit seed extract incorporation in a layer-by-layer electrostatic deposition coating of alginate and chitosan on shrimp (*Litopenaeus vannamei*) stored at 4 °C for 15 days. The resultant coatings inhibited the growth of mesophilic bacteria and psychotropic bacteria with low melanosis thus extended the shelf-life of shrimp.

Starch was also used in edible food packaging with the thin coatings of natural antimicrobial agents. Cassava starch extracted from fresh roots or its non-edible parts (stems, peels, and leaves) has been tested for clean label packaging material (Tumwesigye et al. 2016). Starches acted as a channel for polyphenols releasing from polycaprolactone and pomegranate rind without the extraction of antimicrobial compounds from the fruit rind (Khalid et al. 2018). Berti et al. (2020) suggested the antimicrobial slurry (0.0204 g/100 g natamycin and 0.081 g/100 g nisin) fortified with rice bran added into starch-based edible films/coatings for cheese (Argentinian Port Salut). The subsequent films/coatings reduced yeast and bacterial population below 2 log units after 48 h along with the prevention of *L. innocua* and *S. cerevisiae*. Furthermore, these films released natamycin and nisin, which hindered inoculated microbial growth on the cheese surface.

The edible films based on corn starch containing micro-emulsified beeswax (1% w/w), lauric arginate (2000 mg/L), and natamycin (400 mg/L) used as natural antimicrobial coatings against *Rhizopus stolonifer*, *Colletotrichum gloeosporioides*, *Botrytis cinerea*, and *Salmonella Saintpaul* (Ochoa et al. 2017). Sánchez-Ortega et al. (2016) constructed an oleic acid nano-emulsion (starch-based edible coating suspensions), combined with a mixture of clean label antimicrobial agents [lactic acid (17.5 mg/mL), nisin (3.75 μg/mL), and lauric arginate (0.0625 mg/mL)] having antimicrobial activity against *Brochothrix thermosphacta*, *L. monocytogenes* and *Micrococcus luteus*. Similarly, the edible coating based on starch and bovine gelatin with turmeric as a substitute for synthetic casing in frankfurter sausage (stored at 5–10 °C) was useful to maintain the adequate physicochemical quality 10-day extension of shelf life (Tosati et al. 2017).

Fresh-cut fruits are highly susceptible to microbial spoilage, enzymatic browning, and quality defects such as tissue softening, thus commercial marketing of these fruits is facing a huge challenge in terms of sustainable packaging material. Paper-based food packaging is renewable and biodegradable, but the porous structure is ineffective hostile to grease, water vapor, water, and volatile organic compounds. Kopacic et al. (2018) upgraded the industrially produced paperboard with clean label biopolymers (alginate and chitosan) for functional bio-based coatings to achieve wettability, water absorptiveness, and grease resistance. In this context, bio-based edible coatings with active clean label antimicrobial agents play a vital role to be used for active food packaging. Khan et al. (2019) observed lowered total aerobic count (3.83 log units) for chitosan monomethyl fumaric acid edible coatings on fresh strawberries as compared to uncoated (5.31 log units) stored at 10 °C. While the combination of chitosan coating with clean label antimicrobial extracts (nisin, natamycin, pomegranate, and grape seed) for active packaging of fresh strawberries reduced the mesophilic bacteria (below the limit of 10^8 CFU/g) and yeast and mold (below the limit 10^5 CFU/g) along with the increased the shelf life up to 30 days (Duran et al. 2016). Similarly, liquid edible alginate coatings with the addition of carvacrol and methyl cinnamate (natural antimicrobials) were electrostatically sprayed on fresh strawberries to inhibit the decay (only 5.6% of infected fruits) with a delay of 11 days (7 days for controls) as compared to 16.6% microbial decay was observed in uncoated controls (Peretto et al. 2017).

Elabd (2018) developed an antimicrobial edible coating (potato starch, chitosan, ascorbic acid, and potassium sorbate) for preserving guava slices storage at 4 °C and 70–75% relative humidity for 18 days. The coatings contributed to reducing the microbial growth of guava slices and increasing their shelf-life. While the combination of the pectin-based edible coating [modified with 500 IU/mL nisin (antimicrobial) and 10 g/kg citric acid and calcium chloride (anti-browning and firming agents)] and active low O_2 MAP (5 kPa O_2 for 9 days at 5 °C) maintained the overall sensory and microbiological quality of fresh-cut persimmon (*Rojo Brillante*) slices (Sanchís et al. 2017). However, the combination of pectin films with the incorporation of copaiba oil nano-emulsions proved to have the potential for providing relevant functional quality for food packaging along with the clean label alternative to conventional plastic packaging (Norcino et al. 2020).

Falagán et al. (2018)) added natural antimicrobial agents (vanillin and *trans*-cinnamic acid) in the alginate-based edible coating to inhibit the psychrophilic, mold, and yeast growth during the eighth-day shelf life of fresh-cut peaches (*Flordastar*) as well as to control wedges lost that was 5% in the treated fruits as compared to untreated 25%. Biobased edible films made with sodium alginate/castor oil (Abdel Aziz et al. 2018), alginates/sunflower oil (Dewi and Husni 2020), and alginates/virgin coconut oil (Marismandani and Husni 2020) could be considered clean label alternative to plastic packaging with improved thermal, mechanical, and antibacterial effect against *S. aureus* and *B. subtilis*. Cellulose nanofibril/oil composite films have also been investigated for active edible barriers in food packaging for nutrient delivery or preventing oxidative degradation and microbial growth (Valencia et al. 2019).

Lignocellulosic materials represent a reasonable share as raw materials in biobased food packaging (Ballesteros et al. 2018) while other clean label materials include cellulose-ether derivatives (carboxymethylcellulose, hydroxypropyl cellulose, methylcellulose, and hydroxypropyl methylcellulose. Cellulose nanostructures (nano-cellulose) can be used as a coating on different fruits and vegetables and retained the original color/flavor and texture because of the reduced gas exchange rate between the environment and the food commodity (Azeredo et al. 2016). Designing integrated biopolymer material having antimicrobial properties with favorable mechanical quality remains a great challenge. New developments are mainly presented in Li et al. (2019a, b) work, which is based on antibacterial clean label gelatin nanocomposites. The tough, hydrophobic, active films were created via multiple linking between dopamine-grafted cellulose nanofibrils and montmorillonite in a covalent polymer network.

Conventional packaging films that are safe to be in direct contact with food such as polylactic acid (PLA) can also be part of clean label food packaging, however, the addition/incorporation or surface coatings of natural antimicrobial agents can further increase its advantages (Valerini et al. 2018). The PLA films filled with active agents (thymol, kesum, and curry) exhibited antimicrobial activity against *Staph. aureus* (Mohamad et al. 2020) along with its higher susceptibility to biodegradation as compared to synthetic counterparts (Tawakkal et al. 2016), thus could be used in clean label food packaging of meats, fruits, and vegetables. The most recent work in this area centers around the pH-sensitive and active films consist of cellulose-reinforced PLA having clean label plasticizers and biobased fillers (Khosravi et al. 2020). Similarly, polystyrene sulfonate/ polyvinyl alcohol/curcumin films incorporated with xylitol (6–8%) showed antimicrobial activity versus *Pseudomonas aeruginosa*, *Candida albicans*, and *Staph. aureus* (Ibrahim et al. 2017). Demircan and Zhang (2017) obtained natural cellulosic materials (soluble cellulose-grafted hyperbranched polymers) and Ding et al. (2018) grafted polypropylene nonwoven fabric with aloe-emodin (natural antimicrobial) against *Staph. aureus*, *E. coli*, and *Candida albicans*. More recent work by Thanakkasaranee et al. (2020) used calcined corals composite films (having calcium oxide) to prepare nylon/linear low-density polyethylene (LDPE) films, which delayed the *E. coli* and *Staph. aureus* lag time and growth cycle in milk during storage.

The use of clean label biomaterials such as proteinaceous substances is cost-effective and environment-friendly as compared to the traditional antimicrobial packaging materials including silver nanoparticles, synthetic polymers, or peptides that harm the environment as well as health. However, the performance optimization of the antimicrobial surface coatings having functional molecules is urgently required to control the pathogenic microorganisms. Gu et al. (2017) extracted protein from daily foods (mere hen egg-white lysozyme) and used it for developing clean label nanofilm having antimicrobial/antibiofilm activity against Gram-positive/negative and fungi. The resultant coatings proved to be effective against antifouling agents without adding any other harmful compound. Similarly, Bugatti et al. (2018) composed bio-based polyamide membranes filled with lysozyme (50 wt%) as antimicrobial food packaging.

Many recent studies have shown that the naturally derived plant extracts incorporation into food packaging in the form of coatings/films provided a clean label alternative for antibacterial properties via aromatic and phenolic compounds. For example, pomelo peel extract coatings on rice straw paper (Chollakup et al. 2020a), sea buckthorn (*Hippophaë rhamnoides L.*) leaves, and inner bark of pine trees (*Pinus silvestris*) extract on paper (Brobbey et al. 2017), garlic/lime extract coatings on gelatin-chitosan films (Thaker et al. 2017), geraniol and pomegranate extract coatings on gellan-based films (Tomadoni et al. 2018) showed resistance towards pathogenic microbes.

6.2.2 Antioxidative Food Packaging

Environmental, economic, and food & health safety aspects encouraged towards new bioactive packaging strategies such as the addition of biobased components (having polyphenols) into food packaging enhanced antioxidant activity and radical scavenging activity via the strong interactions with film matrix. Nowadays, biodegradable, or clean label food packaging (i.e., lignin, cellulose, chitosan, starch, gelatin, alginate, etc.) as an alternative to the petroleum-based products, is gaining attention for a well-added value biopolymer due to radical scavenging and antioxidative properties to protect food from spoilage. The clean label antioxidative materials incorporation into biobased packaging are effective in preventing the lipid oxidation of fatty food, enzymatic browning of fresh food, reducing the need for synthetic additives and plastic waste, and extending the shelf-life of fresh-cut or minimally processed packaged food. From the literature, several approaches were used to incorporate clean label agents (can be seen in Table 6.3) from food by-products (Han et al. 2018; Abdelhedi et al. 2018), natural extracts (Rambabu et al. 2019; Piñeros-Hernandez et al. 2017), essential oils (Souza et al. 2017; Wieczyńska and Cavoski 2018), flours and peel powders (de Moraes Crizel et al. 2018; Hanani et al. 2019), and organic acids (Liu et al. 2017) in the packaging to enhance its antioxidative quality as well as to contribute in better protection of food matrix.

Food industries' by-products such as soy protein isolate, fish gelatin, and mango kernel extracts have been increased recognition due to their rich source of nutrients, thus utilized as active agents in food packaging materials. With their advantage of providing a suitable substitute for packaging to direct contact with food, they also increased antioxidant activity (89%), and free radicals scavenging activities (Adilah et al. 2019). According to Barbosa-Pereira et al. (2014), the active films produced by utilizing the agro-industrial residual sources instead of synthetic additives are likely to be more acceptable to consumers to extend the shelf-life of meat products.

Rambabu et al. (2019) used chitosan films for cashew nuts packaging (preserved for 28 days) with the addition of 5% mango leaf extract to enhance their oxidation resistance by 56% as compared to commercial polyamide/polyethylene film. The animal byproducts such as bovine gelatin in addition to sulfated polysaccharides and smooth-hound viscera proteins have also been used for blend or bilayer films to enhance the antioxidant quality (Abdelhedi et al. 2018). Similarly, the bioconversion of fish waste/byproducts (gelatin and chitosan) into biodegradable or edible food packaging can be used in combination with other clean label active agents or natural extracts to prevent ecological problems related to synthetic packaging due to non-biodegradability.

Emulsion films containing carboxymethyl cellulose, potato starch (oxidized), soy protein isolate, gelatin, and ethanolic hop extract (0.1–0.50%) displayed an increased antiradical activity, metal chelating power, and ferric reducing antioxidant power. The films susceptible to solubilization (due to carboxymethyl cellulose and soy protein isolate), had better antioxidant activity compared to water-swollen films (oxidized potato starch and gelatin) because of ethanolic hop extract (Liu et al.

Table 6.3 Clean label antioxidative agents in food packaging

Food packaging material	Active agent	References
Cassava starch films	Coffee extract (2%)/cocoa powder (46%)	Veiga-Santos et al. (2018)
	Green tea and basil extract (5%)	dos Santos Caetano et al. (2018)
	Rosemary extracts (5–20 g/100 g)	Piñeros-Hernandez et al. (2017)
Gelatin-based films	Durian leaf extract (0.5%)	Joanne Kam et al. (2018)
	Pomegranate (Punica granatum L.) peel powder	Hanani et al. (2019)
	Caesalpinia decapetala and Caesalpinia spinosa "Tara" extracts (0.2%)	Gallego et al. (2016)
Gelatin/chitosan films	Gallic acid (0.15/100 g)	Limpisophon and Schleining (2017)
	Procyanidin (1 mg/mL)	Ramziia et al. (2018)
	Epigallocatechin gallate (0–0.20% w/v)	Nilsuwan et al. (2018)
Chitosan films	Mango leaf extract (1–5%)	Rambabu et al. (2019)
	Chinese chive root extract (1–5% w/w)	Riaz et al. (2020)
	Protocatechuic acid (0.5% or 1.0%)	Jun Liu et al. (2017)
	Thinned young apple polyphenols	Sun et al. (2017), Nisar et al. (2019)
	Olive residue flour (10%)	de Moraes Crizel et al. (2018)
	Proanthocyanidins (0–20 wt.%)	Bi et al. (2019)
	Essential oils (ginger, rosemary, sage, tea tree, and thyme) and hydro-alcoholic extracts (from ginger, green and black tea, kenaf leaves, rosemary, and sage plants)	Souza et al. (2017), Wieczyńska and Cavoski (2018)
Chitosan/starch films	Thyme extract	Talón et al. (2017)
Chitosan/gelatin films	Cinnamon, guarana, rosemary, and boldo-do-chile ethanolic extracts	Bonilla and Sobral (2016)
Chitosan/carboxymethyl cellulose films	Blueberry and red grape skin pomace extracts (1–4% w/v)	Kurek et al. (2019a, b)
Soy protein concentrate films	Red grape extract (0–10% w/w)	Ciannamea et al. (2016)
	Licorice residue extract (0–70 g/kg)	Han et al. (2018)
	Chestnut bur extract (20–100 g/kg)	Wang et al. (2016)

(continued)

Table 6.3 (continued)

Food packaging material	Active agent	References
Soy protein isolate/gelatin films	Mango kernel extracts (1–5%)	Adilah et al. (2018)
Sodium alginate films	Tea polyphenol (5–20%, w/w of pomelo peel flour)	Wu et al. (2019a, b)
Methylcellulose/glutaraldehyde films	Maqui (Aristotelia chilensis) berry fruit extract	López De Dicastillo et al. (2016)
PLA films	Origanum vulgare essential oils (20–100 g/kg)	Llana-Ruíz-Cabello et al. (2016)
	Merkén, an aboriginal Chilean spice (3–5 wt.%)	López de Dicastillo et al. (2017)
PVA films	Seaweed extracted cellulose nanocrystal (5% w/v) and basil leaves extract (5–20% w/v)	Singh et al. (2018)
	Green tea extract (0.5–2%)	Chen et al. (2018a)
PVA/clay nanocomposite films	Tea polyphenols (0.5–4%)	Chenwei et al. (2018)
Cellulose acetate/ LDPE bilayer film	Cucumis metuliferus extract (1–5 wt.%)	Arrieta et al. (2020)
Mung bean protein films	Pomegranate peel (2.5–25% w/w)	Moghadam et al. (2020)
Semi-refined carrageenan	Persicaria minor extract, a natural Malaysian herb (0.4–2%)	Yahaya et al. (2019)

2017; Kowalczyk and Biendl 2016). Similarly, emulsified edible films were developed by Fabra et al. (2018) with lipids into alginate films in addition to green tea extract and grape seed extract (14.79 and 3.50 mM Trolox/0.75 g extract, respectively) to enhance the antioxidant capacity of active films.

Veiga-Santos et al. (2018) compared the antioxidant capacity of coffee/cocoa/cassava starch polymer packaging (for palm oil stored at 30 °C for 45-days) with commercial polymer packaging. The packaging increased peroxide index (6.09 times), hexanal production (60.4 times), and conjugated dienoic acid production (6.88 times) in addition to decreased water/light permeability to prevent catalytic oxidation. However, 0.5% incorporation of durian leaf extract into gelatin-based film increases DPPH scavenging activity (17.6 times) and reduced the oxidation rate (3 times) of palm oil as compared to control (Joanne Kam et al. 2018).

High-fat foods such as butter and high moisture content in soft cheeses make them vulnerable to lipid oxidation and mold attacks. Arshad et al. (2020) effectively enhanced their storage stability by using different types of edible coatings incorporated with natural essential oils i.e., active whey powder-based edible coatings for soft cheese (0.5–1% coated clove oil and 1.5–2.5% coated peppermint oil) during 30 days of storage, cornstarch-based edible coatings for butter (0.2–0.4% coated black cumin oil and 1.5–2.5% coated ginger oil) during 90 days of storage at 2–5 °C.

6.2.3 Oxygen Scavengers

In food packaging, O_2 is very important for maintaining the quality and extending the shelf life of products such as ready-to-cook garlic. But the amount of O_2 accelerates the rate of biochemical, microbiological, and physiological reactions, consequently, deteriorate the quality (Singh et al. 2019). O_2 in the headspace of a package is the main cause of food spoilage including oxidation/rancidity, flavor/color or texture alteration, microbial propagation, nutritional quality degradation, and shelf life reduction. It also depends on the nature of food commodities as many food products are sensitive to O_2 and easily degraded upon prolonged exposure.

Prevention of these effects is necessary for longer shelf life by the O_2 depleted environment via food packages. But the removal/reduction of O_2 could be the cause of anaerobic growth of pathogens in the package stored at above 3 °C, which may favor health issues (Remya et al. 2017). There are different technologies used to tackle this problem including vacuum packaging, modified atmosphere packaging (MAP), and the addition of O_2 absorber and scavengers. The addition of O_2 scavengers works as an active packaging technique, which absorbs the dissolved O_2 and O_2 in the headspace and reduces the contact between O_2 and food products (Dey and Neogi 2019). Commercially available O_2 absorbers consist of sulfites, boron/boric acid, sugar alcohols and glycols, unsaturated fatty acids and hydrocarbons, palladium catalysts, enzymes, ascorbic acid, ferrous iron, and antioxidants such as *t*-butyl hydroquinone and butylated hydroxytoluene. Iron powder and ascorbic acid are commonly used as O_2 scavengers but iron-based O_2 scavenging sachets cannot be used in microwaves ovens and liquid foods such as juices and could be a cause of accidental intake by children.

Future researchers are focusing on new strategies or clean label solutions for functional packaging such as incorporating organic or non-metallic O_2 scavengers, synthetic, and natural antioxidants to protect the bioactive polyunsaturated fatty acids and vitamins against oxidation. But synthetic antioxidants are not much appealing to health-conscious consumers (Roman et al. 2016). Therefore, alternative clean label/natural O_2 scavengers are incorporated into the conventional food packaging including reducing the amount of packaged O_2 and inhibit oxidation and nutrient degradation (Johnson et al. 2018). Similarly, Gaikwad et al. (2017a) developed an O_2 scavenging low-density polyethylene (LDPE) film with different pyrogallol (natural phenolic compounds) coatings (1–20%). The resultant films displayed excellent O_2 scavenging performances (0.443 mL/cm^2 at 23 °C and high relative humidity) and were suitable for high moisture food packaging applications

Clean label active packaging solution to food additives reduces O_2 levels by >98% to prevent oxidative spoilage such as lipid oxidation of fish oil-in-water emulsions (Johnson et al. 2018). The addition of O_2 scavenger in food packaging not only inhibits lipid oxidation and nutrient degradation such as essential vitamins but also provides a clean label alternative to the food manufacturers and consumers for extending the shelf life and nutritional quality especially in the foods having sensitive components such as ascorbic acid and α-tocopherol.

Clean label polyphenols such as pyrogallol is an active component that can be used in O_2 scavenging system (containing sodium carbonate) for low water activity food packaging applications (Gaikwad et al. 2016). Gaikwad et al. (2017b) investigates the antioxidant functionality of soybean oil packaged in pyrogallol coated LDPE film (5–20% at 5–60 °C and 95 relative humidity) and confirmed that O_2 scavenging active packaging could be an efficient substitute for direct addition of artificial antioxidants to the stabilization of soybean oil. Moreover, polyphenol coated polypropylene active food packaging showed better oxidative stability as compared to soluble chelators including metal chelating (39.3 ± 2.5 nmol Fe^{3+} cm^{-2}, pH 4.0) and radical scavenging (up to 52.9 ± 1.8 nmol Trolox eq cm^{-2}) capacity, which preventing the lycopene degradation and lipid or ferric iron oxidation via separating iron from the emulsion droplet (Roman et al. 2016). Similarly, the use of gallic acid for food packaging (with a high water activity stored at room temperature) applications proved to be a potential clean label strategy for optimum O_2 scavenger performance (Pant et al. 2018). However, the combination of gallic acid and sodium carbonate, incorporated in a bio-based multilayered thermoplastic starch films having ascorbic acid, acted as a barrier to O_2 and water vapor diffusion and reduced the rate of O_2 absorption (Pant et al. 2017).

Active chitosan PVA films with ginger essential oil as an antimicrobial O_2 scavenger can inhibit aerobic *Pseudomonas* spp., lactic acid bacteria, and *Brochothrix thermosphacta* and in fresh cobia fish (*Rachycentron canadum*) steaks stored at 2 °C. The resultant film increased the storage stability up to 30 days compared to the control pouch (15 days) without affecting organoleptic characteristics. Besides, the combination of O_2 scavenger and edible chitosan coatings showed good potential as a natural antimicrobial MAP packaging technique for sliced sucuk in retaining the color quality (Şahin et al. 2017).

Mahieu et al. (2015) obtained active wheat starch-based films with ascorbic acid (15% w/w) and iron powder (1.5% w/w) as O_2 scavenging agents, which can absorb 13.5 mL of O_2/g of the dry film in 15 days at 80% relative humidity as well as inhibiting aerobic microorganisms and enhancing the shelf-life of packaged food. Starch-based films have a high quantity of absorbed water, which activates iron and ascorbic acid oxidation and triggers retrogradation of plasticized starch crystalline structure and improved the mechanical properties such as tensile strength. While the release of carbon dioxide due to ascorbic acid oxidation produce an antimicrobial effect.

However, the iron-based O_2 scavenging agents need 1–2 days to be activated by the water penetrating from food to the innermost layer of film. The headspace air also hinders O_2 to dissolve into the food which causes an additional delay in the scavenging system. Another limitation of this O_2 reducing packaging system is the opaque and flexible nature of the O_2 scavenging film that might be incompatible for all types of food and packaging lines. O_2 scavengers could be incorporated on the lid, label, or bottom of the container to create innovative packaging systems, where it provides visualization of the food as well as protection to the food (Johnson et al. 2018).

6.2.4 Moisture Absorber

Moisture in food products is useful for marinating the food quality of high water activity-based products to prevent them from drying but the accumulation of excessive moisture/relative humidity and high-water activity of food (in the case of fresh fruits and vegetables, and raw meat & fish) is adverse for the integrity of the package in addition to food spoilage due to accelerated growth of bacteria and mold (Yildirim et al. 2018). The dry food products are most sensitive to high moisture contents within the food package during the storage that may lead to changes in texture/appearance, thus reducing the shelf life of food (Murmu and Mishra 2018). The control of high relative humidity inside the food package is important to preserve the nutritional and sensorial quality of the food.

The simplest approach is to add moisture absorbers in food packages (Gaikwad et al. 2018). Moisture absorbers in the form of sachets, films, tray and pads, control water and microbial of the product, maintain fluid exudates from food products, retain the relative humidity inside the package (Choi and Lee 2013). The potential solutions for moisture regulation are to use MAP or vacuum packaging for reducing moisture, use high barrier packaging for preventing moisture, and use moisture-absorbing/scavenging agents for removing moisture as well as maintaining relative humidity levels inside packages (Bovi et al. 2018). Moisture absorbing/scavenging agents including sheets, blankets, and pads are working on the principle of physical adsorption to control fluid exudates from food products such as drip loss (common for raw fish/meat) due to muscular degradation and storage temperature fluctuation accumulates excessive liquid within the package that reduces the attractiveness of a packed product and makes it less required by consumers (Rux et al. 2016).

Commercially available moisture absorbers include natural clays, calcium oxide, calcium chloride, starch copolymers, carboxymethyl cellulose, and polyacrylate salts are used in the forms of sachet-absorbing pads or molecular sieves. These desiccants, bulky blankets, and sheets have a strong affinity to absorb extra fluids from muscular foods and melted ice, thus placed under packaged chilled foods during shipping (Dobrucka and Cierpiszewski 2014). Previously, other clean label alternatives for moisture absorbers have been used which include fructose, a hygroscopic material/ humectant (White 2014); and sorbitol, a food-grade ingredient (Aday and Caner 2013) for a variety of food packaging applications but organic-based moisture absorbers have the lower-absorbing capacity and relatively expensive as compared to inorganic absorbers (Gaikwad et al. 2018). Therefore, the parallel application of both inorganic and organic-based moisture absorber in active food packaging is a favorable future direction to replace artificial additives (preservatives and polymers) and for obtaining quality packaging for high-water activity foods (Gaikwad et al. 2018).

New developments are mainly presented in Bovi et al. (2018) as they used FruitPad embedded with 30% fructose for fresh strawberries packaging. Results showed less in-package condensation and high moisture absorption (0.94 g of water/g of the pad at 20 °C and 100% relative humidity) as well as less weight loss

(<0.9%) during 5 days of storage. Besides, Gaona-Forero et al. (2018) got satisfactory results by using a mixture of sodium polyacrylate in sachets of porous vegetable paper (70% w/w) for 'Hass' avocado samples that were stored in rigid perforated polyethylene terephthalate (PET) packages for 12 days at 12 °C.

6.3 Clean Label Intelligent Packaging

Advances in food packaging technology have led to intelligent technologies based on indicators, and sensors to monitor or regulate the quality of perishable food during the supply chain. Intelligent food packaging gives reliable micro-level information about the unwanted physical and chemical changes (i.e., alteration of humidity temperature, and pH) that have critically affected food safety (the recent findings of freshness indicators are presented in Table 6.4). Such as the incorporation of bioactive natural safe compounds like α-costic acid and ungeremine into eco-friendly intelligent biofilm was used as bio-fungicide against *Penicillium roqueforti* and *Aspergillus niger* (Valerio et al. 2018). Similarly, quercetin was incorporated into carboxymethyl chitosan to make antioxidant active packaging as well as intelligent Al^{3+}-sensing films that released antioxidant agents into packed fatty foods (Bai et al. 2019). However, the high cost of intelligent packaging related to large-scale manufacturing or commercializing can be reduced by applying advanced cost-effective printing electronics (Wang et al. 2019a). Intelligent packaging further can be classified into different types of sensors or indicators as mentioned below.

6.3.1 Gas Indicators/Ink Sensor Packaging

Visual gas indicators incorporation into intelligent food packaging systems have been developed to monitor packaged product quality economically and simply by colorimetric changes induced via pH-sensitive dyes due to the amount of gas production during food spoilage (Vu and Won 2014). These intelligent indicators provide information about the freshness quality of food without destroying the package. However, incorporation of these strongly hydrophilic dyes into the commercial packaging system faces some difficulties as leaching of these dyes into food. In contrast, Ma et al. (2017) developed tara gum/polyvinyl alcohol (PVA) based colorimetric NH_3 indicator intelligent films incorporating with curcumin (0.1–0.5%) while Zhai et al. (2019) used gellan gum-based colorimetric hydrogen sulfide (H_2S) sensor (0.81 μM detection limits), which shifted color from yellow to colorless resulting non-destructive and real-time monitoring of meat spoilage.

Gas sensors including carbon dioxide (CO_2), O_2, and nano-biosensor are used to estimate the freshness/quality of food products by real-time monitoring the presence of gaseous analytes in the package (Ma et al. 2017). A standard traditional titration method spends hours to examine volatile organic compounds and gas

Table 6.4 Freshness indicators/intelligent packaging

Indicator	Advantage	References
Natural dyes, curcumin E100, beetroot extract E162, grape peel extract E163	Detect food spoilage in raw cod flesh. The response comparable to an artificial dye methyl red.	Tichoniuk et al. (2017)
Natural dyes, B-carotene 1% E160a, chlorophyll 10% E141, curcumin 8% E100 and lutein 0.9% E161b	Pro-ecological and fully biodegradable films which changed color under the influence of external factors and gave the materials the characteristics of intelligent packaging.	Latos-Brozio and Masek (2020)
Natural dyes extracted from butterfly pea, red cabbage, turmeric, and beetroot	All visual indicators changed their colors towards the spoilage of meat in 24 h monitoring at room temperature. The presence of methylamine as the main volatile compound from the spoiled meat, which contributed to the color change of the visual indicators.	Rahim et al. (2020)
Naphthoquinone dyes extracted from *Arnebia euchroma* onto cellulose films	The sensing film changed from rose-red to purple, then to bluish violet providing a good indication of spoilage, which correlated well with TVB-N and total viable count in shrimp and pork under 20 °C, 4 °C and −20 °C.	Dong et al. (2020)
Naphthoquinone dyes extracted from the root of gromwell (*Lithospermum erythrorhizon*) onto cellulose paper	The indicator changed the color from red to dark blue, depending on the pH of buffer solutions. Also, the indicator showed high stability after 4 months of storage and maintained high sensitivity to pH changes in fish and pork.	Ezati et al (2021)
Lycium ruthenicum Murr. (2.5%) onto κ-carrageenan films	Film displayed colorimetric response at a wide range of pH (2–10) and the color change was reversible in milk and shrimps. Also, it has good antioxidant activity.	Jingrong Liu et al. (2019)
Curcumin (3%) onto k-carrageenan based films	Used as a pH colorimetric indicator, which showed a strong redness-shift under alkaline condition in pork and shrimp.	Liu et al. (2018)
Curcumin (0.01% w/v) onto chitosan/oxidized chitin nanocrystal matrix	The color of films varied with the change in pH, and this characteristic was used to monitor the spoilage of hairtail and shrimp during storage.	Wu et al. (2019a, b)
Jambolan fruit (*Syzygium cumini*) anthocyanins (0.1–0.3%) onto chitosan, and PVA films	Films containing anthocyanins showed visible changes from a red color to a blue color when used to monitor shrimp freshness at several temperatures (between −20 °C and 20 °C).	Merz et al. (2020)

(continued)

Table 6.4 (continued)

Indicator	Advantage	References
Mulberry anthocyanin extract (5%) onto sodium carboxymethyl starch and κ-carrageenan films	The red film changed to dark-blue as TVB-N of the fish reached 19.12 mg/100 g. The film showed different colors at different pH.	Zhang et al. (2020)
Mulberry (0.4 g) anthocyanin onto ethylene-vinyl alcohol copolymer-montmorillonite multilayer films	The colorimetric response to volatile ammonia and pH was sensitive, and the color change could be easily distinguished by the naked eye in packaged shrimps.	Kang et al. (2018)
Blueberry-derived anthocyanins onto PLA films	Showed good color stability at 4–25 °C. As the ammonia concentration increased, the sensor color changed from pink to light pink and then to colorless. The sensor could effectively monitor mutton freshness in real-time.	Sun et al. (2020)
Purple sweet potato anthocyanins onto carboxymethyl-cellulose/starch films	The color was changed from red to blue and green when the fish became spoiled while exposed to different pH or ammonia.	Jiang et al. (2020)
Roselle (*Hibiseus sabdariffa L.*) anthocyanins 20–120 mg onto starch/ PVA films	The colorimetric films presented visible color changes over time due to a variety of primary volatile amines known as TVB-N to monitor the freshness of silver carp (*Hypophthalmichthys molitrix*) at 4 °C and 25 °C.	Zhai et al. (2017)
Roselle anthocyanins (0.18% w/v) onto Hydroxypropyl methylcellulose PVA	The films could be applied as intelligent packaging for real-time shrimp freshness monitoring. The film color changed from rose-red to light green at 4 days, corresponding to the onset of spoilage, and then turned into yellow at 8-day, when the shrimp was severely spoilage.	Huang et al. (2020)
Rose anthocyanins onto PVA/ okra mucilage polysaccharide composite film	Films could effectively monitor shrimp freshness in real-time and the color changes at pH 2–12 and was highly sensitive to volatile ammonia, suggesting its potential in intelligent packaging for freshness monitoring of aquatic products and meat foods.	Kang et al. (2020)
Black rice bran anthocyanins into oxidized chitin nanocrystals/ gelatin films	Films were pH-sensitive and showed remarkable color changes in different buffer solutions, used to monitor the freshness of shrimp and hairtail by visible color changes.	Ge et al. (2020)

(continued)

Table 6.4 (continued)

Indicator	Advantage	References
Alizarin (1% w/v) onto cellulose-chitosan film	Freshness monitoring of minced beef through perceptible color changes. Indicator showed a clear color response from yellow to brown and purple at pH 2–11 and had more color stability at 4 °C as compared with 25 °C during storage for 2 months. The indicator revealed color variation from brown to purple when TVB-N reached the necessary amount of 20.53 mg N/100 g.	Ezati et al. (2019)
Methyl red indicator (0.05–1%) film sensor label	Showed the most rapid response to changing the color of the label. Detecting the freshness of tilapia fillets over a shelf life of 10 h at 25 °C.	Ayu Sariningsih et al. (2019)
Sensor unit contains bromothymol blue-phenol red as pH-sensitive dye organized with polymer matrix solution	It monitors visible color changes from yellow to purple and determines the suitability of freshness indicator formulations in fish products, especially skate (*Raja kenojei*).	Lee and Shin (2016)
Bromothymol blue was fixed on the cationic cellulose fibers and added (3%) onto the semen cassiae gum	The visual–olfactory film was sensitive to ammonia with a noticeable color changed from blue to pale yellow for raw milk spoilage while light yellow to blue-green for the freshwater shrimp spoilage.	Cao et al. (2019)
Hydroxyethyl cellulose-based indicator labels using two different mixed indicators (bromothymol blue/phenol red, cresol red/thymol blue) formulations	Change color with spoilage of anchovy (*Engraulis encrasicolus*) stored at 2 °C and 10 °C and can be used as freshness indicator in the fish industry.	Şen et al. (2020)
Indicator labels made by bromothymol blue alone, methyl red and bromothymol blue solutions (3:2)	Changed from yellow-green to orange by increased carbon dioxide concentrations in the package due to food spoilage. The label made with a mixture of methyl red and bromothymol blue can be applied as an easy-to-use and promising indicator for freshness monitoring of packaged fresh-cut green bell pepper.	Chen et al. (2018b)

(continued)

Table 6.4 (continued)

Indicator	Advantage	References
Polydiacetylene/agarose sensor	First detection tool for the freshness of almond milk. The sensor discriminates between triglycerides and FFA by a visible color change and can therefore be used to track fat metabolism during food spoilage. Upon exposure to spoiled almond milk, a visible blue-to-red color change is induced in the film, which is correlated with FFA concentration in the samples.	Weston et al. (2020) communication
Blend of inner poly(ether-block-amide) film, a color-changing layer of eight polymer-immobilized pH dyes, and an outer poly (ethylene terephthalate) film	The optimized colorimetric array freshness indicator was applied to chicken breast packaging and digital images for its color response were acquired using a smartphone camera over storage time at 4 °C and 10 °C.	Kaeun Lee et al. (2019)
Sugarcane bagasse nanocellulose-based hydrogel	Hydrogel worked as a carrier for pH-responsive dyes (bromothymol blue/ methyl red), which changed color according to the freshness of the chicken sample. CO_2 levels were found to increase with the spoilage of chicken, relative to rising levels of microorganisms. The indicator hydrogel optical color changed from green to red on the third day, as the log CFU/g passed the limit of acceptability for human consumption and emission of complicated volatiles.	Lu et al. (2020)

molecules such as H_2, CO, NO_2, O_2, H_2S, NH_3, CO_2, CH_4, etc., which consequently leads to food spoilage and package leakage (Dalmoro et al. 2017). While emerging gas indicators evaluate the on-site freshness or product quality and package integrity (Chang et al. 2017). For this purpose, an ultrasensitive porous-electrode-capped organic gas sensor system was built for fast (takes 1 min) and accurate detection of volatile amine gas (TVB-N as 28–35 mg/100 g freshness limit) from the raw fish meat (Chang et al. 2017). Besides, a wireless O_2 sensor using FeII-poly(4-vinyl pyridine)-single-walled carbon nanotube composites were developed for commercial passive near-field communication tags, and MAP, which can detect O_2 inside a package with a conventional smartphone. The resultant heavy-metal-free and inexpensive sensors were used for non-line-of-sight quality monitoring of nitrogen-filled vegetable packages (Zhu et al. 2017).

Colorimetric O_2 indicators are an important part of the intelligent food packaging system as they change color in the presence of O_2 gas involved in food spoilage. Conventional O_2 indicators include synthetic and toxic redox dyes and strong reducing agents, but their activity is reduced in the air, thus need anaerobic conditions for proper working. However, a natural component-based O_2 indicator consists of laccase, guaiacol, and cysteine was used as a clean label alternative to synthetic and artificial components. Won et al. (2016) redesigned conventional O_2 indicator for in-pack activation, which contributes greatly to intelligent packaging for healthier and safer foods.

Similarly, Ultraviolet (UV)-activated O_2 indicators have the advantages of in-pack activation and irreversibility; however, these dye-based O_2 indicator films suffer from dye leaching upon contact with water. However, Vu and Won (2014) used carrageenan (natural sulfated polysaccharides) to build UV-activated colorimetric O_2 indicator films. Such films lightened upon UV irradiation, while recovered color very rapidly in the presence of O_2 and thus helped to confirm the complete removal of O_2 by absorbers/O_2 scavengers. These types of film can also be used as a seal and leak detector with MAP compared to the other waterproof O_2 indicator film (Vu and Won 2014). Comparably, sodium alginate, sodium caseinate, and carrageenan UV-activated intelligent O_2 indicating films were designed to estimate O_2 concentration (1–10% detection limit), which can be potentially applied for direct contact (vacuum) or non-direct contact (MAP) food-packaging applications (Deshwal et al. 2018).

The accumulation of CO_2 within the package indicates microbial spoilage of that food. Therefore, CO_2 Sensors are used as intelligent labels for estimating the CO_2 level to monitor the quality & freshness of food packaged under MAP conditions. Lee and Ko (2014) used a whey protein isolate based indicator in LDPE film sachets for sensing CO_2-related spoilage and over-ripeness. While recent clean label approach used naturally occurring dyes (anthocyanins) and food-grade material based on amino acid (L-Lysine) and polypeptide (poly-L-lysine) for sensing CO_2 performance. These real-time sensors can detect small changes in gaseous CO_2 (up to 2.5%) in cold preserved food stored at 0–5 °C (Saliu and Della Pergola 2018). Similarly, CO_2-sensitive intelligent packaging film based on the Coomassie brilliant blue dye/chitosan (0.3% w/v) (Meng et al. 2015), and bromothymol blue/tetrabutylammonium ion-paired dye/ multilayered films (Lyu et al. 2019) were used to check the irreversible ripeness in kimchi linked to the gas levels within the package.

Printing clean label functional labels or gas sensors onto biodegradable packaging materials provided a flexible and low-cost smart food packaging system that is the demand for innovative packaging applications (Hrytsenko et al. 2018). Several natural colorants have been used as intelligent indicators in biodegradable polymers/films to estimate the shelf life of packed food products (Latos-Brozio and Masek 2020). Similarly, clean label pH-sensitive edible Inks (betaine dye) obtained from red beet (Ozcan and Arman Kandirmaz 2020) and carboxymethyl chitosan (Ding et al. 2020) has been used in smart packaging, which can be used as a deterioration marker and antimicrobial preservative against *Staph. aureus* and *E. coli*. Besides, the pectin-based biodegradable film (modified with nano clay and methylene blue) was used as a colorimeter sensor that changes color from blue to colorless in the presence of vitamin C. The resultant films can measure vitamin C (10 mg/100 g detection limit) in orange, kiwi, and tangerine juice (Pirsa 2020).

The O_2-sensitive photo-reversible intelligent ink-based packaging system is in ongoing demand in the modern era. Recently, Volison-Klimentiev et al. (2020) developed an oil-based composite consists of O_2-sensitive printable ink to monitor the O_2 level inside food packaging. The high levels of unsaturated fatty acids induced shrinkage upon exposure to atmospheric O_2, thus estimating the integrity of packaged food. Linseed oil for perishable foods (dairy products and seafood) whereas olive oil for nonperishable food (dry grains, cereals, etc.) could be used in intelligent ink sensors in the form of barcode labels or passive RFID tags (Volison-Klimentiev et al. 2020). The existing clean label smart packaging techniques based on RFID tags have been demonstrated by several researchers (Fernández-Salmerón et al. 2015; Barge et al. 2017; Bibi et al. 2017). Weigel et al. (2015) presented a new disposable O_2 sensing RFID system incorporated with linseed oil, which can track O_2 contamination without external energy. However, the most recent RFID smart packaging consists of a wireless sensor network, which builds an intelligent network on secondary packages to identify changes in the packaged environment (Athauda and Chandra Karmakar 2019).

6.3.2 *Time/Temperature Indicator*

As the food industry grows, time-temperature control is one of the main factors to develop an effective cold chain system. Nowadays, intelligent packaging includes thin foldable wrappings with a flexible sensing device attached to it. There are many standardized temperature-sensitive labels used for real-time monitoring of food's condition changes via temperature fluctuation inside the smart packages. Time-temperature indicators (TTI) provide product quality information that is communicated through changes in label color when exposed to temperature over a specified time. The TTI label consists of functional dyes, matrices, however, enzymatic, and microbial TTIs innovations are the recent intervention of TTI system connected to the food label to examine their time/temperature history.

Enzymatic TTI with pH indicator changes color which identifies a pH decrease due to temperature-sensitive enzymatic hydrolysis of a lipid substrate (Anbukarasu et al. 2017). While lactic acid bacteria-based pH-dependent sensors are mostly used for microbial TTI food packaging (Lim et al. 2014; Hsiao and Chang 2017). The microbial TTI with pH indicator takes a long time to respond while Esfahani et al. (2017) used microbial TTI with the α-amylase activity of *Bacillus amyloliquefaciens* vegetative cells to fasten the procedure. Similarly, Tsironi et al. (2017) used enzymatic TTI smart labels to predict the growth of *Vibrio parahaemolyticus* and *Vibrio vulnificus* in oysters and microbial TTI (*Janthinobacterium* sp.) based on the violacein formation to validate the minced beef meat spoilage.

Clean label interventions include TTI system incorporated with xylose or a glycine solution (Yamamoto and Isshiki 2012), glucose-fructose-glycine solutions (Choachamnan and Boonsupthip 2019), glucose oxidase (Rahman et al. 2018), natural tyrosinase (Tsironi et al. 2017), betacyanin dye (Rahimah et al. 2020) and anthocyanidin (extracted from strawberries) (Jang et al. 2018). However, Pereira et al. (2015), used anthocyanins (extracted from red cabbage *Brassica oleracea* var. *capitata*) on a PVA/chitosan polymeric packaging as a clean label TTI system. Besides, Rachmelia and Imawan (2018) indicated that the black corn extract can be used as a smart label on chitosan films because it changed color from purple to blue to yellow at 10–40 °C. Furthermore, the combination of TTI system with RFID tag could be used for the new generation of smart intelligent food packaging (Kim et al. 2016).

6.3.3 pH Indicator

Intelligent packaging not only protects the packaged food but also emits an electric or colorimetric signal in real-time response to any fluctuation in packaged food quality. Food spoilage often leads to pH variations which can be easily evidenced through a colorimetric pH indicator. pH indicators/sensors (solid base and dyes) indicate changes in food pH at the time of purchase or before consumption (Balbinot-Alfaro et al. 2019). Currently available colorimetric sensors are consisting of some chemical dyes to indicate the fluctuation in packed food quality while the innovative clean label pH sensors include natural dyes/pigments or polymers for sensor applications. Such as hibiscus extract incorporated in natural gelatin or chitosan polymeric films were used as a natural pH indicator (from 2.2 to 8.0 and 13.0) that showed a visible color variation on quality fluctuations (Peralta et al. 2019). Similarly, anthocyanins (1.5 g extract/100 g film solution) from eggplant were incorporated into chitosan and starch-based pH indicators to indicate food spoilage (Bilgiç et al. 2019). Additionally, blueberry and red grape skin pomace extracts were added into chitosan and carboxymethyl cellulose bio-based pH indicators (from 4 to 7) to evaluate the quality shift in fresh chicken meat (Kurek et al. 2019b). Moreover, anthocyanins have been extracted from black carrots (Ebrahimi Tirtashi et al. 2019; Moradi et al. 2019), red cabbage (Pourjavaher et al. 2017; Kuswandi et al. 2020; Zia et al. 2021) purple basil leaves (*Ocimum sanctum L.*) (Suthar and Saran 2020).

Purple sweet potato (a root crop) is rich in anthocyanins, thus used in smart packaging as a pH-dependent clean label dye. Antioxidant-rich biocompatible pH-indicating intelligent packaging was developed by incorporating purple potato extracts into the chitosan matrix (Yong et al. 2019; Li et al. 2019a) and agar/potato starch films (Choi et al. 2017). While the addition of purple potato (as a natural dye) with oregano essential oil (as an antimicrobial agent against *E. coli* and *L. monocytogenes*) onto cellulose nanofibers films showed a color shift from red to yellow when pH changed from 2 to 12 (Chen et al. 2020). Besides, the composite films having purple potato with gellan gum shifted color from purple to blue (6 h) and yellow-green (24 h) with the pH changes caused by *E. coli* growth and may be served as an easy-to-use indicator system to detect the spoilage of protein-rich foods (Wei et al. 2017).

The addition of different types of natural plant extracts into a carbohydrate-based film (carboxymethyl cellulose, carrageenan, and pectin) has recently become a very popular area. For this purpose butterfly pea (*Clitoria ternatea L.*) flower extract (Netramai et al. 2020), hibiscus extract in starch films (Othman et al. 2018), blueberry extract in starch films (Andretta et al. 2019) such as corn starch (Luchese et al. 2017), and cassava starch (Luchese et al. 2018), blueberry and blackberry pomace extracts in chitosan films (Kurek et al. 2018; Halász and Csóka 2018), mulberry extracts and chitosan nanoparticles in PVA films (Ma et al. 2018), Jambolão (*Syzygium cumini*) skins extract in methylcellulose films (da Silva Filipini et al. 2020) and *Lycium ruthenicum Murr* in cassava starch films (Qin et al. 2019) were used to develop potential biodegradable and intelligent films as an alternative to synthetic plastic nonbiodegradable films in pH sensors.

Furthermore, pH-sensitive Sago (*Metroxylon sagu*) polymeric films (Ahmad et al. 2020) and sago starch film with torch ginger extract as a potential colorimetric indicator responded well according to pH variation during the safe monitoring of packaged food (Xue Mei et al. 2020). However, green tea and basil extracts (5 wt.%) were incorporated in thermoplastic biodegradable cassava starch films, that changes color due to pH-sensitive chlorophyll and carotenoids in the extracts. The films presented antioxidant activity due to phenolic compounds of the extracts (Medina et al. 2017). Such films are rich in pH-sensitive anthocyanins that changed color by altering pH. These films not only showed improved mechanical and barrier quality but also monitor the freshness of packed food products that are sensitive to lipid oxidation and pH modification.

A new pH-sensing film was developed by using tamarind seed polysaccharides and natural dye extracted from litmus lichen (2.5%) changed color from orange to blue in the pH range of 4–10 (Liang and Wang 2017). While black soybean seed coat extract (containing 21.38% cyanidin-3-*O*-galactoside, 63.30% cyanidin-3-*O*-glucoside, and 11.92% peonidin-3-*O*-glucoside) have been added to chitosan-based films (Wang et al. 2019b) as an intelligent pH indicator. Besides, smart color-changing paper packaging sensors with pH-sensitive chromophores (anthraquinone and azo) have been suggested by (Zhang et al. 2019a) for clean label intelligent food packaging.

6.4 Conclusion

Food packaging is an essential part of the food manufacturing process since it preserves the quality of food products during storage. Consumer interests focus on the originality and naturalness of the package and its eco-advantages (other than economic and safety concerns) motivated scientists and food manufacturers to seek alternative ways to substitute traditional packaging. Additionally, natural products are being incorporated into food packaging as active antimicrobial and antioxidant substitutes for plastic polymers. For this purpose, low-cost biodegradable polymers-based food packaging films have been developed to replace the synthetic chemical packaging. However, the combination of biodegradable packaging material with clean label bioactive components such as the addition of essential plant oils or plant extracts in active and intelligent packaging gives extra information about the improvement of food quality aspects. To improve packaged food quality, investigations are currently looking to bring clean label biopolymer to industrial level food packaging applications.

References

Abdel Aziz MS, Salama HE, Sabaa MW (2018) Biobased alginate/castor oil edible films for active food packaging. LWT Food Sci Technol 96:455–460. https://doi.org/10.1016/j.lwt.2018.05.049

Abdelhedi O, Nasri R, Jridi M, Kchaou H, Nasreddine B, Karbowiak T, Debeaufort F, Nasri M (2018) Composite bioactive films based on smooth-hound viscera proteins and gelatin: physicochemical characterization and antioxidant properties. Food Hydrocoll 74:176–186. https://doi.org/10.1016/j.foodhyd.2017.08.006

Aday MS, Caner C (2013) The shelf life extension of fresh strawberries using an oxygen absorber in the biobased package. LWT- Food Sci Technol 52(2):102–109. https://doi.org/10.1016/j.lwt.2012.06.006

Adilah ZAM, Jamilah B, Hanani ZAN (2019) Storage stability of soy protein isolate films incorporated with mango kernel extract at different temperature. Food Hydrocoll 87:541–549

Adilah ZAM, Jamilah B, Hanani ZAN (2018) Functionaland antioxidant properties of protein-based films incorporated with mango kernel extract for active packaging. Food Hydrocoll 74:207–218. https://doi.org/10.1016/j.foodhyd.2017.08.017

Ahmad AN, Abdullah Lim S, Navaranjan N (2020) Development of sago (Metroxylon sagu)-based colorimetric indicator incorporated with butterfly pea (Clitoria ternatea) anthocyanin for intelligent food packaging. J Food Saf 40(4). https://doi.org/10.1111/jfs.12807

Ahmed I, Lin H, Zou L, Brody AL, Li Z, Qazi IM, Pavase TR, Lv L (2017) A comprehensive review on the application of active packaging technologies to muscle foods. Food Control 82:163–178. https://doi.org/10.1016/j.foodcont.2017.06.009

Alehosseini A, Gómez-Mascaraque LG, Martínez-Sanz M, López-Rubio A (2019) Electrospun curcumin-loaded protein nanofiber mats as active/bioactive coatings for food packaging applications. Food Hydrocoll 87:758–771. https://doi.org/10.1016/j.foodhyd.2018.08.056

Alkan Tas B, Sehit E, Erdinc Tas C, Unal S, Cebeci FC, Menceloglu YZ, Unal H (2019) Carvacrol loaded halloysite coatings for antimicrobial food packaging applications. Food Packag Shelf Life 20:100300. https://doi.org/10.1016/j.fpsl.2019.01.004

Al-Tayyar NA, Youssef AM, Al-hindi R (2020) Antimicrobial food packaging based on sustainable Bio-based materials for reducing foodborne Pathogens: A review. Food Chem 310:125915. https://doi.org/10.1016/j.foodchem.2019.125915

Anbukarasu P, Sauvageau D, Elias AL (2017) Time-temperature indicator based on enzymatic degradation of dye-loaded polyhydroxybutyrate. Biotechnol J 12(9):1–9. https://doi.org/10.1002/biot.201700050

Andretta R, Luchese CL, Tessaro IC, Spada JC (2019) Development and characterization of pH-indicator films based on cassava starch and blueberry residue by thermocompression. Food Hydrocoll 93:317–324. https://doi.org/10.1016/j.foodhyd.2019.02.019

Araújo A, Galvão A, Filho CS, Mendes F, Oliveira M, Barbosa F, Filho MS, Bastos M (2018) Okra mucilage and corn starch bio-based film to be applied in food. Polym Test 71:352–361. https://doi.org/10.1016/j.polymertesting.2018.09.010

Araújo, A., Galvão, A., Silva Filho, C., Mendes, F., Oliveira, M., Barbosa, F., ... & Bastos, M. (2018). Okra mucilage and corn starch bio-based film to be applied in food. Polymer Testing, 71, 352–361.

Arrieta MP, Garrido L, Faba S, Guarda A, Galotto MJ, de Dicastillo CL (2020) Cucumis metuliferus fruit extract loaded acetate cellulose coatings for antioxidant active packaging. Polymers 12(6):1–19. https://doi.org/10.3390/POLYM12061248

Arshad R, Sameen A, Huma N, Zia MA (2020) Exploring the potential of active edible coating on the shelf stability of dairy products. Pak J Agric Sci 57(1):237–244. https://doi.org/10.21162/PAKJAS/20.8182

Ashrafi A, Jokar M, Mohammadi Nafchi A (2018) Preparation and characterization of biocomposite film based on chitosan and kombucha tea as active food packaging. Int J Biol Macromol 108:444–454. https://doi.org/10.1016/j.ijbiomac.2017.12.028

Assis RQ, Rios PDA, de OliveriaRios A, Olivera FC (2020) Biodegradable packaging of cellulose acetate incorporated with norbixin, lycopene or zeaxanthin. Ind Crop Prod 147:112212. https://doi.org/10.1016/j.indcrop.2020.112212

Athauda T, Chandra Karmakar N (2019) Review of RFID-based sensing in monitoring physical stimuli in smart packaging for food-freshness applications. In: Wireless power transfer. Cambridge University Press, Cambridge, pp 161–174. https://doi.org/10.1017/wpt.2019.6

Aydemir Sezer U, Sanko V, Yuksekdag ZN, Uzundağ D, Sezer S (2016) Use of oxidized regenerated cellulose as bactericidal filler for food packaging applications. Cellulose 23(5):3209–3219. https://doi.org/10.1007/s10570-016-1000-4

Azeredo HMC, Rosa MF, Mattoso LHC (2016) Nanocellulose in bio-based food packaging applications. Ind Crop Prod 1–8. https://doi.org/10.1016/j.indcrop.2016.03.013

Azeredo HM, Rosa MF, Mattoso LHC (2017) Nanocellulose in bio-based food packaging applications. Ind Crops Prod 97:664–671

Bai R, Zhang X, Yong H, Wang X, Liu Y, Liu J (2019) Development and characterization of antioxidant active packaging and intelligent Al 3+ -sensing films based on carboxymethyl chitosan and quercetin. Int J Biol Macromol 126:1074–1084. https://doi.org/10.1016/j.ijbiomac.2018.12.264

Balbinot-Alfaro E, Craveiro DV, Lima KO, Costa HLG, Lopes DR, Prentice C (2019) Intelligent packaging with pH indicator potential. Food Eng Rev 11(4):235–244. https://doi.org/10.1007/s12393-019-09198-9

Ballesteros LF, Michelin M, Vicente AA, Teixeira JA, Cerqueira MÂ (2018) Use of lignocellulosic materials in bio-based packaging. In Lignocellulosic materials and their use in bio-based packaging (pp. 65–85). Springer, Cham.

Barbosa-Pereira L, Angulo I, Lagarón JM, Paseiro-Losada P, Cruz JM (2014) Development of new active packaging films containing bioactive nanocomposites. Innov Food Sci Emerg Technol 26:310–318. https://doi.org/10.1016/j.ifset.2014.06.002

Barge P, Biglia A, Comba L, Gay P, Ricauda Aimonino D, Tortia C (2017) Temperature and position effect on readability of passive UHF RFID labels for beverage packaging. Chem Eng Trans 58:169–174. https://doi.org/10.3303/CET1758029

Benbettaïeb N, O'Connell C, Viaux A-S, Bou-Maroun E, Seuvre AM, Brachais CH, Debeaufort F (2019) Sorption kinetic of aroma compounds by edible bio-based films from marine-by product macromolecules: effect of relative humidity conditions. Food Chem 298:125064. https://doi.org/10.1016/j.foodchem.2019.125064

Berti S, Flores SK, Jagus RJ (2020) Improvement of the microbiological quality of Argentinian Port Salut cheese by applying starch-based films and coatings reinforced with rice bran and containing natural antimicrobials. J Food Process Preserv 44. https://doi.org/10.1111/jfpp.14827

Bi F, Zhang X, Bai R, Liu Y, Liu J, Liu J (2019) Preparation and characterization of antioxidant and antimicrobial packaging films based on chitosan and proanthocyanidins. Int J Biol Macromol 134:11–19. https://doi.org/10.1016/j.ijbiomac.2019.05.042

Bibi F, Guillaume C, Gontard N, Sorli B (2017) A review: RFID technology having sensing aptitudes for food industry and their contribution to tracking and monitoring of food products. Trends Food Sci Technol 62:91–103. https://doi.org/10.1016/j.tifs.2017.01.013

Bierhalz ACK, Da Silva MA, Kieckbusch TG (2012) Natamycin release from alginate/pectin films for food packaging applications. J Food Eng 110(1):18–25. https://doi.org/10.1016/j.jfoodeng.2011.12.016

Bilgiç S, Söğüt E, Seydim AC (2019) Chitosan and starch based intelligent films with anthocyanins from eggplant to monitor pH variations. Turk J Agric Food Sci Technol 7(sp1):61. https://doi.org/10.24925/turjaf.v7isp1.61-66.2705

Bonilla J, Sobral PJA (2016) Investigation of the physicochemical, antimicrobial and antioxidant properties of gelatin-chitosan edible film mixed with plant ethanolic extracts. Food Biosci 16:17–25. https://doi.org/10.1016/j.fbio.2016.07.003

Bovi GG, Caleb OJ, Klaus E, Tintchev F, Rauh C, Mahajan PV (2018) Moisture absorption kinetics of FruitPad for packaging of fresh strawberry. J Food Eng 223:248–254. https://doi.org/10.1016/j.jfoodeng.2017.10.012

Brobbey KJ, Saarinen J, Alakomi H, Yang B, Toivakka M (2017) Efficacy of natural plant extracts in antimicrobial packaging systems. J Appl Packag Res 9(1):6

Bugatti V, Vertuccio L, Viscusi G, Gorrasi G (2018) Antimicrobial membranes of bio-based pa 11 and hnts filled with lysozyme obtained by an electrospinning process. Nanomaterials 8(3):139. https://doi.org/10.3390/nano8030139

Burgos N, Armentano I, Fortunati E, Dominici F, Luzi F, Fiori S, Cristofaro F, Visai L, Jiménez A, Kenny JM (2017) Functional properties of plasticized bio-based poly(lactic acid)_poly(hydroxybutyrate) (PLA_PHB) films for active food packaging. Food Bioproc Tech 10(4):770–780. https://doi.org/10.1007/s11947-016-1846-3

Cao L, Ma Q, Liang T, Sun G, Chi W, Zhang C, Li J, Wang L (2019) A semen cassia gum-based film with visual–olfactory function for indicating the freshness change of animal protein-rich food. Int J Biol Macromol 133:243–252. https://doi.org/10.1016/j.ijbiomac.2019.04.045

Caro N, Medina E, Díaz-Dosque M, López L, Abugoch L, Tapia C (2016) Novel active packaging based on films of chitosan and chitosan/quinoa protein printed with chitosan-tripolyphosphate-thymol nanoparticles via thermal ink-jet printing. Food Hydrocoll 52:520–532. https://doi.org/10.1016/j.foodhyd.2015.07.028

Ceballos RL, Ochoa-Yepes O, Goyanes S, Bernal C, Famá L (2020) Effect of yerba mate extract on the performance of starch films obtained by extrusion and compression molding as active and smart packaging. Carbohydr Polym 244:116495. https://doi.org/10.1016/j.carbpol.2020.116495

Chang LY, Chuang MY, Zan HW, Meng HF, Lu CJ, Yeh PH, Chen JN (2017) One-minute fish freshness evaluation by testing the volatile amine gas with an ultrasensitive porous-electrode-capped organic gas sensor system. ACS Sens 2(4):531–539. https://doi.org/10.1021/acssensors.6b00829

Chen CW, Xie J, Yang FX, Zhang HL, Xu ZW, Liu JL, Chen YJ (2018a) Development of moisture-absorbing and antioxidant active packaging film based on poly(vinyl alcohol) incorporated with green tea extract and its effect on the quality of dried eel. J Food Process Preserv 42(1):1–11. https://doi.org/10.1111/jfpp.13374

Chen HZ, Zhang M, Bhandari B, Guo Z (2018b) Applicability of a colorimetric indicator label for monitoring freshness of fresh-cut green bell pepper. Postharv Biol Technol 140:85–92. https://doi.org/10.1016/j.postharvbio.2018.02.011

Chen S, Wu M, Lu P, Gao L, Yan S, Wang S (2020) Development of pH indicator and antimicrobial cellulose nanofibre packaging film based on purple sweet potato anthocyanin and oregano essential oil. Int J Biol Macromol 149:271–280. https://doi.org/10.1016/j.ijbiomac.2020.01.231

Cheng J, Wang H, Kang S, Xia L, Jiang S, Chen M, Jiang S (2019) An active packaging film based on yam starch with eugenol and its application for pork preservation. Food Hydrocoll 96:546–554. https://doi.org/10.1016/j.foodhyd.2019.06.007

Chenwei C, Zhipeng T, Yarui M, Weiqiang Q, Fuxin Y, Jun M, Jing X (2018) Physicochemical, microstructural, antioxidant and antimicrobial properties of active packaging films based on poly(vinyl alcohol)/clay nanocomposite incorporated with tea polyphenols. Prog Org Coat 123:176–184. https://doi.org/10.1016/j.porgcoat.2018.07.001

Choachamnan J, Boonsupthip W (2019) Glucose-fructose-glycine time–temperature integrator and its potential application in heating process control of food softness. J Food Process Eng 42(8):1–9. https://doi.org/10.1111/jfpe.13287

Choi HY, Lee YS (2013) Characteristics of moisture-absorbing film impregnated with synthesized attapulgite with acrylamide and its effect on the quality of seasoned laver during storage. J Food Eng 116(4):829–839. https://doi.org/10.1016/j.jfoodeng.2013.01.023

Choi I, Lee JY, Lacroix M, Han J (2017) Intelligent pH indicator film composed of agar/potato starch and anthocyanin extracts from purple sweet potato. Food Chem 218:122–128. https://doi.org/10.1016/j.foodchem.2016.09.050

Chollakup R, Kongtud W, Sukatta U, Piriyasatits K, Premchookiat M, Jarerat A (2020a) Development of rice straw paper coated with pomelo peel extract for bio-based and antibacterial packaging. Key Eng Mater 847 KEM:141–146. https://doi.org/10.4028/www.scientific.net/KEM.847.141

Chollakup R, Pongburoos S, Boonsong W, Khanoonkon N, Kongsin K, Sothornvit R, Sukyai P, Sukatta U, Harnkarnsujarit N (2020b) Antioxidant and antibacterial activities of cassava starch and whey protein blend films containing rambutan peel extract and cinnamon oil for active packaging. LWT Food Sci Technol 130:109573. https://doi.org/10.1016/j.lwt.2020.109573

Ciannamea EM, Stefani PM, Ruseckaite RA (2016) Properties and antioxidant activity of soy protein concentrate films incorporated with red grape extract processed by casting and compression molding. LWT Food Sci Technol 74:353–362. https://doi.org/10.1016/j.lwt.2016.07.073

Dalmoro V, Zimnoch dos Santos JH, Pires M, Simanke A, Baldino GB, Oliveira L (2017) Encapsulation of sensors for intelligent packaging. In: Food packaging. Elsevier, Amsterdam. https://doi.org/10.1016/b978-0-12-804302-8.00004-2

Demircan D, Zhang B (2017) Facile synthesis of novel soluble cellulose-grafted hyperbranched polymers as potential natural antimicrobial materials. Carbohydr Polym 157:1913–1921. https://doi.org/10.1016/j.carbpol.2016.11.076

Deshwal GK, Panjagari NR, Badola R, Singh AK, Minz PS, Ganguly S, Alam T (2018) Characterization of biopolymer-based UV-activated intelligent oxygen indicator for food-packaging applications. J Packag Technol Res 2(1):29–43. https://doi.org/10.1007/s41783-018-0029-2

Dewi MY, Husni A (2020) Characterization of biobased alginate/glycerol/sunflower oil as biodegradable packaging. In: E3S web of conferences, 147. https://doi.org/10.1051/e3sconf/202014703004

Dey A, Neogi S (2019) Oxygen scavengers for food packaging applications: a review. Trends Food Sci Technol 90:26–34. https://doi.org/10.1016/j.tifs.2019.05.013

Dey D, Dharini V, Periyar Selvam S, Rotimi Sadiku E, Mahesh Kumar M, Jayaramudu J, Nath Gupta U (2020) Physical, antifungal, and biodegradable properties of cellulose nanocrystals and chitosan nanoparticles for food packaging application. Mater Today Proc. https://doi.org/10.1016/j.matpr.2020.04.885

Di Filippo MF, Dolci LS, Liccardo L, Bigi A, Bonvicini F, Gentilomi GA, Passerini N, Panzavolta S, Albertini B (2021) Cellulose derivatives-snail slime films: new disposable eco-friendly materials for food packaging. Food Hydrocoll 111:106247. https://doi.org/10.1016/j.foodhyd.2020.106247

Diana K, Marlina SS, Helwati H (2019) Effect of natural ingredients addition as antimicrobial agents in Dioscoreahispida Dennst starch-based biofilm. IOP Conf Ser Earth Environ Sci 364(1). https://doi.org/10.1088/1755-1315/364/1/012006

Ding L, Wang H, Liu D, Zheng Z (2018) Surface attachment of natural antimicrobial coatings onto conventional polypropylene nonwoven fabric and its antimicrobial performance assessment. J Food Prot 81(2):172–177. https://doi.org/10.4315/0362-028X.JFP-17-177

Ding F, Hu B, Lan S, Wang H (2020) Flexographic and screen printing of carboxymethyl chitosan based edible inks for food packaging applications. Food Packag Shelf Life 26:100559. https://doi.org/10.1016/j.fpsl.2020.100559

Dobrucka R, Cierpiszewski R (2014) Active and intelligent packaging food – research and development – a review. Polish J Food Nutr Sci 64(1):7–15. https://doi.org/10.2478/v10222

Dong H, Ling Z, Zhang X, Zhang X, Ramaswamy S, Xu F (2020) Smart colorimetric sensing films with high mechanical strength and hydrophobic properties for visual monitoring of shrimp and pork freshness. Sens Actuators B 309:127752. https://doi.org/10.1016/j.snb.2020.127752

dos Santos Caetano K, Almeida Lopes N, Haas Costa TM, Brandelli A, Rodrigues E, Hickmann Flôres S, Cladera-Olivera F (2018) Characterization of active biodegradable films based on cassava starch and natural compounds. Food Packag Shelf Life 16:138–147. https://doi.org/10.1016/j.fpsl.2018.03.006

Duran M, Aday MS, Zorba NND, Temizkan R, Büyükcan MB, Caner C (2016) Potential of antimicrobial active packaging "containing natamycin, nisin, pomegranate and grape seed extract in chitosan coating" to extend shelf life of fresh strawberry. Food Bioprod Process 98:354–363. https://doi.org/10.1016/j.fbp.2016.01.007

Ebrahimi Tirtashi F, Moradi M, Tajik H, Forough M, Ezati P, Kuswandi B (2019) Cellulose/chitosan pH-responsive indicator incorporated with carrot anthocyanins for intelligent food packaging. Int J Biol Macromol 136:920–926. https://doi.org/10.1016/j.ijbiomac.2019.06.148

Elabd MA (2018) Effect of antimicrobial edible coatings on quality and shelf life of minimal processed guava (Psidium guajava). Alex J Food Sci Technol 15(1):65–76. http://ajfs.journals.ekb.eg/article_16349_67ccdb5c84a55a99f12a0a8c0ea6dcf6.pdf

El-Fattah MA, El Saeed AM, Azzam AM, Abdul-Raheim ARM, Hefni HHH (2016) Improvement of corrosion resistance, antimicrobial activity, mechanical and chemical properties of epoxy coating by loading chitosan as a natural renewable resource. Prog Org Coat 101:288–296. https://doi.org/10.1016/j.porgcoat.2016.09.002

Esfahani SS, Emtiazi G, Rabbani M (2017) Preparation of a microbial time-temperature indicator by using the vegetative form of Bacillus amyloliquefaciens for monitoring the quality of chilled food products. Appl Food Biotechnol 4(2):79–84. https://doi.org/10.22037/afb.v4i2.15809

Ezati P, Tajik H, Moradi M (2019) Fabrication and characterization of alizarin colorimetric indicator based on cellulose-chitosan to monitor the freshness of minced beef. Sens Actuat B Chem 285:519–528. https://doi.org/10.1016/j.snb.2019.01.089

Ezati P, Bang YJ, Rhim JW (2021) Preparation of a shikonin-based pH-sensitive color indicator for monitoring the freshness of fish and pork. Food Chem 337:127995. https://doi.org/10.1016/j.foodchem.2020.127995

Fabra MJ, Falcó I, Randazzo W, Sánchez G, López-Rubio A (2018) Antiviral and antioxidant properties of active alginate edible films containing phenolic extracts. Food Hydrocoll 81:96–103. https://doi.org/10.1016/j.foodhyd.2018.02.026

Falagán N, Artés F, Aguayo E (2018) Antimicrobial properties of edible coatings enriched with natural additives on fresh-cut peaches. Acta Hortic 1194:519–525. https://doi.org/10.17660/ActaHortic.2018.1194.76

Fernández-Salmerón J, Rivadeneyra A, Martínez-Martí F, Capitán-Vallvey LF, Palma AJ, Carvajal MA (2015) Passive UHF RFID tag with multiple sensing capabilities. Sensors (Switzerland) 15(10):26769–26782. https://doi.org/10.3390/s151026769

Fonseca-García A, Jiménez-Regalado EJ, Aguirre-Loredo RY (2021) Preparation of a novel biodegradable packaging film based on corn starch-chitosan and poloxamers. Carbohydr Polym 251:117009. https://doi.org/10.1016/j.carbpol.2020.117009

Gaikwad KK, Lee JY, Lee YS (2016) Development of polyvinyl alcohol and apple pomace biocomposite film with antioxidant properties for active food packaging application. J Food Sci Technol 53(3):1608–1619. https://doi.org/10.1007/s13197-015-2104-9

Gaikwad KK, Singh S, Lee YS (2017a) A pyrogallol-coated modified LDPE film as an oxygen scavenging film for active packaging materials. Prog Org Coat 111:186–195. https://doi.org/10.1016/j.porgcoat.2017.05.016

Gaikwad KK, Singh S, Suk Y (2017b) A new pyrogallol coated oxygen scavenging film and their effect on oxidative stability of soybean oil under different storage conditions. Food Sci Biotechnol 26(6):1535–1543. https://doi.org/10.1007/s10068-017-0232-x

Gaikwad KK, Singh S, Ajji A (2018) Moisture absorbers for food packaging applications. Environ Chem Lett 17(2):609–628. https://doi.org/10.1007/s10311-018-0810-z

Gallego MG, Gordon MH, Segovia F, Almajano Pablos MP (2016) Gelatine-based antioxidant packaging containing Caesalpinia decapetala and tara as a coating for ground beef patties. Antioxidants 5(2):10. https://doi.org/10.3390/antiox5020010

Gao Q, Lei M, Zhou K, Liu X, Wang S, Li H (2020) Preparation of a microfibrillated cellulose/chitosan/polypyrrole film for active food packaging. Prog Org Coat 149:105907. https://doi.org/10.1016/j.porgcoat.2020.105907

Gaona-Forero A, Agudelo-Rodríguez G, Herrera AO, Castellanos DA (2018) Modeling and simulation of an active packaging system with moisture adsorption for fresh produce. Application in 'Hass' avocado. Food Packag Shelf Life 17:187–195. https://doi.org/10.1016/j.fpsl.2018.07.005

Ge Y, Li Y, Bai Y, Yuan C, Wu C, Hu Y (2020) Intelligent gelatin/oxidized chitin nanocrystals nanocomposite films containing black rice bran anthocyanins for fish freshness monitorings. Int J Biol Macromol 155:1296–1306. https://doi.org/10.1016/j.ijbiomac.2019.11.101

Gu J, Su Y, Liu P, Li P, Yang P (2017) An environmentally benign antimicrobial coating based on a protein supramolecular assembly. ACS Appl Mater Interfaces 9(1):198–210. https://doi.org/10.1021/acsami.6b13552

Gu R, Yun H, Chen L, Wang Q, Huang X (2020) Regenerated cellulose films with amino-terminated hyperbranched polyamic anchored nanosilver for active food packaging. ACS Appl Biomater 3(1):602–610. https://doi.org/10.1021/acsabm.9b00992

Guo Y, Tian D, Shen F, Yang G, Long L, He J, Song C, Zhang J, Zhu Y, Huang C, Deng S (2019) Transparent cellulose/technical lignin composite films for advanced packaging. Polymers 11(9):1–11. https://doi.org/10.3390/polym11091455

Halász K, Csóka L (2018) Black chokeberry (Aronia melanocarpa) pomace extract immobilized in chitosan for colorimetric pH indicator film application. Food Packag Shelf Life 16:185–193. https://doi.org/10.1016/j.fpsl.2018.03.002

Han Y, Yu M, Wang L (2018) Preparation and characterization of antioxidant soy protein isolate films incorporating licorice residue extract. Food Hydrocoll 75:13–21. https://doi.org/10.1016/j.foodhyd.2017.09.020

Hanani ZAN, Yee FC, Nor-Khaizura MAR (2019) Effect of pomegranate (Punica granatum L.) peel powder on the antioxidant and antimicrobial properties of fish gelatin films as active packaging. Food Hydrocoll 89:253–259. https://doi.org/10.1016/j.foodhyd.2018.10.007

Herbes C, Beuthner C, Ramme I (2018) Consumer attitudes towards biobased packaging – a cross-cultural comparative study. J Clean Prod 194:203–218. https://doi.org/10.1016/j.jclepro.2018.05.106

Hrytsenko O, Hrytsenko D, Shvalagin V, Grodziuk G, Kompanets M (2018) The use of carbon nanoparticles for inkjet-printed functional labels for smart packaging. J Nanomater 2018:1–10. https://doi.org/10.1155/2018/6485654

Hsiao HI, Chang JN (2017) Developing a microbial time–temperature indicator to monitor total volatile basic nitrogen change in chilled vacuum-packed grouper fillets. J Food Process Preserv 41(5):1–9. https://doi.org/10.1111/jfpp.13158

Huang J, Chen M, Zhou Y, Li Y, Hu Y (2020) Functional characteristics improvement by structural modification of hydroxypropyl methylcellulose modified polyvinyl alcohol films incorporating roselle anthocyanins for shrimp freshness monitoring. Int J Biol Macromol 162:1250–1261. https://doi.org/10.1016/j.ijbiomac.2020.06.156

Ibrahim S, Soliman O, Sultan M, Abdel-Aziz MS (2017) Synergistic antimicrobial effect of xylitol with curcumin: water vapor barrier, mechanical and thermal properties of PSS/PVA packaging films. Int J Appl Eng Res 12(20):10360–10366

Iordanskii A, Zhulkina A, Olkhov A, Fomin S, Burkov A, Stilman M (2018) Characterization and evaluation of controlled antimicrobial release from petrochemical (PU) and biodegradable (PHB) packaging. Polymers 10(8):817. https://doi.org/10.3390/polym10080817

Jang HD, Yang JH, Kim DH, Ahn MH, Han SH, Lee SJ (2018) A novel printable time-temperature integrator with anthocyanidin, a natural coloring matter. Korean J Packag Sci Technol 24(2):49–56. https://doi.org/10.20909/kopast.2018.24.2.49

Jha P (2020) Effect of grapefruit seed extract ratios on functional properties of corn starch-chitosan bionanocomposite films for active packaging. Int J Biol Macromol 163:1546–1556. https://doi.org/10.1016/j.ijbiomac.2020.07.251

Jiang G, Hou X, Zeng X, Zhang C, Wu H, Shen G, Li S, Luo Q, Li M, Liu X, Chen A, Wang Z, Zhang Z (2020) Preparation and characterization of indicator films from carboxymethyl-cellulose/starch and purple sweet potato (Ipomoea batatas (L.) lam) anthocyanins for monitoring fish freshness. Int J Biol Macromol 143:359–372. https://doi.org/10.1016/j.ijbiomac.2019.12.024

Joanne Kam WY, Mirhosseini H, Abas F, Hussain N, Hedayatnia S, Florence Chong HL (2018) Antioxidant activity enhancement of biodegradable film as active packaging utilizing crude extract from durian leaf waste. Food Control 90:66–72. https://doi.org/10.1016/j.foodcont.2018.02.036

Johnson DR, Inchingolo R, Decker EA (2018) The ability of oxygen scavenging packaging to inhibit vitamin degradation and lipid oxidation in fish oil-in-water emulsions. Innov Food Sci Emerg Technol 47:467–475. https://doi.org/10.1016/j.ifset.2018.04.021

Kang S, Wang H, Guo M, Zhang L, Chen M, Jiang S, Li X, Jiang S (2018) Ethylene-vinyl alcohol copolymer-montmorillonite multilayer barrier film coated with mulberry anthocyanin for freshness monitoring. J Agric Food Chem 66(50):13268–13276. https://doi.org/10.1021/acs.jafc.8b05189

Kang S, Wang H, Xia L, Chen M, Li L, Cheng J, Li X, Jiang S (2020) Colorimetric film based on polyvinyl alcohol/okra mucilage polysaccharide incorporated with rose anthocyanins for shrimp freshness monitoring. Carbohydr Polym 229:115402. https://doi.org/10.1016/j.carbpol.2019.115402

Khalid S, Yu L, Feng M, Meng L, Bai Y, Ali A, Liu H, Chen L (2018) Development and characterization of biodegradable antimicrobial packaging films based on polycaprolactone, starch and pomegranate rind hybrids. Food Packag Shelf Life 18:71–79. https://doi.org/10.1016/j.fpsl.2018.08.008

Khan I, Tango CN, Chelliah R, Oh DH (2019) Development of antimicrobial edible coating based on modified chitosan for the improvement of strawberries shelf life. Food Sci Biotechnol 28(4):1257–1264. https://doi.org/10.1007/s10068-018-00554-9

Khosravi A, Fereidoon A, Khorasani MM, Naderi G, Ganjali MR, Zarrintaj P, Saeb MR, Gutiérrez TJ (2020) Soft and hard sections from cellulose-reinforced poly(lactic acid)-based food packaging films: a critical review. Food Packag Shelf Life 23:100429. https://doi.org/10.1016/j.fpsl.2019.100429

Khumkomgool A, Saneluksana T, Harnkarnsujarit N (2020) Active meat packaging from thermoplastic cassava starch containing sappan and cinnamon herbal extracts via LLDPE blown-film extrusion. Food Packag Shelf Life 26:100557. https://doi.org/10.1016/j.fpsl.2020.100557

Kim MJ, Shin HW, Lee SJ (2016) A novel self-powered time-temperature integrator (TTI) using modified biofuel cell for food quality monitoring. Food Control 70:167–173. https://doi.org/10.1016/j.foodcont.2016.05.031

Kim JH, Hong W, sung, & Oh, S. W. (2018) Effect of layer-by-layer antimicrobial edible coating of alginate and chitosan with grapefruit seed extract for shelf-life extension of shrimp (Litopenaeus vannamei) stored at 4 °C. Int J Biol Macromol 120:1468–1473. https://doi.org/10.1016/j.ijbiomac.2018.09.160

Kopacic S, Walzl A, Zankel A, Leitner E, Bauer W (2018) Alginate and chitosan as a functional barrier for paper-based packaging materials. MDPI Coat 8(7). https://doi.org/10.3390/coatings8070235

Kowalczyk D, Biendl M (2016) Physicochemical and antioxidant properties of biopolymer/candelilla wax emulsion films containing hop extract - a comparative study. Food Hydrocoll 60:384–392. https://doi.org/10.1016/j.foodhyd.2016.04.010

Kurek M, Garofulić IE, Bakić MT, Ščetar M, Uzelac VD, Galić K (2018) Development and evaluation of a novel antioxidant and pH indicator film based on chitosan and food waste sources of antioxidants. Food Hydrocoll 84:238–246. https://doi.org/10.1016/j.foodhyd.2018.05.050

Kurek M, Hlupić L, Elez Garofulić I, Descours E, Ščetar M, Galić K (2019a) Comparison of protective supports and antioxidative capacity of two bio-based films with revalorised fruit pomaces extracted from blueberry and red grape skin. Food Packaging and Shelf Life 20. https://doi.org/10.1016/j.fpsl.2019.100315

Kurek M, Hlupić L, Ščetar M, Bosiljkov T, Galić K (2019b) Comparison of two pH responsive color changing bio-based films containing wasted fruit pomace as a source of colorants. J Food Sci 84(9):2490–2498. https://doi.org/10.1111/1750-3841.14716

Kuswandi B, Asih NPN, Pratoko DK, Kristiningrum N, Moradi M (2020) Edible pH sensor based on immobilized red cabbage anthocyanins into bacterial cellulose membrane for intelligent food packaging. Packag Technol Sci 33(8):321–332. https://doi.org/10.1002/pts.2507

Latos-Brozio M, Masek A (2020) The application of natural food colorants as indicator substances in intelligent biodegradable packaging materials. Food Chem Toxicol 135:110975. https://doi.org/10.1016/j.fct.2019.110975

Lee K, Ko S (2014) Proof-of-concept study of a whey protein isolate based carbon dioxide indicator to measure the shelf-life of packaged foods. Food Sci Biotechnol 23(1):115–120. https://doi.org/10.1007/s10068-014-0015-6

Lee GY, Shin HS (2016) Development of freshness indicator for quality of skate (Raja kenojei) during storage. Food Sci Biotechnol 25(5):1485–1489. https://doi.org/10.1007/s10068-016-0230-4

Lee K, Park H, Baek S, Han S, Kim D, Chung S, Yoon JY, Seo J (2019) Colorimetric array freshness indicator and digital color processing for monitoring the freshness of packaged chicken breast. Food Packag Shelf Life 22:100408. https://doi.org/10.1016/j.fpsl.2019.100408

Li K, Jin S, Chen H, Li J (2019a) Bioinspired interface engineering of gelatin/cellulose nanofibrils nanocomposites with high mechanical performance and antibacterial properties for active packaging. Compos Part B Eng 171:222–234. https://doi.org/10.1016/j.compositesb.2019.04.043

Li Y, Ying Y, Zhou Y, Ge Y, Yuan C, Wu C, Hu Y (2019b) A pH-indicating intelligent packaging composed of chitosan-purple potato extractions strength by surface-deacetylated chitin nanofibers. Int J Biol Macromol 127:376–384. https://doi.org/10.1016/j.ijbiomac.2019.01.060

Liang T, Wang L (2017) A pH-sensing film from tamarind seed polysaccharide with litmus lichen extract as an indicator. Polymers 10(1):13. https://doi.org/10.3390/polym10010013

Lim SH, Choe WY, Son BH, Hong KW (2014) Development of a microbial time-temperature integrator system using lactic acid bacteria. Food Sci Biotechnol 23(2):483–487. https://doi.org/10.1007/s10068-014-0066-8

Limpisophon K, Schleining G (2017) Use of gallic acid to enhance the antioxidant and mechanical properties of active fish gelatin film. J Food Sci 82(1):80–89. https://doi.org/10.1111/1750-3841.13578

Liu J, Liu S, Wu Q, Gu Y, Kan J, Jin C (2017) Effect of protocatechuic acid incorporation on the physical, mechanical, structural and antioxidant properties of chitosan film. Food Hydrocoll 73:90–100. https://doi.org/10.1016/j.foodhyd.2017.06.035

Liu J, Wang H, Wang P, Guo M, Jiang S, Li X, Jiang S (2018) Films based on κ-carrageenan incorporated with curcumin for freshness monitoring. Food Hydrocoll 83:134–142. https://doi.org/10.1016/j.foodhyd.2018.05.012

Liu J, Wang H, Guo M, Li L, Chen M, Jiang S, Li X, Jiang S (2019) Extract from Lycium ruthenicum Murr. Incorporating κ-carrageenan colorimetric film with a wide pH–sensing range for food freshness monitoring. Food Hydrocoll 94:1–10. https://doi.org/10.1016/j.foodhyd.2019.03.008

Liu X, Xu Y, Zhan X, Xie W, Yang X, Cui SW, Xia W (2020) Development and properties of new kojic acid and chitosan composite biodegradable films for active packaging materials. Int J Biol Macromol 144:483–490. https://doi.org/10.1016/j.ijbiomac.2019.12.126

Llana-Ruíz-Cabello M, Pichardo S, Jiménez-Morillo NT, Bermúdez JM, Aucejo S, González-Vila FJ, Cameán AM, González-Pérez JA (2016) Molecular characterisation of a bio-based active packaging containing Origanum vulgare L. essential oil using pyrolysis gas chromatography-mass spectrometry. J Sci Food Agric 96(9):3207–3212. https://doi.org/10.1002/jsfa.7502

López De Dicastillo C, Rodríguez F, Guarda A, Galotto MJ (2016) Antioxidant films based on cross-linked methyl cellulose and native Chilean berry for food packaging applications. Carbohydr Polym 136:1052–1060. https://doi.org/10.1016/j.carbpol.2015.10.013

López de Dicastillo C, Bruna J, Torres A, Alvarado N, Guarda A, Galotto MJ (2017) A traditional aboriginal condiment as an antioxidant agent in the development of biodegradable active packaging. J Appl Polym Sci 134(15):1–10. https://doi.org/10.1002/app.44692

Lu P, Yang Y, Liu R, Liu X, Ma J, Wu M, Wang S (2020) Preparation of sugarcane bagasse nanocellulose hydrogel as a colourimetric freshness indicator for intelligent food packaging. Carbohydr Polym 249:116831. https://doi.org/10.1016/j.carbpol.2020.116831

Luchese CL, Sperotto N, Spada JC, Tessaro IC (2017) Effect of blueberry agro-industrial waste addition to corn starch-based films for the production of a pH-indicator film. Int J Biol Macromol 104:11–18. https://doi.org/10.1016/j.ijbiomac.2017.05.149

Luchese CL, Abdalla VF, Spada JC, Tessaro IC (2018) Evaluation of blueberry residue incorporated cassava starch film as pH indicator in different simulants and foodstuffs. Food Hydrocoll 82:209–218. https://doi.org/10.1016/j.foodhyd.2018.04.010

Lyu JS, Choi I, Hwang KS, Lee JY, Seo J, Kim SY, Han J (2019) Development of a BTB−/TBA+ ion-paired dye-based CO_2 indicator and its application in a multilayered intelligent packaging system. Sens Actuators B 282:359–365. https://doi.org/10.1016/j.snb.2018.11.073

Ma Q, Du L, Wang L (2017) Tara gum/polyvinyl alcohol-based colorimetric NH_3 indicator films incorporating curcumin for intelligent packaging. Sens Actuators B 244:759–766. https://doi.org/10.1016/j.snb.2017.01.035

Ma Q, Liang T, Cao L, Wang L (2018) Intelligent poly (vinyl alcohol)-chitosan nanoparticles-mulberry extracts films capable of monitoring pH variations. Int J Biol Macromol 108:576–584. https://doi.org/10.1016/j.ijbiomac.2017.12.049

Mahieu A, Terrié C, Youssef B (2015) Thermoplastic starch films and thermoplastic starch/polycaprolactone blends with oxygen-scavenging properties: influence of water content. Ind Crop Prod 72:192–199. https://doi.org/10.1016/j.indcrop.2014.11.037

Maraveas C (2020) Production of sustainable and biodegradable polymers from agricultural waste. Polymers 12(5):1127. https://doi.org/10.3390/POLYM12051127

Marismandani ADP, Husni A (2020) Development and characterization of biobased alginate/glycerol/virgin coconut oil as biodegradable packaging. In: E3S Web of Conferences, 147. https://doi.org/10.1051/e3sconf/202014703016

Martínez-Sanz M, Martínez-Abad A, López-Rubio A (2019) Cost-efficient bio-based food packaging films from unpurified agar-based extracts. Food Packag Shelf Life 21 100367. https://doi.org/10.1016/j.fpsl.2019.100367

Medina-Jaramillo C, Ochoa-Yepes O, Bernal C, Famá L (2017) Active and smart biodegradable packaging based on starch and natural extracts. Carbohydr Polym 176:187–194. https://doi.org/10.1016/j.carbpol.2017.08.079

Meng X, Lee K, Kang TY, Ko S (2015) An irreversible ripeness indicator to monitor the CO_2 concentration in the headspace of packaged kimchi during storage. Food Sci Biotechnol 24(1):91–97. https://doi.org/10.1007/s10068-015-0014-2

Meng W, Shi J, Zhang X, Lian H, Wang Q, Peng Y (2020) Effects of peanut shell and skin extracts on the antioxidant ability, physical and structure properties of starch-chitosan active packaging films. Int J Biol Macromol 152:137–146. https://doi.org/10.1016/j.ijbiomac.2020.02.235

Menzel C (2020) Improvement of starch films for food packaging through a three-principle approach: antioxidants, cross-linking and reinforcement. Carbohydr Polym 250:116828. https://doi.org/10.1016/j.carbpol.2020.116828

Merz B, Capello C, Leandro GC, Moritz DE, Monteiro AR, Valencia GA (2020) A novel colorimetric indicator film based on chitosan, polyvinyl alcohol and anthocyanins from jambolan

(Syzygium cumini) fruit for monitoring shrimp freshness. Int J Biol Macromol 153:625–632. https://doi.org/10.1016/j.ijbiomac.2020.03.048

Moghadam M, Salami M, Mohammadian M, Khodadadi M, Emam-Djomeh Z (2020) Development of antioxidant edible films based on mung bean protein enriched with pomegranate peel. Food Hydrocoll 104:105735. https://doi.org/10.1016/j.foodhyd.2020.105735

Mohamad N, Mazlan MM, Tawakkal ISMA, Talib RA, Kian LK, Fouad H, Jawaid M (2020) Development of active agents filled polylactic acid films for food packaging application. Int J Biol Macromol 163:1451–1457. https://doi.org/10.1016/j.ijbiomac.2020.07.209

Moradi M, Tajik H, Almasi H, Forough M, Ezati P (2019) A novel pH-sensing indicator based on bacterial cellulose nanofibers and black carrot anthocyanins for monitoring fish freshness. Carbohydr Polym 222:115030. https://doi.org/10.1016/j.carbpol.2019.115030

de Moraes Crizel T, de Oliveira Rios A, Alves VD, Bandarra N, Moldão-Martins M, Hickmann Flôres S (2018) Active food packaging prepared with chitosan and olive pomace. Food Hydrocoll 74:139–150. https://doi.org/10.1016/j.foodhyd.2017.08.007

Müller PS, Carpiné D, Yamashita F (2020) Influence of pinhão starch and natural extracts on the performance of thermoplastic cassava starch/PBAT extruded blown films as a technological approach for bio-based packaging material. J Food Sci 85:2832. https://doi.org/10.1111/1750-3841.15355

Murmu SB, Mishra HN (2018) Selection of the best active modified atmosphere packaging with ethylene and moisture scavengers to maintain quality of guava during low-temperature storage. Food Chem 253:55–62. https://doi.org/10.1016/j.foodchem.2018.01.134

Netramai S, Kijchavengkul T, Kham-ngam C, Sirinupong P, Kwanmuang S, Samsudin H, Lertsiri S (2020) Development of colorimetric film with butterfly pea (Clitoria ternatea L.) extract for application in intelligent packaging. Development, 18, 20.

Nilsuwan K, Benjakul S, Prodpran T (2018) Properties and antioxidative activity of fish gelatin-based film incorporated with epigallocatechin gallate. Food Hydrocoll 80:212–221. https://doi.org/10.1016/j.foodhyd.2018.01.033

Nisar T, Wang ZC, Alim A, Iqbal M, Yang X, Sun L, Guo Y (2019) Citrus pectin films enriched with thinned young apple polyphenols for potential use as bio-based active packaging. CYTA J Food 17(1):695–705. https://doi.org/10.1080/19476337.2019.1640798

Norcino LB, Mendes JF, Natarelli CVL, Manrich A, Oliveira JE, Mattoso LHC (2020) Pectin films loaded with copaiba oil nanoemulsions for potential use as bio-based active packaging. Food Hydrocoll 106:105862. https://doi.org/10.1016/j.foodhyd.2020.105862

Ochoa TA, García-Almendárez BE, Reyes AA, Pastrana DMR, López GFG, Belloso OM, González CR (2017) Design and characterization of corn starch edible films including beeswax and natural antimicrobials. Food Bioproc Tech 10(1):103–114. https://doi.org/10.1007/s11947-016-1800-4

de Oliveira Filho JG, Rodrigues JM, Valadares ACF, de Almeida AB, de Lima TM, Takeuchi KP, Alves CCF, de Sousa HAF, da Silva ER, Dyszy FH, Egea MB (2019) Active food packaging: alginate films with cottonseed protein hydrolysates. Food Hydrocoll 92:267–275. https://doi.org/10.1016/j.foodhyd.2019.01.052

Ortiz CM, Salgado PR, Dufresne A, Mauri AN (2018) Microfibrillated cellulose addition improved the physicochemical and bioactive properties of biodegradable films based on soy protein and clove essential oil. Food Hydrocoll 79:416–427. https://doi.org/10.1016/j.foodhyd.2018.01.011

Othman M, Yusup AA, Zakaria N, Khalid K (2018) Bio-polymer chitosan and corn starch with extract of hibiscus rosa-sinensis (hibiscus) as PH indicator for visually-smart food packaging. In: AIP Conference Proceedings, 1985. https://doi.org/10.1063/1.5047198

Otoni CG, Demoura MR, Aouada FA, Camilloto GP, Cruz RS, Lorevice MV, Soares N d FF, Mattoso LHC (2014) Antimicrobial and physical-mechanical properties of pectin/papaya puree/cinnamaldehyde nanoemulsion edible composite films. Food Hydrocoll 41:188–194. https://doi.org/10.1016/j.foodhyd.2014.04.013

Ozcan A, Arman Kandirmaz E (2020) Natural ink production and printability studies for smart food packaging. Color Res Appl 45(3):495–502. https://doi.org/10.1002/col.22488

Pant AF, Sangerlaub S, Muller K (2017) Gallic acid as an oxygen scavenger in bio-based multi-layer packaging films. Materials 10(5):489. https://doi.org/10.3390/ma10050489
Pant AF, Dorn J, Reinelt M (2018) Effect of temperature and relative humidity on the reaction kinetics of an oxygen scavenger based on Gallic acid. Frontiers. Chemistry 6(NOV):1–6. https://doi.org/10.3389/fchem.2018.00587
Peralta J, Bitencourt-Cervi CM, Maciel VBV, Yoshida CMP, Carvalho RA (2019) Aqueous hibiscus extract as a potential natural pH indicator incorporated in natural polymeric films. Food Packag Shelf Life 19:47–55. https://doi.org/10.1016/j.fpsl.2018.11.017
Pereira VA, de Arruda INQ, Stefani R (2015) Active chitosan/PVA films with anthocyanins from Brassica oleraceae (Red Cabbage) as time-temperature indicators for application in intelligent food packaging. Food Hydrocoll 43:180–188. https://doi.org/10.1016/j.foodhyd.2014.05.014
Peretto G, Du WX, Avena-Bustillos RJ, Berrios DJ, Sambo P, McHugh TH (2017) Electrostatic and conventional spraying of alginate-based edible coating with natural antimicrobials for preserving fresh strawberry quality. Food Bioproc Tech 10(1):165–174. https://doi.org/10.1007/s11947-016-1808-9
Piñeros-Hernandez D, Medina-Jaramillo C, López-Córdoba A, Goyanes S (2017) Edible cassava starch films carrying rosemary antioxidant extracts for potential use as active food packaging. Food Hydrocoll 63:488–495. https://doi.org/10.1016/j.foodhyd.2016.09.034
Pirsa S (2020) Biodegradable film based on pectin/nano-clay/methylene blue: structural and physical properties and sensing ability for measurement of vitamin C. Int J Biol Macromol 163. https://doi.org/10.1016/j.ijbiomac.2020.07.041
Pourjavaher S, Almasi H, Meshkini S, Pirsa S, Parandi E (2017) Development of a colorimetric pH indicator based on bacterial cellulose nanofibers and red cabbage (Brassica oleraceae) extract. Carbohydr Polym 156:193–201. https://doi.org/10.1016/j.carbpol.2016.09.027
Priyadarshi R, Rhim JW (2020) Chitosan-based biodegradable functional films for food packaging applications. Innov Food Sci Emerg Technol 62:102346. https://doi.org/10.1016/j.ifset.2020.102346
Qin Y, Liu Y, Yong H, Liu J, Zhang X, Liu J (2019) Preparation and characterization of active and intelligent packaging films based on cassava starch and anthocyanins from Lycium ruthenicum Murr. Int J Biol Macromol 134:80–90. https://doi.org/10.1016/j.ijbiomac.2019.05.029
Qin Y, Liu Y, Zhang X, Liu J (2020) Development of active and intelligent packaging by incorporating betalains from red pitaya (Hylocereus polyrhizus) peel into starch/polyvinyl alcohol films. Food Hydrocoll 100:105410. https://doi.org/10.1016/j.foodhyd.2019.105410
Rachmelia D, Imawan C (2018) Time temperature indicator label using black corn extract and chitosan matrix. J Phys Conf Ser 1120(1). https://doi.org/10.1088/1742-6596/1120/1/012041
Rahim MZA, Husin N, Mohd Noor MA, Yet ZR, Ismail-Fitry MR (2020) Screening of natural colours from various natural resources as potential reusable visual indicators for monitoring food freshness. Malaysian J Anal Sci 24(2):288–299
Rahimah S, Malinda W, Zaida, Sukri N, Salma JK, Tallei TE, Idroes R (2020) Betacyanin as bio-indicator using time-temperature integrator for smart packaging of fresh goat milk. Scientific World Journal 2020:4303140. https://doi.org/10.1155/2020/4303140
Rahman ATMM, Kim DH, Jang HD, Yang JH, Lee SJ (2018) Preliminary study on biosensor-type time-temperature integrator for intelligent food packaging. Sensors 18(6):1949. https://doi.org/10.3390/s18061949
Rajeswari A, Christy EJS, Swathi E, Pius A (2020) Fabrication of improved cellulose acetate-based biodegradable films for food packaging applications. Environ Chem Ecotoxicol 2:107–114. https://doi.org/10.1016/j.enceco.2020.07.003
Rambabu K, Bharath G, Banat F, Show PL, Cocoletzi HH (2019) Mango leaf extract incorporated chitosan antioxidant film for active food packaging. Int J Biol Macromol 126:1234–1243. https://doi.org/10.1016/j.ijbiomac.2018.12.196
Ramziia S, Ma H, Yao Y, Wei K, Huang Y (2018) Enhanced antioxidant activity of fish gelatin–chitosan edible films incorporated with procyanidin. J Appl Polym Sci 135(10):1–10. https://doi.org/10.1002/app.45781

Razavi R, Tajik H, Moradi M, Molaei R, Ezati P (2020) Antimicrobial, microscopic and spectroscopic properties of cellulose paper coated with chitosan sol-gel solution formulated by epsilon-poly-L-lysine and its application in active food packaging. Carbohydr Res 489:107912. https://doi.org/10.1016/j.carres.2020.107912

Reichert CL, Bugnicourt E, Coltelli MB, Cinelli P, Lazzeri A, Canesi I, Braca F, Martínez BM, Alonso R, Agostinis L, Verstichel S, Six L, De Mets S, Gómez EC, Ißbrücker C, Geerinck R, Nettleton DF, Campos I, Sauter E et al (2020) Bio-based packaging: materials, modifications, industrial applications and sustainability. Polymers 12(7). https://doi.org/10.3390/polym12071558

Remya S, Mohan CO, Venkateshwarlu G, Sivaraman GK, Ravishankar CN (2017) Combined effect of O2 scavenger and antimicrobial film on shelf life of fresh cobia (Rachycentron canadum) fish steaks stored at 2 °C. Food Control 71:71–78. https://doi.org/10.1016/j.foodcont.2016.05.038

Rezaee M, Askari G, EmamDjomeh Z, Salami M (2020) UV-irradiated gelatin-chitosan bio-based composite film, physiochemical features and release properties for packaging applications. Int J Biol Macromol 147:990–996. https://doi.org/10.1016/j.ijbiomac.2019.10.066

Riaz A, Lagnika C, Luo H, Dai Z, Nie M, Hashim MM, Liu C, Song J, Li D (2020) Chitosan-based biodegradable active food packaging film containing Chinese chive (Allium tuberosum) root extract for food application. Int J Biol Macromol 150:595–604. https://doi.org/10.1016/j.ijbiomac.2020.02.078

Ribeiro Sanches MA, Camelo-Silva C, da Silva Carvalho C, Rafael de Mello J, Barroso NG, da Silva L, Barros E, Silva PP, Pertuzatti PB (2021) Active packaging with starch, red cabbage extract and sweet whey: characterization and application in meat. LWT Food sci Technol 135. https://doi.org/10.1016/j.lwt.2020.110275

Rokbani H, Daigle F, Ajji A (2019) Long- and short-term antibacterial properties of low-density polyethylene-based films coated with zinc oxide nanoparticles for potential use in food packaging. J Plast Film Sheet 35(2):117–134. https://doi.org/10.1177/8756087918822677

Rollini M, Mascheroni E, Capretti G, Coma V, Musatti A, Piergiovanni L (2017) Propolis and chitosan as antimicrobial and polyphenols retainer for the development of paper based active packaging materials. Food Packag Shelf Life 14:75–82. https://doi.org/10.1016/j.fpsl.2017.08.011

Roman MJ, Decker EA, Goddard JM (2016) Retaining oxidative stability of emulsified foods by novel nonmigratory polyphenol coated active packaging. J Agric Food Chem 64(27):5574–5582. https://doi.org/10.1021/acs.jafc.6b01933

Romani VP, Prentice-Hernández C, Martins VG (2017) Active and sustainable materials from rice starch, fish protein and oregano essential oil for food packaging. Ind Crop Prod 97:268–274. https://doi.org/10.1016/j.indcrop.2016.12.026

Rux G, Mahajan PV, Linke M, Pant A, Sängerlaub S, Caleb OJ, Geyer M (2016) Humidity-regulating trays: moisture absorption kinetics and applications for fresh produce packaging. Food Bioproc Tech 9(4):709–716. https://doi.org/10.1007/s11947-015-1671-0

Saberi B, Chockchaisawasdee S, Golding JB, Scarlett CJ, Stathopoulos CE (2017) Characterization of pea starch-guar gum biocomposite edible films enriched by natural antimicrobial agents for active food packaging. Food Bioprod Process 105:51–63. https://doi.org/10.1016/j.fbp.2017.06.003

Şahin A, Çarkcıoğlu E, Demirhan B, Candoğan K (2017) Chitosan edible coating and oxygen scavenger effects on modified atmosphere packaged sliced sucuk. J Food Process Preserv 41(6):1–8. https://doi.org/10.1111/jfpp.13213

Saliu F, Della Pergola R (2018) Carbon dioxide colorimetric indicators for food packaging application: applicability of anthocyanin and poly-lysine mixtures. Sens Actuators B 258:1117–1124. https://doi.org/10.1016/j.snb.2017.12.007

Sánchez-Ortega I, García-Almendárez BE, Santos-López EM, Reyes-González LR, Regalado C (2016) Characterization and antimicrobial effect of starch-based edible coating suspensions. Food Hydrocoll 52:906–913. https://doi.org/10.1016/j.foodhyd.2015.09.004

Sanchís E, Ghidelli C, Sheth CC, Mateos M, Palou L, Pérez-Gago MB (2017) Integration of antimicrobial pectin-based edible coating and active modified atmosphere packaging to preserve

the quality and microbial safety of fresh-cut persimmon (Diospyros kaki Thunb. cv. Rojo Brillante). J Sci Food Agric 97(1):252–260. https://doi.org/10.1002/jsfa.7722
Sariningsih KA, Rostini I, Haetami K (2019) Methyl red indicator on smart packaging as a freshness sensor for tilapia fillets. Asian Food Sci J 13(4):1–9. https://doi.org/10.9734/afsj/2019/v13i430114
Şen F, Uzunsoy İ, Kahraman MV (2020) Hydroxyethyl cellulose-based indicator labels to track freshness of anchovy (Engraulis encrasicolus). Color Res Appl 45(5):962–967. https://doi.org/10.1002/col.22517
Sganzerla WG, Rosa GB, Ferreira ALA, da Rosa CG, Beling PC, Xavier LO, Hansen CM, Ferrareze JP, Nunes MR, Barreto PLM, de Lima Veeck AP (2020) Bioactive food packaging based on starch, citric pectin and functionalized with Acca sellowiana waste by-product: characterization and application in the postharvest conservation of apple. Int J Biol Macromol 147:295–303. https://doi.org/10.1016/j.ijbiomac.2020.01.074
Shahbazi M, Ahmadi SJ, Seif A, Rajabzadeh G (2016) Carboxymethyl cellulose film modification through surface photo-crosslinking and chemical crosslinking for food packaging applications. Food Hydrocoll 61:378–389. https://doi.org/10.1016/j.foodhyd.2016.04.021
Sharma P, Wichaphon J, Klangpetch W (2020) Antimicrobial and antioxidant activities of defatted Moringa oleifera seed meal extract obtained by ultrasound-assisted extraction and application as a natural antimicrobial coating for raw chicken sausages. Int J Food Microbiol 332:108770. https://doi.org/10.1016/j.ijfoodmicro.2020.108770
Shoja Gharehbagh S, Khanjari A, Yeganmohammadi Davaji M, Akhondzadeh Basti A (2017) The use of chitosan and whey protein isolate edible films incorporated with zataria multiflora boiss. essential oil as an active packaging ingredient against some common foodborne bacteria. J Food Biosci Technol 7(1):41–48
da Silva Filipini G, Romani VP, Guimarães Martins V (2020) Biodegradable and active-intelligent films based on methylcellulose and jambolão (Syzygium cumini) skins extract for food packaging. Food Hydrocoll 109:106139. https://doi.org/10.1016/j.foodhyd.2020.106139
Singh S, Gaikwad KK, Lee YS (2018) Antimicrobial and antioxidant properties of polyvinyl alcohol bio composite films containing seaweed extracted cellulose nano-crystal and basil leaves extract. Int J Biol Macromol 107:1879–1887. https://doi.org/10.1016/j.ijbiomac.2017.10.057
Singh S, Gaikwad KK, Suk Y (2019) Development and application of a pyrogallic acid-based oxygen scavenging packaging system for shelf life extension of peeled garlic. Sci Hortic 256:108548. https://doi.org/10.1016/j.scienta.2019.108548
Singh AA, Sharma S, Srivastava M, Majumdar A (2020) Modulating the properties of polylactic acid for packaging applications using biobased plasticizers and naturally obtained fillers. Int J Biol Macromol 153:1165–1175. https://doi.org/10.1016/j.ijbiomac.2019.10.246
Souza VGL, Fernando AL, Pires JRA, Rodrigues PF, Lopes AAS, Fernandes FMB (2017) Physical properties of chitosan films incorporated with natural antioxidants. Ind Crop Prod 107:565–572. https://doi.org/10.1016/j.indcrop.2017.04.056
Stejskal N, Miranda JM, Martucci JF, Ruseckaite RA, Aubourg SP, Barros-Velázquez J (2020) The effect of gelatine packaging film containing a Spirulina platensis protein concentrate on atlantic mackerel shelf life. Molecules 25(14):1–13. https://doi.org/10.3390/molecules25143209
Sun L, Sun J, Chen L, Niu P, Yang X, Guo Y (2017) Preparation and characterization of chitosan film incorporated with thinned young apple polyphenols as an active packaging material. Carbohydr Polym 163:81–91. https://doi.org/10.1016/j.carbpol.2017.01.016
Sun W, Liu Y, Jia L, Saldaña MDA, Dong T, Jin Y, Sun W (2020) A smart nanofibre sensor based on anthocyanin/poly-l-lactic acid for mutton freshness monitoring. Int J Food Sci Technol. https://doi.org/10.1111/ijfs.14648
Suthar MK, Saran PL (2020) Anthocyanins from Ocimum sanctum L., a promising biomolecule for development of cost-effective and widely applicable pH indicator. 3 Biotech 10(9):1–11. https://doi.org/10.1007/s13205-020-02380-5
Talón E, Trifkovic KT, Nedovic VA, Bugarski BM, Vargas M, Chiralt A, González-Martínez C (2017) Antioxidant edible films based on chitosan and starch containing polyphe-

nols from thyme extracts. Carbohydr Polym 157:1153–1161. https://doi.org/10.1016/j.carbpol.2016.10.080
Tawakkal ISMA, Cran MJ, Bigger SW (2016) Release of thymol from poly(lactic acid)-based antimicrobial films containing kenaf fibres as natural filler. LWT Food Sci Technol 66:629–637. https://doi.org/10.1016/j.lwt.2015.11.011
Thaker M, Hanjabam MD, Gudipati V, Kannuchamy N (2017) Protective effect of fish gelatin-based natural antimicrobial coatings on quality of Indian Salmon Fillets during refrigerated storage. J Food Process Eng 40(1):1–10. https://doi.org/10.1111/jfpe.12270
Thanakkasaranee S, Sadeghi K, Lim IJ, Seo J (2020) Effects of incorporating calcined corals as natural antimicrobial agent into active packaging system for milk storage. Mater Sci Eng C 111:110781. https://doi.org/10.1016/j.msec.2020.110781
Tichoniuk M, Radomska N, Cierpiszewski R (2017) The application of natural dyes in food freshness indicators designed for intelligent packaging. Studia Oeconomica Posnaniensia 5(7):19–34. https://doi.org/10.18559/soep.2017.7.2
Tomadoni B, Moreira MR, Pereda M, Ponce AG (2018) Gellan-based coatings incorporated with natural antimicrobials in fresh-cut strawberries: microbiological and sensory evaluation through refrigerated storage. LWT Food Sci Technol 97:384–389. https://doi.org/10.1016/j.lwt.2018.07.029
Tosati JV, Messias VC, Carvalho PIN, Rodrigues Pollonio MA, Meireles MAA, Monteiro AR (2017) Antimicrobial effect of edible coating blend based on turmeric starch residue and gelatin applied onto fresh Frankfurter sausage. Food Bioproc Tech 10(12):2165–2175. https://doi.org/10.1007/s11947-017-1985-1
Tsironi T, Ronnow P, Giannoglou M, Taoukis P (2017) Developing suitable smart TTI labels to match specific monitoring requirements: the case of Vibrio spp. growth during transportation of oysters. Food Control 73:51–56. https://doi.org/10.1016/j.foodcont.2016.06.041
Tumwesigye KS, Oliveira JC, Sousa-Gallagher MJ (2016) Integrated sustainable process design framework for cassava biobased packaging materials: critical review of current challenges, emerging trends and prospects. Trends Food Sci Technol 56:103–114. https://doi.org/10.1016/j.tifs.2016.08.001
Vähä-Nissi M, Koivula HM, Räisänen HM, Vartiainen J, Ragni P, Kenttä E, Kaljunen T, Malm T, Minkkinen H, Harlin A (2017) Cellulose nanofibrils in biobased multilayer films for food packaging. J Appl Polym Sci 134(19):2–9. https://doi.org/10.1002/app.44830
Valencia L, Nomena EM, Mathew AP, Velikov KP (2019) Biobased cellulose nanofibril-oil composite films for active edible barriers [research-article]. ACS Appl Mater Interfaces 11(17):16040–16047. https://doi.org/10.1021/acsami.9b02649
Valerini D, Tammaro L, Di Benedetto F, Vigliotta G, Capodieci L, Terzi R, Rizzo A (2018) Aluminum-doped zinc oxide coatings on polylactic acid films for antimicrobial food packaging. Thin Solid Films 645:187–192. https://doi.org/10.1016/j.tsf.2017.10.038
Valerio F, Masi M, Cimmino A, Moeini SA, Lavermicocca P, Evidente A (2018) Antimould microbial and plant metabolites with potential use in intelligent food packaging. Nat Prod Res 32(13):1605–1610. https://doi.org/10.1080/14786419.2017.1385018
Veiga-Santos P, Silva LT, de Souza CO, da Silva JR, Albuquerque ECC, Druzian JI (2018) Coffee-cocoa additives for bio-based antioxidant packaging. Food Packag Shelf Life 18:37–41. https://doi.org/10.1016/j.fpsl.2018.08.005
Volison-Klimentiev A, Cullari L, Shachar-Michaely G, Dor S, Peretz-Damari S, Afik N, Regev O (2020) Vegetable-oil-based intelligent ink for oxygen sensing. ACS Sens. https://doi.org/10.1021/acssensors.0c01777
Vu CHT, Won K (2014) Leaching-resistant carrageenan-based colorimetric oxygen indicator films for intelligent food packaging. J Agric Food Chem 62(29):7263–7267. https://doi.org/10.1021/jf5014764
Wang H, Wang L (2017) Developing a bio-based packaging fi lm from soya by-products incorporated with valonea tannin. J Clean Prod 143:624. https://doi.org/10.1016/j.jclepro.2016.12.064

Wang H, Hu D, Ma Q, Wang L (2016) Physical and antioxidant properties of flexible soy protein isolate films by incorporating chestnut (Castanea mollissima) bur extracts. LWT- Food Sci Technol 71:33–39. https://doi.org/10.1016/j.lwt.2016.03.025

Wang L, Wu Z, Cao C (2019a) Technologies and fabrication of intelligent packaging for perishable products. Appl Sci 9(22):4858. https://doi.org/10.3390/app9224858

Wang X, Yong H, Gao L, Li L, Jin M, Liu J (2019b) Preparation and characterization of antioxidant and pH-sensitive films based on chitosan and black soybean seed coat extract. Food Hydrocoll 89:56–66. https://doi.org/10.1016/j.foodhyd.2018.10.019

Wei YC, Cheng CH, Ho YC, Tsai ML, Mi FL (2017) Active gellan gum/purple sweet potato composite films capable of monitoring pH variations. Food Hydrocoll 69:491–502. https://doi.org/10.1016/j.foodhyd.2017.03.010

Weigel C, Schneider M, Schmitt J, Hoffmann M, Kahl S, Jurisch R (2015) Self-sufficient sensor for oxygen detection in packaging via radio-frequency identification. J Sens Sens Syst 4(1):179–186. https://doi.org/10.5194/jsss-4-179-2015

Weston M, Kuchel RP, Chandrawati R (2020) A polydiacetylene-based colorimetric sensor as an active use-by date for plant-based milk alternatives. Macromol Rapid Commun 41(18):1–7. https://doi.org/10.1002/marc.202000172

White JS (2014) Sucrose, HFCS, and fructose: history, manufacture, composition, applications, and production. In Fructose, high fructose corn syrup, sucrose and health (pp. 13–33). Humana Press, New York, NY.

Wieczyńska J, Cavoski I (2018) Antimicrobial, antioxidant and sensory features of eugenol, carvacrol and trans-anethole in active packaging for organic ready-to-eat iceberg lettuce. Food Chem 259:251–260. https://doi.org/10.1016/j.foodchem.2018.03.137

Won K, Jang NY, Jeon J (2016) A natural component-based oxygen indicator with in-pack activation for intelligent food packaging. J Agric Food Chem 64(51):9675–9679. https://doi.org/10.1021/acs.jafc.6b04172

Wu C, Sun J, Chen M, Ge Y, Ma J, Hu Y, Pang J, Yan Z (2019a) Effect of oxidized chitin nanocrystals and curcumin into chitosan films for seafood freshness monitoring. Food Hydrocoll 95:308–317. https://doi.org/10.1016/j.foodhyd.2019.04.047

Wu H, Lei Y, Zhu R, Zhao M, Lu J, Xiao D, Jiao C, Zhang Z, Shen G, Li S (2019b) Preparation and characterization of bioactive edible packaging films based on pomelo peel flours incorporating tea polyphenol. Food Hydrocoll 90:41–49. https://doi.org/10.1016/j.foodhyd.2018.12.016

Xue Mei L, Mohammadi Nafchi A, Ghasemipour F, Mat Easa A, Jafarzadeh S, Al-Hassan AA (2020) Characterization of pH sensitive sago starch films enriched with anthocyanin-rich torch ginger extract. Int J Biol Macromol 164:4603–4612. https://doi.org/10.1016/j.ijbiomac.2020.09.082

Yahaya WAW, Almajano MP, Yazid NA, Azman NAM (2019) Antioxidant activities and total phenolic content of Malaysian herbs as components of active packaging film in beef patties. Antioxidants 8(7):204. https://doi.org/10.3390/antiox8070204

Yamamoto T, Isshiki K (2012) Development of the indicator using Maillard reaction to warn against the temperature rise of the chilled food. Japanese Journal of Food Chemistry and Safety, 19(2): 84–87. https://doi.org/10.18891/jjfcs.19.2_84

Yang Y, Liu H, Wu M, Ma J, Lu P (2020) Bio-based antimicrobial packaging from sugarcane bagasse nanocellulose/nisin hybrid films. Int J Biol Macromol 161:627–635. https://doi.org/10.1016/j.ijbiomac.2020.06.081

Yildirim S, Röcker B, Pettersen MK, Nilsen-Nygaard J, Ayhan Z, Rutkaite R, Radusin T, Suminska P, Marcos B, Coma V (2018) Active Packaging Applications for Food. Compr Rev Food Sci Food Safety 17(1):165–199. https://doi.org/10.1111/1541-4337.12322

Yoksan R, Chirachanchai S (2010) Silver nanoparticle-loaded chitosan-starch based films: fabrication and evaluation of tensile, barrier and antimicrobial properties. Mater Sci Eng C 30(6):891–897. https://doi.org/10.1016/j.msec.2010.04.004

Yong H, Wang X, Bai R, Miao Z, Zhang X, Liu J (2019) Development of antioxidant and intelligent pH-sensing packaging films by incorporating purple-fleshed sweet potato extract into chitosan matrix. Food Hydrocoll 90:216–224. https://doi.org/10.1016/j.foodhyd.2018.12.015

Zhai X, Shi J, Zou X, Wang S, Jiang C, Zhang J, Huang X, Zhang W, Holmes M (2017) Novel colorimetric films based on starch/polyvinyl alcohol incorporated with roselle anthocyanins for fish freshness monitoring. Food Hydrocoll 69:308–317. https://doi.org/10.1016/j.foodhyd.2017.02.014

Zhai X, Li Z, Shi J, Huang X, Sun Z, Zhang D, Zou X, Sun Y, Zhang J, Holmes M, Gong Y, Povey M, Wang S (2019) A colorimetric hydrogen sulfide sensor based on gellan gum-silver nanoparticles bionanocomposite for monitoring of meat spoilage in intelligent packaging. Food Chem 290:135–143. https://doi.org/10.1016/j.foodchem.2019.03.138

Zhang H, Hou A, Xie K, Gao A (2019a) Smart color-changing paper packaging sensors with pH sensitive chromophores based on azo-anthraquinone reactive dyes. Sens Actuat B Chem 286:362–369. https://doi.org/10.1016/j.snb.2019.01.165

Zhang T, Yu Q, Fang L, Wang J, Wu T, Song P (2019b) All-organic multilayer coatings for advanced poly(lactic acid) films with high oxygen barrier and excellent antifogging properties. ACS Appl Polym Mater 1(12):3470–3476. https://doi.org/10.1021/acsapm.9b01030

Zhang C, Sun G, Cao L, Wang L (2020) Accurately intelligent film made from sodium carboxymethyl starch/κ-carrageenan reinforced by mulberry anthocyanins as an indicator. Food Hydrocoll 108:106012. https://doi.org/10.1016/j.foodhyd.2020.106012

Zhao G, Lyu X, Lee J, Cui X, Chen WN (2019) Biodegradable and transparent cellulose film prepared eco-friendly from durian rind for packaging application. Food Packag Shelf Life 21:100345. https://doi.org/10.1016/j.fpsl.2019.100345

Zhu R, Desroches M, Yoon B, Swager TM (2017) Wireless oxygen sensors enabled by Fe(II)-polymer wrapped carbon nanotubes. ACS Sens 2(7):1044–1050. https://doi.org/10.1021/acssensors.7b00327

Zia J, Mancini G, Bustreo M, Zych A, Donno R, Athanassiou A, Fragouli D (2021) Porous pH natural indicators for acidic and basic vapor sensing. Chem Eng J 403:126373. https://doi.org/10.1016/j.cej.2020.126373

Chapter 7
Consumer Behavior and Industry Implications

Shahida Anusha Siddiqui, Maximilian Julius Pahmeyer, Mohammad Mehdizadeh, Andrey Ashotovich Nagdalian, Natalya Pavlovna Oboturova, and Ahmed Taha

7.1 The Trend of the Clean Label in Different Countries

The need for transparency in health and food safety always confronts customers with questions such as product content, the origin of production, production processes, additives used, and the status of ethical factors in the production process. These factors seem to be answered transparently in clean-label foods. This is why the clean label is increasingly advancing and developing among food industry trends. Today, clean labels and clean food are recognized as an international

S. A. Siddiqui (✉)
Technical University of Munich Campus Straubing for Biotechnology and Sustainability
Straubing, Germany

DIL e.V.–German Institute of Food Technologies, Quakenbrück, Germany
e-mail: S.Siddiqui@dil-ev.de

M. J. Pahmeyer
DIL e.V.–German Institute of Food Technologies, Quakenbrück, Germany

Faculty of Agricultural Sciences and Landscape Architecture, Osnabrück University of Applied Sciences, Osnabrück, Germany

M. Mehdizadeh
Faculty of Agricultural and Natural Resources, University of Mohaghegh Ardabili, Ardabili, Iran

A. A. Nagdalian · N. P. Oboturova
Faculty of Food Engineering and Biotechnology, North Caucasus Federal University, Stavropol, Russian Federation

A. Taha
Department of Functional Materials and Electronics, Center for Physical Sciences and Technology, Vilnius, Lithuania

Faculty of Agriculture (Saba Basha), Department of Food Science, Alexandria University, Alexandria, Egypt
e-mail: ahmed-taha@alexu.edu.eg

C. M. Galanakis (ed.), *The Age of Clean Label Foods*,
https://doi.org/10.1007/978-3-030-96698-0_7

standard in nutrition. They can influence the decisions of customers and consumers in different regions all over the world. So, increasing demands to purchase and use food with a clean label and free of chemicals in other countries. In general, "clean label" as one of the biggest trends in the food market (Cheung et al. 2016), is a new and widely used term in the food industry, which refers to challenges including the health and safety of food products from the perspective of maintaining environmental sustainability. Park and Kim (2021) reviewed the clean label trend in starch (as an abundant natural ingredient) using blending, physical and enzymatic modification methods. The importance of using clean food has been considered for many years. Recently, due to the power of the media and social media platforms across the world, environmentally friendly organizations and NGOs, as well as critics of the use of chemicals and pesticides in agricultural ecosystems it has become a fundamental issue and a global trend (Cheboi et al. 2021). Replacing salt, sugar, saturated and trans fats, and artificial colors with natural additives with minimum adverse effects on human health is one of the most important aspects of clean food. Milner et al. (2020) evaluated the effects of safe alternatives for sugar in cake products. They reported the effectiveness of apple pomace, whey permeate, oligofructose, polydextrose as sucrose-replacing additives. Yi et al. (2018) reported the significant effect of kiwifruit puree application as a natural additive for preserving apple juice. McDonnell et al. (2013) suggested applying natural-based antimicrobial additives for enhancing the healthiness of meat and poultry products. Table 7.1 represents the policy of clean labels for various food products in different countries worldwide.

7.2 Review of the TOP Ten Undesirable Ingredients for Consumers in the Composition of Products

In general, a range of factors has been identified for creating a negative image for certain food ingredients, especially additives, in the minds of consumers. Asioli et al. (2017) reviewed factors for three food trends, namely organic food, natural food, and "free-from" artificial additives/ingredients, the last of which will be dealt with more closely in this section. They grouped the essential factors into six distinct categories, according to the psychological model of food choice developed by Mojet (Köster 2009):

1. Intrinsic product characteristics include sensory appearance, complexity, and possibly resultant irritation.
2. Biological and physiological factors, like genetics, age and gender, and Oro-gastro-intestinal physiology.
3. Psychological factors are about the consumers' cognition, experiences, and traits.
4. Situational factors are about the surroundings and time of food choice.
5. Socio-cultural factors, related to socio-economic influences on and thus believes of the individual, and trust in industry and government.

Table 7.1 Clean label food policy from different countries

Country	Products	Policy	References
India	Bread	Replacing commercial additives with hard red winter wheat flour for "clean-label" bread products	Rahman and Simsek (2020)
Brazil	Meat	Use the chia (*Salvia hispanica* L.) mucilage in powder or gel format for reducing the phosphate in emulsified meat products	Câmara et al. (2020)
Taiwan	Meat	Removing chemical additives, using simple ingredients, and minimum processing	Huang et al. (2017)
Switzerland	Meat	Replacing nitrite based preservatives in meat products by chitosan orange peel complex	Lent (2019)
Iran	Meat	Use the chitosan for a positive impact on microbial and oxidative activity in chicken meat	Mehdizadeh and Langrood (2019)
Brazil	Salmon fish	Use the pink pepper residue extract for limiting antibacterial activity and enhancing the shelf life of salmon fillet	Merlo et al. (2019)
France	Bakery products	Replacing chemical preservatives by lactic acid bacteria for controlling spoilage fungi	Le Lay et al. (2016)
Finland	Bread	Using *Lactobacillus plantarum* to inhibit rope spoilage in bread	Katina et al. (2002)
Ireland	Bakery products	Using *Lactobacillus amylovorus* for the preservation of bakery products	Ryan et al. (2011)
Germany	Biscuits	Replacing synthetic antioxidant with phenolic-based natural antioxidants	Ua-Arak et al. (2017)
Italy	Corn flour	Using physical modifications on corn flour as a clean label strategy	Carcelli et al. (2020)
Ireland	Cake	Using natural additive derived from *Linum usitatissimum* as a fat alternative	Garvey et al. (2021)
Spain	Chorizo	Using natural phenolic compounds derived from vegetables for replacing synthetic nitrates	Martínez-Zamora et al. (2021)
Portugal	Yogurt	Replacing synthetic antioxidant with natural additives from chamomile and fennel plant	Caleja et al. (2016)

6. Extrinsic product characteristics, like packaging and perceived sustainability and risks.

The review results by Asioli et al. (2017) are graphically summarized in Fig. 7.1. Concerning preferences for products free from additives, they found factors for all categories except for situational factors.

A cursory examination of these results shows that there appear to be more socio-cultural factors and extrinsic product characteristics with influence on preference for free from products than characteristics intrinsic to the product or ingredient itself. Examples include education about the purpose and origin of additives and trust in governments or industries. However, to assess the undesirability of a given ingredient, intrinsic product characteristics, and extrinsic characteristics of the component such as media scandals and other influences on consumer perception of the component are of the most significant concern. In that regard, it can be said that

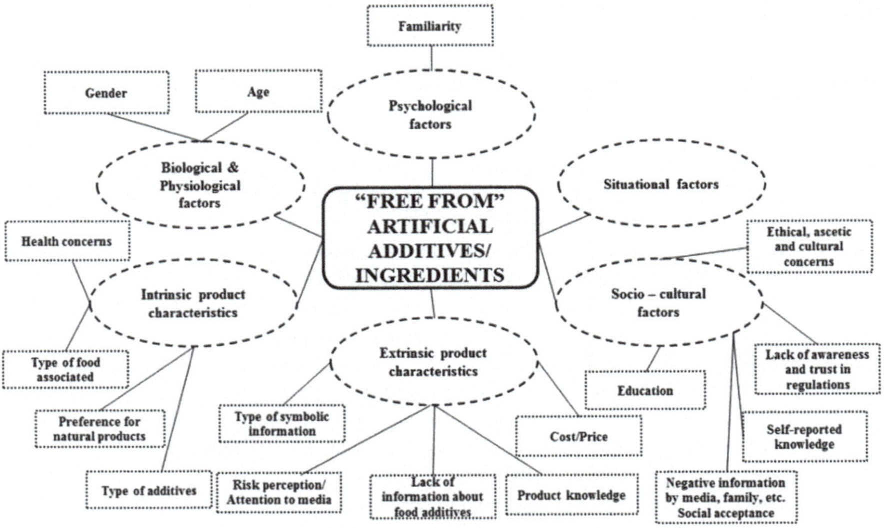

Fig. 7.1 Factors that drive consumer preferences for food that is "free from" artificial ingredients and additives (Asioli et al. 2017)

ingredients perceived as natural are more accepted than those perceived as unnatural (Evans et al. 2010), such as chemicals. Especially the latter is disliked irrespective of their concentration in the product due to insensitivity to dose-response factors (Dickson-Spillmann et al. 2011). For exclusive products, longer ingredient lists and unknown processing methods also negatively influence naturalness perception (Murley and Chambers 2019). However, the mere choice of declaration of the ingredient name can be a factor since e-numbers are perceived as less natural than chemical names for the same ingredient (Siegrist and Sütterlin 2017). Especially additives serving techno-functional processes are met with distrust, in contrast to additives serving health purposes (Bastian et al. 2015). Familiarity is also an essential factor; traditional ingredients or ingredients used in kitchens are commonly perceived better (Asioli et al. 2017; Zimmerman et al. 2021). Asioli et al. (2017) also mention risk perception as an extrinsic factor, while the only product intrinsic risk factor directly associated and they mention are health concerns. In addition, sustainability concerns such as environmental impacts (i.e., climate change or biodiversity loss) or the treatment of people working in the primary production of ingredients have gained importance as a factor for consumer decisions (Zhang et al. 2019a, b; Nagdalian et al. 2021).

In the following, a selection of food ingredients and components will be discussed regarding consumer aversion toward them and possible reasons for that in light of the factors mentioned above. This includes scientifically supported adverse effects of these ingredients and results of consumer surveys concerning their acceptability and influence on purchase choice as well as considerations on their naturalness. However, the attitude-liking (Hemmerling et al. 2016) gap should be kept in mind. Some people may prefer the organoleptic properties of food with ingredients they state an aversion to.

7.2.1 Sugar

High sugar intake in wealthy countries is seen as a significant driver of obesity and diabetes and, as a result, has fallen out of favor for many health-conscious consumers and is recognized as one of the leading health risks in food (Siddiqui et al. 2022a). In some countries, such as Great Britain or Mexico, a tax on sweetened beverages has been introduced and caused consumers to buy fewer sweetened beverages and companies to reduce the sugar content in their products (Arantxa Colchero et al. 2017; Hashem et al. 2019). Regardless of taxes, studies have shown that 32% of consumers prefer to buy products with lower sugar content when the high sugar content is represented by health warning messages or as teaspoons of sugar contained. This could enhance their understanding of the high sugar content. Purely numerical information, which is the only mandatory information on sugar content, is less likely to influence consumer choice (Scapin et al. 2021).

However, it has also been found that reducing the sugar content of conventionally sweetened products reduces liking of the products across consumer and product groups when consumers are unaware of differences in sugar content (Markey et al. 2015). This shows that, even though reductions in sugar content of processed foods have health benefits and motivate consumers to buy products, lower contents, especially in traditionally sweet foods, may lead to less acceptance from a sensory perspective.

7.2.2 Salt

Sodium chloride or table salt is commonly used in processed food to preserve properties and taste enhancement. Excessive salt consumption can cause heightened blood pressure and increase the risk of cardiovascular diseases, and individuals in developed countries exceed the minimum required daily consumption by up to 25 times (Albarracín et al. 2011). Due to this fact, consumers recognize it as one of the leading health hazards in food after sugar (Profeta et al. 2021). According to a survey in Australia, a majority (two thirds) of consumers are aware of these adverse effects and the fact that a large portion of salt consumption stems from processed foods, and 21% indicated they would consider salt content as a factor for food purchases (Webster et al. 2010). Of course, as for sugar, reducing salt content may reduce sensory acceptance due to the familiar organoleptic and taste-enhancing properties of salt. Strategies such as substitution with different salts are under consideration (Khetra et al. 2016; Sinesio et al. 2019).

7.2.3 Fats

In wealthy countries, fat is generally regarded as a macronutrient undesirable in more significant quantities due to its high energy content and resultant promotion of weight gain as well as adverse effects such as heightened cholesterol levels, blood pressure, and cardiovascular diseases, associated with saturated and trans-fatty acids in particular (Te Morenga and Montez 2017). Fat is perceived as the most relevant negatively health impacting food component by some due to a focus on weight reduction in the health perception of many people in industrialized countries. Many consumers stated in surveys that they would generally instead buy products labeled as 'reduced fat' compared to regular fat products, mainly if they were health-conscious and for products perceived as healthy such as yogurt. However, this was not true for some products, i.e., chips (Oostenbach et al. 2019), likely because they can be perceived to have inferior sensory properties (Hamilton et al. 2000). Additionally, certain fatty acids, especially trans fatty acids, are associated with additional health risks like cancer and diabetes (Islam et al. 2019). Hydrogenated vegetable oils are a relevant source of artificial trans fatty acids and should be avoided, even though only a small amount of consumers seem to be aware of this (Kamel and Al Otaibi 2018).

When talking about ingredients, certain fats or oils are commonly added for sensory reasons but are disliked by consumers due to health risks or issues connected to their production impacting the environment or society. One prominent example is palm oil.

Palm oil is a common ingredient with a growing market share being applied in many processed foods due to a competitive price and advantageous physical characteristics. However, it has come under public attack, and many manufacturers nowadays declare their products to be free from palm oil. Indeed, there are health and environmental concerns connected to palm oil use. Palmitic acid, which has a high concentration in palm oil, may increase the risk of obesity, type 2 diabetes mellitus, cardiovascular diseases, and cancer (Magri et al. 2015; Binker-Cosen et al. 2017; Fatima et al. 2019). From an environmental and social perspective, increased palm oil production in tropical areas, such as Indonesia and Malaysia, which account for 80% of the global output (USDA (United States Department of Agriculture) 2018), can lead to environmental degradation in biodiversity hotspots (Reijnders and Huijbregts 2008; Azhar et al. 2017) and aggravation of the economic situation of smallholder farmers (Mertz et al. 2009; Lee et al. 2014) as well as being connected to human rights violations (Corciolani et al. 2019).

On the other hand, owing to the high specific productivity of oil palms, alternative oil plants may lead to even higher land use. In a survey, consumers stated concern over palm oil use due to these health and sustainability factors irrespective of the fact that only 7% of the participants felt informed about the impacts of palm oil. These concerns, in turn, were identified as a factor for buying free from palm oil products, with consumers influenced by current information on palm oil from media being the most palm oil averse (Borrello et al. 2019). Furthermore, even sustainably produced palm oil labels could not lower consumers' risk perception than palm oil-free labels (Vergura et al. 2019).

7.2.4 Colorants

Generally, colorants can be divided into natural and synthetic pigments. Colorants of synthetic origin have been widely used in the past but are being replaced by dyes from natural sources. Negative consumer perceptions and thus phase out by food companies of these compounds have been affected by toxicity, allergic reactions, and adverse effects on behavior and neurocognitive function (Martins et al. 2016). For instance, a study suggested a connection between certain colorants and the prevalence of Attention Deficit Hyperactivity Disorder (ADHD) (Mccann et al. 2007). This may be why colorants are perceived as less natural than flavors (Murley and Chambers 2019) (Table 7.2).

Alternatively, there is a range of natural colorants, derived from, i.e., plants, such as chlorophyll (E140) or sodium copper chlorophyllin (E141) for green color, carotenoids such as β-carotene (E160a) or capsanthin (E160c) for yellow and orange, anthocyanins or betacyanins from beetroot, such as betanin (E162) for Red and purple or C-phycocyanin from cyanobacteria for blue color. However, since extracting some of these compounds is a quite elaborate technical process and some of them can also be synthesized chemically, their naturalness is up for debate, as illustrated by the example of brown colorants based on caramelization of carbohydrates, which may be carried out alone (E150a) or in the presence of ammonia, sulfite, or both (E150b through d) where the ammonia process (E150c) has raised toxicological concerns for higher intakes (Coultate and Blackburn 2018). On the other hand, some naturally sourced colorants, such as Anthocyanins, carotenoids, phenolic compounds, beet derivatives, annatto, and some curcuminoids, have even been shown to have health-promoting effects (Martins et al. 2016). As for consumer perception, colorants declared as E-numbers or with complicated chemical names may present a disincentive to consume a food irrespective of the compounds' origin. In contrast, general health-promoting effects may cause a positive association, offsetting the general dislike against techno-functional ingredients. This only works provided the consumers are in some way informed about the ingredient and its benefits, preferably if the ingredients are indeed perceived as natural, which may not be the case even if they are extracted from natural sources.

7.2.5 Artificial Flavors

Flavoring of foods, often using taste-inducing compounds or compounds mimicking natural likes, either extracted from natural sources or chemically synthesized, is an established method in the food industry to increase the liking of products by

Table 7.2 Synthetic food colorants currently permitted for use by EFSA and FDA

Commercial name	E-number	FD&C	Permitted in EU or USA	CI name	Possible side effects
Tartrazine	E102	Yellow No. 5	Both	CI Food Yellow 4	Allergy, ADHD
Quinoline Yellow WS	E104	None	EU only	CI Food Yellow 13	ADHD
Sunset Yellow FCF	E110	Yellow No. 6	Both	CI Food Yellow 3	ADHD
Carmoisine	E122	None	EU only	CI Food Red 3	ADHD
Ponceau 4R	E124	None	EU only	CI Food Red 7	ADHD
Allura Red AC	E129	Red No. 40	Both	CI Food Red 17	ADHD
Patent Blue V	E131	None	EU only	CI Food Blue 5	
Indigo Carmine	E132	Blue No. 2	Both	CI Food Blue 1	
Brilliant Blue FCF	E133	Blue No. 1	Both	CI Food Blue 2	
Green S	E142	None	EU only	CI Acid Green 50	
Fast Green FCF	E143	Green No. 3	USA only	CI Food Green 3	
Brilliant Black BN	E151	None	EU only	CI Food Black 1	
Brown HT	E155	None	EU only	CI Food Brown 3	
Orange B	None	Orange B	USA only	CI Acid Orange 137	
Citrus Red No. 2	None	Citrus Red No. 2	USA only	CI Solvent Red 80	

Adapted from Coultate and Blackburn (2018)

consumers. Additionally, flavor enhancers are used to strengthen existing flavors in food. However, especially artificial flavors are perceived as unnatural and undesired by many consumers. It has been found that excessive flavoring can even decrease the sensory acceptability of foods. In contrast, information on added flavors decreases acceptance for specific groups which are interested in organic food or exhibit food neophobia (Endrizzi et al. 2019). As for other additives, flavors declared as e-numbers are generally regarded as unnatural and possibly unsafe irrespective of their origin or actual health effects due to general distrust and prevailingly adverse reporting in the media (Van Gunst and Roodenburg 2019).

One well-known such compound is Glutamate, commonly employed as Monosodium Glutamate to enhance the taste. Glutamate comprises 20% of dietary protein intake. Still, it has been widely associated with headache, myalgia, backache, neck pain, nausea, diaphoresis, tingling, flushing, and backache in its artificially added form chest heaviness (Simon 2003). Even though WHO declared it as non-health threatening in appropriate doses, its presence in products is often checked for by consumers due to these controversial effects and a belief that it increases appetite and thus facilitates weight gain (Brother International GmbH 2020). Alternatively, many producers utilize yeast extracts, which also contain glutamic acid and legally do not have to be declared as a flavoring agent or under the chemical name of glutamate. This practice has been criticized as misleading by consumer agencies (Verbraucherzentrale Nordrhein Westfalen 2010) and could cause distrust by consumers.

Sweeteners, commonly employed to retain taste while reducing sugar content, are sometimes regarded as an excellent alternative to sugars. However, due to their low perceived naturalness and suspected health effects, some consumers do not want them in their food. One prominent example is Aspartame, which is low calorie and 180 times sweeter than sucrose but has shown some ambivalent evidence of causing granulomatous panniculitis, urticaria, and headaches (Simon 2003).

7.2.6 *Preservatives*

Preservatives are another category of food additives that consumers tend to be generally concerned about due to a lack of information and understanding of their types and uses (Ibrahim et al. 2021; Shim et al. 2011). There is already a trend in the food industry to declare "free from preservatives" on the packaging of food products. They were found to be the ingredient category most commonly advertised as lacking in a product in market analysis in Germany (Verbraucherzentrale Nordrhein Westfalen 2010).

Nitrates are a much-criticized ingredient, the salts of which are sometimes used to preserve certain meat products. There is a legal ban on it in the EU, but for some meat products, they can be used. However, nitrates and nitrites have come under attack as groundwater and food contaminants (mainly in the context of agricultural effluents, see below) and, alongside known cases of vascular headache, there are concerns that they might be cancerous due to their metabolic product nitrosamine, a known carcinogen (Simon 2003). This has led to negative attention in media and by consumers and illustrates that they should be avoided as ingredients.

Sulfites are commonly used to prevent browning of foods or contamination of fermentations, e.g., in winemaking. They can cause asthmatic reactions (Simon 2003) and, as preservatives, are not accepted by some consumers even though professional winemakers regard their presence in wines as positive (Bastian et al. 2015).

Alternative options for preserving foods are acids and antioxidants, declarable by the e-numbers 300 through 399. Commonly used food acids include vinegar, citric acid, tartaric acid, malic acid, fumaric acid, and lactic acid (Khanum Mirza Safiullah Baig and Sultan Kasim 2018). Even though they technically serve the purpose of lengthening the shelf life of products, they legally do not forfeit the possibility to declare a product as "free from preservatives" in the EU. This, in principle, allows using them in clean label products. However, this practice has been criticized by consumer protection agencies as misleading (Verbraucherzentrale Nordrhein Westfalen 2010) and may cause distrust by some consumers.

7.2.7 *Allergens*

Many allergens from natural and synthetic sources are found in foods. People with the associated allergies or intolerances present a definite reason for not buying products containing them. There is legislation requiring the declaration of many allergens in many countries to ensure the safety of allergic consumers.

In addition, some significant allergens have gained unpopularity even among people who are tolerant of them. Two of the most prominent examples are lactose and gluten. Lactose is the most abundant sugar in many kinds of milk, particularly cow's milk, and can cause adverse digestive reactions for the not insignificant lactose intolerant fraction of the population who suffer from the absence or dysfunction of the protein lactase the small intestine. This can be avoided by either substituting milk or breaking down lactose using supplemented lactase, either in the production process, producing a lactose-free product, or by the consuming person. Many consumers are well aware of these facts (Erika and Ózsvári 2020), and additionally, despite no proven strong adverse effects for tolerant people, sales of lactose-free products have been rising, particularly among passive consumers (Savarese et al. 2021).

Gluten is a protein found in wheat, and thus many bakery products containing wheat meal. It may be associated with celiac disease, wheat allergy, and nonceliac gluten sensitivity and believed to cause bowel irritation and inflammatory symptoms. However, the role of gluten in these symptoms is not clearly established, and sometimes the symptoms do not subside as a result of a gluten-free diet (El-Salhy et al. 2015). Despite the uncertain nature of this evidence, the consumption of gluten-free products has been rising considerably in western countries like the United States of America. Some researchers consider it a socially motivated rather than a medical phenomenon (Malone and Bailey Norwood 2020) which is based on self-diagnosis and a lay understanding of gluten-related disorders (Moore 2014). Nevertheless, the increasing popularity of gluten-free products illustrates that restriction or exclusion of gluten is essential to many consumers and may broaden the potential market for products adhering to this trend.

7.2.8 *Genetically Modified Organisms*

Genetically modified organisms or GMOs are a great discussion over environmental and health concerns. Their use does offer potential benefits such as enhanced climate resistance, the potential for producing pharmaceutically important compounds or improving product quality. Still, they have been mainly used to grow pesticide-resistant crops and thus simplify pest control in the past, a controversial practice from an environmental point of view. Even though their use is not legally prohibited in most countries since no adverse effects could be proven scientifically (sometimes attributed to insufficiently rigorous risk assessment (Hilbeck et al. 2020)), many consumers remain skeptical, likely because of widespread negative advocacy, i.e., by NGOs (Valentinov et al. 2019). Studies have shown that more consumers tend to have a negative perception of GMOs and would usually instead buy non-GMO products, even at a higher price, with health concerns or unpredictable consequences of DNA modification being the most relevant reasons (Popek and Halagarda 2017; Yuan 2020). Labeling GM foods as such is compulsory in the EU (Popek and Halagarda 2017) and to a lesser extent in the USA (Mosier et al. 2020), with stricter regulations upcoming in 2022 (Oselinsky et al. 2021). GMO labels influence a majority of consumers (Wunderlich et al. 2018). Still, there are differences between regions and among demographics and a gap between stated perceptions and purchasing behavior (Oselinsky et al. 2021). But since even retail markets now employ GMO-free labeling (Russo 2015) or even enforce GMO-free production across all their labels, and mainly European countries continue to stifle the use of this technology regardless of a lack of scientific evidence for side effects, their use should be avoided at least on the European market. Other parts of the world, such as Asia or Africa, may be more open to this technology (Paarlberg 2010), but public opinion and market trends should be assessed based on geographic location.

7.2.9 Subsection: Contaminants

Although not technically ingredients and usually not listed on ingredient lists, contaminants mostly from agricultural production will be mentioned here briefly due to their importance in food health discussions. One of the most relevant categories of pollutants is pesticides, such as the total herbicide glyphosate, which has been detected in some foods and humans (Krüger et al. 2014). Glyphosate is controversially being discussed as a possible carcinogen (Torretta et al. 2018), which has led to media attention and discussions on its prohibition (Pardo and Martínez 2019; Villnow et al. 2019). Another source of contaminants may be fertilizers which may lead to heightened concentrations of, e.g., nitrate, dealt with previously regarding preservatives. There are maximum legal concentrations, especially for baby food, where nitrate can lead to infant methemoglobinemia since some nitrate concentrations in plant leaves. Thus vegetables are unavoidable but should be closely monitored (Greer and Shannon 2005). Antibiotics are another contaminant from animal husbandry that may be found in foods. The attention towards contamination through pesticides and fertilizers may also be considered a driving factor of the trend towards organic food (Winter and Davis 2006) since the prohibition of these substances in organic farming forestalls contamination (Mie et al. 2017). Other sources of contaminants may include machine parts and technical fluids from all along the food supply chain and chemical or radioactive effluents or microplastics making their way into the sea or onto agricultural land and then into food sources (Barboza et al. 2018). Especially microplastics have recently received attention due to plastics pollution's general issue (Rhein and Schmid 2020).

7.2.10 Subsection: Animal Products

There is a range of alternative dietary lifestyles that are gaining prevalence, with vegetarian and vegan lifestyles are perhaps the most prominent examples, especially among younger people (Patelakis et al. 2019). Additionally, an increasing number of people are reducing their uptake of animal products due to health, environmental, or animal welfare concerns as well as cultural and religious beliefs, such as adherence to halal or kosher diets (McGirr et al. 2017). This trend has led to increased sales of alternative products replacing or mimicking those from animal sources (Searing et al. 2020). Furthermore, such issues make products from specific production systems unattractive even to consumers who consume meat. Products from production systems with increased animal welfare and animal welfare labels increase consumers' stated interest in buying a product (Cornish et al. 2020).

Sales of meat, eggs, or dairy products may remain strong or even growing in many parts of the world (Searing et al. 2020) and may not be replaced in the mainstream of many products which are primarily made of these materials. However, when considering food ingredients, products, where only small amounts of easily replaceable animal products are contained or used for technical purposes should be given special attention in this regard. Examples include animal-sourced gelatine (Duasa et al. 2021) or soft drinks filtered using animal livers. Especially since these ingredients' purpose or even existence may be unknown to many consumers, their use may appear intransparent to consumers. It may cause distrust, especially if vegetarians or vegans feel like they are sold animal products without their knowledge.

Generally, increasing awareness of environmental and animal welfare issues in the media and by consumers should prompt food manufacturers to ensure good animal welfare in animal product supply chains to forestall possible scandals. The total exclusion of animal products or at least meat can prevent this issue altogether while also making products attractive for a growing group of vegans and vegetarians.

7.3 Modern Methods of Detecting Food Fraud and Identifying Dirty Labels

Food fraud is one of the main challenges of food supply. It can harm public health, cause economic losses for individuals and companies, and cause a significant threat to the integrity of the food supply (Spink 2019). According to the Global Food Safety Initiative (GFSI), food fraud is defined as "*A collective term encompassing the deliberate and intentional substitution, addition, tampering or misrepresentation of food, food ingredients or food packaging, labeling, product information or false or misleading statements made about a product for economic gain that could impact consumer health*" (Global Food Safety Initiative 2018). The US Food and Drug Administration (FDA) defined food fraud as the "*Fraudulent, intentional substitution or addition of a substance in a product for the purpose of increasing the apparent value of the product or reducing the cost of its production, i.e., for economic gain*" (Ryan 2016). The European Commission identified four characteristics in food fraud definitions; (1) violation of EU food law; (2) Intention; (3) Economic gain; (4) consumer deception (Ryan 2016). There are many types of food fraud, including dilution, mislabeling, concealment, unapproved enhancement, substitution, and more (Fig. 7.2) (Spink et al. 2019).

The main types of food fraud are defined as follows (Manning 2016; Global Food Safety Initiative 2018; Robson et al. 2020);

- **Dilution:** Adding a low-price solution to a high-price ingredient, diluting it for economic gain purposes.
- **Substitution:** To replace an ingredient or a part of the product (with high value) with a similar ingredient (with a low value).
- **Counterfeiting:** To copy a trading name, packaging method, recipe, processing technology, etc., for economic gain.
- **Concealment:** To hide the bad quality of a food product or an ingredient.
- **Unapproved enhancement:** To add strange and unapproved ingredients to food products to improve the quality characteristics.

Fig. 7.2 Types of food fraud

- **Grey market:** The unofficial trade of food products outside the authorized channel of the manufacturer.
- **Mislabeling:** Place inaccurate claims on packaging labels for economic gain.

7.3.1 Food Labels and Mislabeling

According to the Codex Alimentarius Commission, "*Labelling includes any written, printed or graphic matter that is present on the label, accompanies the food, or is displayed near the food, including that for the purpose of promoting its sale or disposal*" (FAO/WHO 2007). Food labels are essential to provide all necessary information about food products. The following details should be displayed on the food labels (FAO/WHO 2007; Berryman 2014):

- The food name
- Ingredients' list
- The net content and weight
- The manufacturer's name and address
- Country of origin
- Lot identification
- Durability indication (Production and expiry dates)
- Instructions for storage and use
- Nutrition information

Mislabeling means displaying inaccurate claims on the labels for economic gain. A claim is "*any representation which states, suggests or implies that a food has particular characteristics relating to its origin, nutritional properties, nature, production, processing, composition or any other quality*" (FAO/WHO 2007). People and policymakers should be aware of food fraud, particularly mislabeling and misleading claims. Legislators should prohibit several claims, including (FAO/WHO 2007):

- Claims saying that a food product can provide all essential nutrients to the human body, except if a Codex standard approved such claims.
- Claims stating that ordinary food or a balanced natural diet is not able to supply a sufficient amount of all nutrients.
- Claims which cannot be validated.
- Claims imply that a food product can prevent or treat a disease or a physiological disorder without official scientific approval.
- Claims that doubt other products' safety or arouse fear in the consumer.

Mislabeling can cause significant economic losses to consumers and governments. Moreover, mislabeling can affect natural resources. For example, mislabeling some fish species under protection causes resource losses (Jacquet and Pauly 2008). Therefore, policymakers are always willing to update and improve the detection methods of food fraud organizations to protect society and governments.

7.3.2 *Modern Methods to Detect Food Fraud and Identifying Dirty Labels (Mislabeling)*

The detection of food fraud has become difficult because of many alternative substances that can be used for adulterating food products (McGrath et al. 2018a, b). The targeted and non-targeted analysis methods are the main approaches used for detecting food fraud. The targeted analysis includes the methods used to identify specific marker substances and compare the results with known limits to confirm if the tested substance is in excess (McGrath et al. 2018a, b). The non-targeted methods include measuring the sample's complete profile and comparing the result with old (control) samples in the library of the instruments' software. Appropriate statistical analysis is conducted to confirm the difference between new and aged samples (McGrath et al. 2018a, b). The advantage of non-targeted methods is that they widen the range of detected components to overcome the complexities of food adulteration (Cavanna et al. 2018). The main advantage of the targeted methods is that the results are assessed component-by-component using univariate statistics. At the same time, multivariate statistical models are needed to evaluate the data obtained from non-targeted methods (Cavanna et al. 2018). Several non-targeted methods have been used for detecting food fraud including, near-infrared (NIR), nuclear magnetic resonance (NMR), Fourier transform infrared (FT-IR), chromatography, and spectrometry-based techniques (McGrath et al. 2018a, b). In addition, DNA barcoding-based identification methods have been employed as a fast technique to detect mislabeling of food products, particularly meat and fish samples (Carvalho et al. 2015). We will discuss the main techniques used to detect food fraud and mislabeling in the following context.

Food fraud, meaning the intentional misleading of consumers or non-compliance with food laws with the goal of financial gain, has been an issue throughout history. It can, in severe cases, pose a significant health risk, such as in the prominent case of melamine added to baby formula, endangering the health of numerous Chinese children. Food fraud can also be related to many other types of food, such as fish and seafood, where cheaper and more food poisoning-prone species may be used. Conventionally produced ingredients illegally used in organic foods can also be an example of food fraud. Honey and other sweeteners, meat and grain-based foods, pulses, juices, dairy, coffee, and some processed foods are further examples of reported cases of food fraud (Johnson 2014; Hong et al. 2017). Many food adulterations may not pose a significant health risk, but they are fraud and undermine consumer trust in the food industry.

A range of analytical methods exists to detect this sort of impurity and will be dealt with in this chapter.

7.3.3 Chromatography

Chromatography serves the separation of mixtures and is carried out by dissolving the mixture in a solvent (mobile phase) and letting this mixture run through a stationary phase, i.e., a column, where the components of the original mixture travel at different apparent velocities, thus creating other bands each of which contains, optimally, just one component of the original mixture. This enables assessing their relative contents in the mix using various methods (Poole 2003). In this manner, dilution of valuable ingredients with lower-value alternatives can be identified, such as expensive juices of fruits with purported antioxidant properties diluted with cheaper alternative juices. The components which serve as identifiers can range from common macronutrients to specific metabolomic products, which can even allow (Vaclavik et al. 2012). Out of the range of chromatographic methods available, the methods often used for detection of food adulteration are:

- Thin-layer chromatography (TLC)
- High-performance liquid chromatography (HPLC)
- Gas chromatography (GC)

HPLC is primarily used for fruits, fruit juices, and sweeteners, and LC is the third most commonly employed technique overall (Hong et al. 2017).

7.3.4 Mass Spectrometry-Based

Mass spectrometry (MS) involves ionizing a sample and measuring the mass-to-charge ratios of the formed ions. These mass spectra can then be used to identify, characterize and quantify proteins. A wide range of different techniques in terms of ionization techniques and spectrometers exist (Domon and Aebersold 2006), and the commonly used techniques for food can be divided into liquid chromatography (LC)-MS/MS and gas chromatography (GC/MSD). MS is the most commonly used analytical technique in food applications, accounting for 20.6% of all analyses in scientific literature, most commonly for the analysis of spices, extracts, cereals, grains, and pulses, and LC for herbs, oils, and organic foods (Hong et al. 2017).

7.3.5 Electrophoretic Methods

Electrophoresis is a technique that enables the separation of DNA, RNA, or protein fragments by charge and molecular weight, by applying an electric field to a sample in a gel, through which molecules with lower molecular weight or higher charge travel more quickly, thus ending up at a different point in the gel after electrophoresis. This produces a pattern that enables the classification of a sample according to a different protein or DNA fragments contained within, often coupled with different isolation, multiplication, or analysis techniques.

A commonly used protein-based electrophoretic method is Sodium dodecyl sulfate-polyacrylamide gel electrophoresis (SDS-PAGE), which separates proteins based on molecular weight and enables characterization on protein fractions.

The most common DNA-based method is Polymerase chain reaction (PCR). It can be used to quickly multiply gene samples and, when using specific markers, quickly identify if DNA is from a given species using gel electrophoresis. The specific markers make DNA synthesis stop at a defined point in the sequence, and thus PCR produces DNA segments of differing lengths. These then create a characteristic pattern using gel electrophoresis, where DNA segments with higher molecular weight travel slower along an electric field through a gel. This pattern, in turn, identifies if a DNA sample is from a species and is mainly used for meat and meat products, fish and seafood, and milk and milk products, where it can detect whether an animal product is from the species claimed or from a different, cheaper species. Overall, PCR is the second most common technique used for foods, at 18.5% of all literature accounts (Hong et al. 2017).

Somewhat less commonly employed variants of DNA-based electrophoretic methods are (Hong et al. 2017):

- PCR-restriction fragment length polymorphism (RFLP)
- random amplified polymorphic DNA (RAPD)
- amplified fragment length polymorphism (AFLP)
- simple sequence repeats (SSR)

7.3.5.1 DNA Barcoding

The DNA sequence is different from a cellular organism to another. Even in closely related individuals such as family members, differences at the DNA sequence level are sufficient to be distinguished (Primrose 2019). The process starts with DNA extraction and purification from the food sample. Then, a specific genetic region is amplified using PCR techniques. After amplification, the DNA fragments are analyzed using electrophoresis techniques (Primrose 2019). DNA barcoding was developed by the United States Food and Drug Administration (FDA) to detect fish products fraud, but it is currently used for dairy and plant samples as well (Galimberti et al. 2013). It uses a selected genetic target (region) of approximately 650 base pairs (bp) of the mitochondrial gene cytochrome c oxidase subunit I (COI) to identify the sample at the species level (Carvalho et al. 2015). Recently, Xing et al. (2020a, b) used 16S rRNA (~220 bp) and cytochrome C oxidase I (COI) gene (~650 bp) to detect the mislabeling of 52 meat samples in the Chinese market. They found that 23% of the tested samples were mislabeled and contained undeclared species (Xing et al. 2020b). In the US, Scales et al. (2021) investigated the mislabeling potential of the American bison meat using PCR-satellite fragment length polymorphism (PCR-SFLP). They concluded that 6.7% of 45 American bison meat were mislabeled and contained domestic cattle meats (Scales et al. 2021). Kalogianni (2018) reviewed several DNA barcoding methods which have been used to detect milk adulteration (Kalogianni 2018). DNA barcoding is also used for detecting the potential mislabeling of fish products (Xiong et al. 2020; Xing et al. 2020a), poultry products (Hellberg et al. 2017), fruits (Wu et al. 2018), and honey products (Soares et al. 2018). DNA barcoding techniques can be considered practical tools for detecting food product mislabeling.

7.3.6 Spectroscopic Methods

Spectroscopic methods involve measurements of the reflection (and thus absorbance) of electromagnetic waves at different wavelengths. Since other organic compounds have different absorbances, spectroscopy allows rapid and non-destructive classification with little need for sample preparation. This, in turn, enables in-factory classification. However, they are not as commonly used as the previously mentioned methods. Infrared spectroscopy (IR) has been used to classify all manner of foods, and near-infrared (NIR) and mid-infrared (MIR) are also suitable for food classification. Fourier transform infrared (FTIR) enables analysis across a broader spectrum of wavelengths. Nuclear magnetic resonance (NMR) has also seen some applications for foods. Furthermore, other spectroscopic methods sometimes applied for foods are (Nagdalian et al. 2021; Hong et al. 2017):

- Raman
- Site-specific natural isotope fractionation (SNIF)-NMR
- Inductively coupled plasma-atomic emission spectrometer (ICP-AES)

7.3.7 Biochemical Methods

The enzyme-linked immunosorbent assay (ELISA) technique has been employed in food on rare occasions. It can be used to detect contamination with a grain such as wheat, rye, barley, and oats in gluten-free foods, which is a type of fraud that may be a risk to consumers with severe celiac disease (Osorio et al. 2019).

Biosensors are another possible but rarely employed method to detect food fraud. In a biosensor, a biological analyte reacts with some chemical component of the biosensor, producing a small electric current that indicates the presence of the analyte (Nature 2021).

Analytical methods for detecting food fraud remain a dynamic area of research. The remaining challenges include the move from targeted to non-targeted techniques, which would allow detection of many types of food fraud with a single method (McGrath et al. 2018a, b). Furthermore, to ensure food fraud detection from the legislative side, early warning systems and competent authorities must be implemented and data analytics leveraged (Ulberth 2020).

7.3.8 Nuclear Magnetic Resonance (NMR)

NMR is one of the most widely used analytical methods for detecting food fraud. It is used to verify the origin and species of many animal and plant products. Any components containing hydrogen exhibit NMR signals, and the food components then can be identified based on these signals. Thus, almost all substances of a food mixture can be placed in one spectrum. Therefore, a fingerprint of food samples can be obtained from the NMR spectra (Monakhova et al. 2013). Recently, NMR techniques have been used to detect adulteration and authentication in different food products. For example, NMR spectroscopy has been successfully used to confirm the existence of Robusta and Arabica species in coffee (Monakhova et al. 2015). Moreover, NMR spectroscopy has been used for detecting the fraud of different food products such as edible oils (Guyader et al. 2018), wine (Godelmann et al. 2013), spices (Petrakis et al. 2015), and fruit juice (Spraul et al. 2009). NMR spectroscopy can provide a unique fingerprint of food products and is considered an essential tool for detecting food fraud.

7.4 Research of Consumer Preferences When Choosing a Product

Consumer preference is defined as examining the processes that affect the selection of products needed by different people. Understanding consumer preferences could help the producers to detect patterns of consumers' behavior in decision-making to purchase or ignore a specific product. Numerous factors play a role in consumer preference for product selection. An accurate understanding of consumer preference in choosing the product is a complex process in which there are generally many differences in consumer preference from person to person. One of the most important strategies of producers of various products has been increasing production level and meeting the growing demand of consumers. However, recently, paying attention to consumer preference and attracting people's attention in the community in product selection is of particular importance for producers (Skreli et al. 2017). Labels, packaging, price, quality, durability, production process, and brand name are the most important factors influencing consumer preference in product selection (Venkatraman et al. 2012). Isik and Yasar (2015) evaluate the effects of brand name on consumers' preferences and found positive correlations between brand name and consumer preferences. Table 7.3 present the most important factors affecting consumer preference for choosing products.

Table 7.3 Factors affecting consumer preference for choosing products

Country	Products	Affecting factors	References
Iran	Dairy products	Price and family cost	Ahmadi Kaliji et al. (2019)
Taiwan	Traditional vegetable markets	Price level and products labeling	Li and Houston (2001)
India	Organic food products	Food label (organic food)	Sivathanu (2015)
Indonesia	Imported cosmetics	Brand name, packaging, location, and store environment	Kawa et al. (2013)
Vietnam	Shrimp product	Product labeling with eco-certifications	Xuan (2021)
Philippine	Rice	Aroma, big grains, and cohesiveness	Del-Mundo and Juliano (1981)
USA	Locally grown products	Age, gender, income, and product quality	Carpio and Isengildina-Massa (2009)
Scotland	Organic products	Attitude, ethical lifestyle, and price	Michaelidou and Hassan (2010)
South Africa	Different foods	Price, distribution, and promotion	Radder and Roux (2005)
Vietnam	Food products	Convenience and price	Dang et al. (2018)
USA	Apples	Eco-labeling	Loureiro et al. (2001)
Sweden	Organic foods	Healthism and trust	Anisimova (2016)
Bangladesh	Fast food	Brand reputation, accessibility, previous experience, cost, and quality of the food	Islam and Ullah (2010)
Iran	Dairy products	Functional, social, emotional, and epistemic values	Rahnama and Rajabpour (2017)
Ireland	Organic meat	Food safety	O'Donovan and McCarthy (2002)
USA	Different foods	Cultural, psychological, lifestyle, and food trends	Asp (1999)
Malaysia	Rice	Flavor, the taste of cooking, price, and location	Musa et al. (2011)

In recent years, the requirements and expectations of customers in purchasing food products have changed widely. Typically, customers pay special attention to the ingredients and product label's reliability and understandability in choosing the required food products. Seventy-six percent of those surveyed in a statistical population of Asia, North America, and Europe were eager to buy and use products that contained known and reliable ingredients. Fifty-two percent of these people were willing to pay 10% extra to purchase these products (Cassiday 2017). Brunner et al. (2018) found that customers are often more inclined to understand the technology used to produce food products. Hartmann et al. (2018) reported that consumers' priority when buying food products was to have a short list of ingredients, followed by knowledge of the product's ingredients. However, Busken (2013) listed these customer preferences as chemical-free ingredients and additives, recognizing the ingredients and minimum processing.

Usually, customers' trust in food of natural origin and the result of local and farm products is much higher than the food of chemical origin and the creation of chemical processes in the laboratory. According to Lent (2019), customers' prioritization of decision-making factors for purchasing the food products was ranked as recognition of ingredients, natural with no additive, clarity of nutritional data, and branding. One of the most significant modifications for promoting the health of food products is the use of phenolic compounds of natural origin to increase food products' shelf life (Xie et al. 2019). The 'health' label of food products is the most critical factor for customers. However, the labels such as 'free from chemicals' and 'organic' are of particular importance (Asioli et al. 2017). Other preferences, including no genetically modified organisms, pesticide-free, and organic food, are among the significant labels considered by consumers (Basha et al. 2015 Ditlevsen et al. 2019). Organic products free of recombinant bovine somatotropin (rBST) with no antibiotic are among the most important labels for milk products (Bernard and Bernard 2009). Another label that customers pay attention to is making food products. So, hand-made food products have a higher priority than machine-made ones (Abouab and Gomez 2015).

7.5 Use of Non-traditional Raw Materials in Food Products

In the face of increasing demands for nutritious food by a generally growing population, science, governments, and industry are increasingly looking for alternative raw materials for food products. A sufficient and sustainable supply of protein is most prominently an excellent topic of discussion considering the environmental (Aiking 2014) and health (Ayivi et al. 2021; Ekmekcioglu et al. 2018) impacts associated with many traditional protein sources, particularly of animal origin. However, as outlined above, non-traditional and unfamiliar ingredients are often met with distrust by consumers, especially if perceived as unnatural. This issue has to be accounted for when considering non-traditional materials for food production.

7.5.1 Cultured Meat

One alternative to meat from animals is meat grown from animal cells in bioreactors, called cultured or in-vitro meat. This has the advantage of overcoming the myriad ethical concerns associated with animal rearing if animal slaughter can be avoided altogether in its production. In addition, it is hoped to present a more sustainable production pathway, and healthier products (Aiking 2011), but challenges due to the high degree of processing required needing to be solved (van der Weele et al. 2019). However, this form of meat, also often called "lab meat" or "lab-grown meat" inlay media (Carrington 2020), can be perceived as unnatural by consumers, presenting a barrier to acceptance, especially under the influence of lack of information and food neophobia (Post et al. 2020). This issue may be overcome by better information of consumers. It has been found that arguing that conventional meat is unnatural increases acceptance of cultured meat, whereas discussing that cultured meat itself is natural or that naturalness is not an essential factor is less persuasive or even has a negative impact on acceptance (Bryant et al. 2019).

7.5.2 *Insects*

Insects are a sustainable potential animal protein source receiving increasing attention from scientists and entrepreneurs. However, since they are not traditionally consumed in western countries, in contrast to many Asian and African regions, acceptance by consumers is an issue alongside unclear legislation. Insects are often associated with being dirty or unhygienic (Siddiqui et al. 2022b, c; Delsignore & Siddiqui, 2022; Van Huis 2013), among other culturally entrenched prejudices (Kauppi et al. 2019). There are many differences in insect acceptance between regions, as consumers are less skeptical about them, i.e., in northern Europe than elsewhere on the continent and between individuals. For instance, males are more likely to consume insects than females. Being younger and intending to reduce meat consumption are additional factors increasing the likelihood of trying insects (Verbeke 2015). This is likely related to greater environmental and health concerns and lower food neophobia in these groups (Kauppi et al. 2019). Product properties and purchasing context also have a considerable influence on product acceptability; more highly processed insect products where the insects themselves are not recognizable cause less aversion toward them, significantly if they are associated with known flavors (Schösler et al. 2012; Caparros Megido et al. 2014, 2016; Lensvelt and Steenbekkers 2014). Of course, unfavorable tastes also reduce the likeliness of repurchase of insect products (Tan et al. 2016), a problem aggravated by the choice of insects whose production is easier to scale over better-tasting ones (House 2018). Marketing of the products, i.e., as meat replacements and in comparison with different alternatives, also influences selection criteria (House 2016). Additionally, the fact that consumers prefer plant-based alternative proteins over animal-based alternative proteins dampens the acceptance of insects (Onwezen et al. 2021).

Strategies in this domain can be, again, informing consumers about benefits or marketing products with abstract representations (Kauppi 2016) and under less negatively connotated names (in the case of insects, this can mean lesser-known scientific or indigenous names rather than the often negatively connotated common names, which is a contrast to chemical ingredients) (Looy et al. 2014) Other measures listed by Kauppi et al. (2019) pertain to product design, with products not recognizable as insects such as meals being preferable (Kauppi et al. 2019).

7.5.3 *Algae*

Microalgae, in particular, are an alternative resource receiving much attention due to their favorable nutrient content and potential for highly efficient growth if challenges related to technological complexity can be overcome (Pahmeyer et al. 2022; van der Weele et al. 2019). Macroalgae, such as seaweed, have long been employed in Asian diets as a good nutrient source (Afonso et al. 2019). Some algae, such as spirulina, have a history of being used in food supplements by different consumer groups (Moons et al. 2018). In recent years, their adoption as main or minor components in food and beverages has been rising as whole biomass and source of purified coloring or other functional agents such as, by far most commonly, carrageenans as thickening agents (Boukid and Castellari 2021). To a lesser extent, health and environmental reasons are a common motivator for their consumption across different consumer groups, while food neophobia for this product is ambiguous but has to be considered (Onwezen et al. 2021). The purchasing of these products can be enhanced by health and environmental claims and disguising algae-based ingredients to increase familiarity and independent promotors to increase trust. Consumers' attitudes also play a relevant factor, and the products are more accepted the more environmentally friendly they are (Onwezen et al. 2021).

7.6 Principles of Creating Attractive Products for the Consumer

A complex interplay between various factors influences purchasing decisions of consumers. A wellspring of scientific research has been conducted on this topic, with mathematical consumer choice models created to assess how different pricing and quality strategies may influence other consumer choices (Wang et al. 2021). When talking specifically about food, besides price and quality, many more factors come into play. As can be seen in the discussion of undesirable ingredients above text, health is a very important factor for many consumers and sustainability factors such as environmental friendliness or social impacts. However, the influence of taste and other sensory properties and purchasing price should not be underestimated, and often, food producers must compromise between different factors. For instance, as illustrated above, reducing or excluding health detrimental ingredients may reduce sensory liking. Likewise, some ingredients (to substitute others or add some health or sustainability benefit) may increase the price or hurt sensory properties. Furthermore, stated preferences or intentions from surveys may deviate from actual purchasing behavior for subconscious or unstated reasons.

The selection of ingredients has an excellent influence on the abovementioned factors for making a product attractive.

7.6.1 Appearance and Taste

Appearance and taste remain among the most important factors for product attractiveness. However, it is hard to find generalized heuristics as to which ingredients are generally positive for taste since each meal has unique properties and each consumer's individual preferences. It is hard for producers to estimate the importance of consumers, and they are often not fully met. In general, when developing a product, sensory tests should be carried out by a large and diverse assessing group as possible to gain insight into how the interplay of taste properties affects the attractiveness of a product. Some institutions offer sensory panels where a group of consumers tests new products. When taking a more advanced approach, the liking of different assessors can be mathematically modeled as a function of factors such as different tastes and colors, enabling to move toward most appreciated product characteristics in a directed, evolutionary approach (Halagarda and Suwała 2018).

7.6.2 Health

Health awareness is the most important predictor of whether consumers inform themselves about ingredients in foods (Bornkessel et al. 2014). This leads to the conclusion that health-conscious individuals are the most critical ones when it comes to a product's ingredients, and changes made to make a product more attractive can be effective if they address this point. Well-known health-promoting nutrients are, for instance, fiber and antioxidants. For many athletes or dieters, high protein content may also increase product attractiveness.

One relatively new development in that realm is the French Nutri-score, which has by now also found application in different countries. It is a front-of-pack label that states the health of a product on a five-point scale from A (healthy) to E (unhealthy), according to the so-called Food Standards Agency nutrient profiling system (FSA-NPS). It is believed to adequately describe the nutritional quality and thus provide a decision aid to consumers (Julia and Hercberg 2017). The score gives negative points for some and positive points for other nutrients. The scoring scheme is shown in Fig. 7.3. The left table on the top left shows ingredients that give negative points, meaning energy, sugars, saturated fat, and Sodium (which depends on salt addition). Positive points are given for fruits, vegetables, fiber, and protein. This is in line with general health recommendations and a good guideline regardless of whether the nutriscore is declared on the product in the end or not. However, using the nutriscore is an effective measure to communicate the healthiness of a product and positively influence purchase intentions (De Temmerman et al. 2021).

Many novel ingredients with possible or even proven health-promoting effects, including them in products, may help toward that effect. However, if the ingredient is not yet widely known and has no common but rather only a scientific name, including a health claim is substantial since unknown products with scientific names

1. Attribution of points, based on the content of nutrients and other elements per 100 g of a food/beverage

			Specific cut-offs: beverages			Specific cut-offs: fats	
Points	Energy (kJ)	Sugars (g)	Energy (kJ)	Sugars (g)	Saturated fat (g)	Saturated fat/ Lipids (%)	Sodium (mg)
0	≤ 335	≤ 4.5	≤ 0	0	≤ 1	< 10	< 90
1	> 335	> 4.5	≤ 30	≤ 1.5	> 1	< 16	> 90
2	> 670	> 9	≤ 60	≤ 3	> 2	< 22	> 180
3	> 1,005	> 13.5	≤ 90	≤ 4.5	> 3	< 28	> 270
4	> 1,340	> 18	≤ 120	≤ 6	> 4	< 34	> 360
5	> 1,675	> 22.5	≤ 150	≤ 7.5	> 5	< 40	> 450
6	> 2,010	> 27	≤ 180	≤ 9	> 6	< 46	> 540
7	> 2,345	> 31	≤ 210	≤ 10.5	> 7	< 52	> 630
8	> 2,680	> 36	≤ 240	≤ 12	> 8	< 58	> 720
9	> 3,015	> 40	≤ 270	≤ 13.5	> 9	< 64	> 810
10	> 3,350	> 45	≤ 270	> 13.5	> 10	≥ 64	> 900
	0–10 (a)	0–10 (b)	0–10 (a)	0–10 (b)	0–10 (c)	0–10 (c)	0–10 (d)
Total	Points A = (a) + (b) + (c) + (d) [0 – 40]						

		Specific cut-offs: beverages		
Points	Fruits, veg (%)	Fruits, veg (%)	Fiber (g)	Protein (g)
0	≤ 40	≤ 40	≤ 0.7	≤ 1.6
1	< 40		> 0.7	> 1.6
2	> 60	> 40	> 1.4	> 3.2
3	–		> 2.1	> 4.8
4	–	> 60	> 2.8	> 6.4
5	> 80		> 3.5	> 8.0
6				
7				
8				
9				
10		> 80		
	0–5 (a)	0–10 (a)	0–5 (b)	0–5 (c)
Total	Points C = (a) + (b) + (c) [0 – 15]			

2. Final score: -15 to 40 points

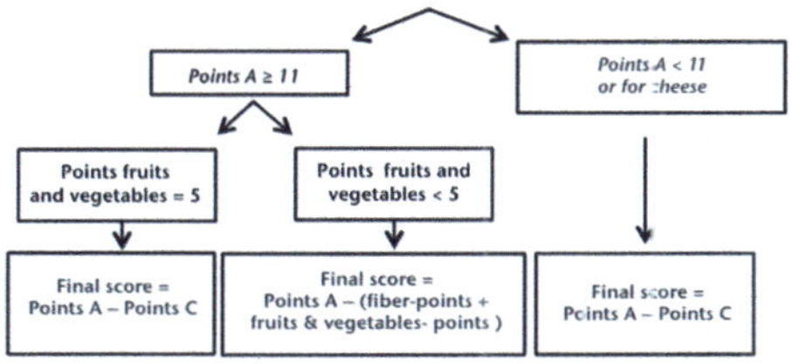

3. Attribution of colors

Foods (points)	Beverages (points)	Color
min to -1	water	green
0 to 2	min to 1	light green
3 to 10	2–5	yellow
11 to 18	6–9	orange
19 to max	10 to max	dark orange

NUTRI-SCORE

A B C D E

Green: highest quality

Red: lowest quality

Fig. 7.3 Attribution of points and colors for the Nutri-score. (Taken from Julia and Hercberg 2017)

do not awaken trust in consumers. However, when an enhanced function or reduced disease risk is claimed, it can increase positive associations to the ingredients and attractiveness of the product; this works exceptionally well for women and older people, with younger people focusing more on disease prevention whereas elderly are more favorable toward short term effects (Ares et al. 2009).

7.6.3 *Sustainability*

There are many possible definitions for the concept of sustainability. In a broad sense, a sustainable approach enables carrying out an activity without degrading any other systems for an, ideally, indefinite amount of time. The united nations' sustainable development goals (SDGs) are the most internationally agreed-upon guideline for implementing sustainability. Food consumers most commonly identify sustainability as "preservation of natural resources," followed by "decent working conditions" and "accessibility for everyone to healthy and safe food." Thus, sustainability has social, environmental, and economic dimensions from a food producers' perspective. The policy dimensions, of course, have to be taken into account as well but increasing the sustainability of a product compared to those adhering to the legal minimum can give especially early adopters a competitive edge (Peano et al. 2019). However, sustainability as a concept is not yet well known by all consumers everywhere (Rejman et al. 2019). Generally, consumers set different priorities in sustainability; they may be focused on a human-nature balance, setting value on products that avoid negative impacts on the environment. The same group also sets the value in good working conditions along the supply chain and a secure and healthy food supply for all. Some ingredients associated with negative impacts were presented above the text. Others may put more value in social welfare and thus inclusive economic environment, short and local supply chains, or environmentally enhancements associated with the product, or yet others (a smaller group) to preserve local ecosystems (Peano et al. 2019). This garners some broad guidelines when developing a sustainable product for consumers: Avoid ingredients associated with negative environmental impacts; ensure compliance to social welfare and working conditions standards across the supply chain; wherever possible, focus on locally sourced ingredients with short transportation routes; source ingredients commonly associated with better environmental performance, such as organic foods (Tobi et al. 2019).

Packaging can also be an essential factor for the actual environmental friendliness of a product, with lower amounts of packaging used and especially avoidance of single-use plastics as a good measure (Rhein and Schmid 2020); however, consumer knowledge in that regard is lacking, and price and product quality, as well as visual cues such as colors and pictures, are often more important determinants of their behavior (Ketelsen et al. 2020).

7.6.4 Presentation and Labeling

The importance of consumer information on any benefits or avoided negative aspects is evident if a product should be made more attractive by these attributes—a common strategy to ensure this is easily understandable manner labels. Many official and private labels exist, and their relevance to consumers depends on their credibility and consumers' knowledge of the issues involved (Kumar et al. 2021).

Furthermore, the placement and presentation of a product significantly influence purchasing behavior; it is essential to increase product visibility by placing it where consumers expect it. For instance, when producing an alternative product (i.e., a vegetarian meat alternative), it is beneficial to place it next to the products it is meant to substitute (Vandenbroele et al. 2021).

Numerous types of non-traditional raw materials are utilized in the food industry to improve the quality of the final product, increase its shelf life and increase consumers' acceptance of the products. We will discuss colorants and sweeteners as the leading natural non-traditional food additives in the following sections.

7.6.5 Natural Colorants

The sources of colors in the food products are pigments naturally present in the food or the added colorants during the food production process. The added colorants can be from synthesized or natural colorants. The main aim of adding colorants to the food products is to improve the final product's appearance and make it attractive for the consumers (Francis 1992). Most consumers and policymakers became aware of the negative impacts of synthetic colorants. Thus, natural pigments and colorants have attracted much attention due to their health benefits, antioxidant activities, and the growing concern about the safety of synthetic colorants (Wissgott and Bortlik 1996). Antioxidants, such as bioactive compounds, vitamins, and phytochemicals, play a vital role in promoting humans health. By quenching free radicals and singlet oxygen in the human body, antioxidants can reduce oxidative stress and prevent many diseases such as cancer and cardiovascular disease (CVD) (Leong et al. 2018). Carotenoids, anthocyanins, and betalains are the primary natural pigments used in many food applications (Rodriguez-Amaya 2019). These pigments are obtained from many plant and animal sources. Carotenoids are the primary source of orange, red, and yellow plant pigments. Carotenoids are classified into two classes based on their structure; carotenes (such as lycopene, b-carotene, etc.) are included in the first category and consisted of hydrocarbons without oxygen molecules. The second category comprises xanthophylls (including astaxanthin, zeaxanthin, fucoxanthin, and lutein) which contain oxygen molecules (Octavianti et al. 2011). Carrots, spinach, broccoli, tomato, orange, apricot, and mango are the primary natural sources of carotenoids (Octavianti et al. 2011; Mezzomo and Ferreira 2016). Anthocyanins are water-soluble polyphenolic pigments (red, purple, and blue colors) that play a vital role as a defense agent against stresses and attract pollinators to the flowers. These polyphenols positively impact human health due to their antioxidant, anti-inflammatory, and antithrombotic activities (Zhang et al. 2019a, b). In addition, studies showed that anthocyanins have the potential to protect against CVD (Krga and Milenkovic 2019). Apple, all berry species, eggplant, and lentil are the primary natural rich sources of anthocyanins (Zhang et al. 2019a, b). Betalains are water-soluble pigments, which contain nitrogen in their structure. They are synthesized naturally from tyrosine. They are structurally classified into two groups: the yellow-orange betaxanthins and the red-violet betacyanins. Betalains have high antioxidant activity and therefore promote human health. The colored Swiss chard (Beta vulgaris L. ssp. cicla) and yellow and red beetroot (Beta vulgaris L. ssp. Vulgaris), and leafy or grain amaranth (Amaranthus sp.) are the primary natural sources of betalains (Azeredo 2009). Natural pigments and their sources are recommended to consumers due to their health benefits.

7.6.6 *Sweeteners*

Sweeteners, both artificial and natural, are essential to the human diet and the food industry. Recently, dietitians recommended that the consumption of simple sugars should not exceed 10% of the daily caloric intake in reducing the risk of obesity (Mooradian et al. 2017). There are several plant sources of highly sweet compounds, including glycosides, steroids, proteins, dihydroisocoumarins, amino acids, proanthocyanidins, flavonoids, terpenoids, and phenylpropanoids (Pawar et al. 2013). Stevia (Stevia rebaudiana) has recently used a natural a non-caloric sweetener in many food industries. Besides, clinical studies showed that using stevia as a sweetener could reduce glucose concentration in plasma (Curi et al. 1986). Moreover, some bioactive compounds in stevia (steviol and stevioside) could stimulate the pancreas to release insulin. Thus, this plant could be used to manage type 2 diabetes (Jeppesen et al. 2000). However, the safety of using such substances should be investigated, and their effect on human health in the long term needs to be studied.

7.7 Conclusion

Investigating the relationship between consumers' behavior and the clean label of food products is crucial for both consumers and food producers. Policymakers in different countries impose different policies of clean labels for food products. Several ingredients are undesirable in food products such as salts, sugars, synthetic colorants, allergens, trans fats and synthetic preservatives. Therefore, food producers and researchers always seek to develop novel food products with natural and healthy ingredients. The utilization of natural colorants and sweeteners attracted the attention of food developers towards sustainable food production. With the intensive population growth, the high demand for food and the seek of illegal economic gain, food fraud started to be the main concern of food safety authorities. Different types of food fraud including dilution, substitution, mislabeling, counterfeiting and grey market were identified. Advanced food fraud detection methods have been used recently to improve the trust in food supply chains. Spectroscopic, chromatographic and DNA-based methods are the main methods recently employed for food fraud detection. With the increasing demand for food towards achieving the food security goals, several initiatives have been launched to use some non-traditional food sources in food production. Insects, algae and cultured meat are among those non-traditional food sources. Extensive research still needed to improve the nutritional properties and safety of these food sources.

References

Abouab N, Gomez P (2015) Human contact imagined during the production process increases food naturalness perceptions. Appetite 91:273–277. https://doi.org/10.1016/j.appet.2015.04.002

Afonso NC et al (2019) Brown macroalgae as valuable food ingredients. Antioxidants 8:365. https://doi.org/10.3390/antiox8090365

Ahmadi Kaliji S, Mojaverian SM, Amirnejad H, Canavari M (2019) Factors affecting consumers' dairy products preferences. AGRIS On-Line Pap Econ Inform 11(2):3–11. https://doi.org/10.7160/aol.2019.110201

Aiking H (2011) Future protein supply. Trends Food Sci Technol 22:112–120. https://doi.org/10.1016/j.tifs.2010.04.005

Aiking H (2014) Protein production: planet, profit, plus people? Am J Clin Nutr 100:483S. https://doi.org/10.3945/ajcn.113.071209

Albarracín W et al (2011) Salt in food processing; usage and reduction: a review. Int J Food Sci Technol 46:1329–1336. https://doi.org/10.1111/j.1365-2621.2010.02492.x

Anisimova T (2016) Integrating multiple factors affecting consumer behavior toward organic foods: the role of healthism, hedonism, and trust in consumer purchase intentions of organic foods. J Food Prod Mark 22(7):809–823. https://doi.org/10.1080/10454446.2015.1121429

Arantxa Colchero M et al (2017) In Mexico, evidence of sustained consumer response two years after implementing a sugar-sweetened beverage tax. Health Aff 36(3):564–571. https://doi.org/10.1377/hlthaff.2016.1231

Ares G, Giménez A, Gámbaro A (2009) Consumer perceived healthiness and willingness to try functional milk desserts. Influence of ingredient, ingredient name and health claim. Food Qual Prefer 20(1):50–56. https://doi.org/10.1016/j.foodqual.2008.07.002

Asioli D, Aschemann-Witzel J, Caputo V, Vecchio R, Annunziata A, Naes T, Varela P (2017) Making sense of the "clean label" trends: a review of consumer food choice behavior and discussion of industry implications. Food Res Int 99:58–71. https://doi.org/10.1016/j.foodres.2017.07.022

Asp EH (1999) Factors affecting food decisions made by individual consumers. Food Policy 24(2–3):287–294. https://doi.org/10.1016/S0306-9192(99)00024-X

Ayivi R, Ibrahim S, Colleran H, Silva R, Williams L, Galanakis C, Fidan H, Tomovska J, Siddiqui S (2021) COVID-19: human immune response and the influence of food ingredients and active compounds. Bioactive Compounds in Health and Disease. 4(6):100. https://doi.org/10.31989/bchd.v4i6.802

Azeredo HMC (2009) Betalains: properties, sources, applications, and stability - a review. Int J Food Sci Technol 44:2365–2376. https://doi.org/10.1111/j.1365-2621.2007.01668.x

Azhar B et al (2017) The global palm oil sector must change to save biodiversity and improve food security in the tropics. J Environ Manag 203:457–466. https://doi.org/10.1016/j.jenvman.2017.08.021

Barboza LGA et al (2018) Marine microplastic debris: an emerging issue for food security, food safety and human health. Mar Pollut Bull 133:336–348. https://doi.org/10.1016/j.marpolbul.2018.05.047

Basha MB, Mason C, Shamsudin MF, Hussain HI, Salem MA (2015) Consumers attitude towards organic food. Proc Econ Finance 31:444–452. https://doi.org/10.1016/S2212-5671(15)01219-8

Bastian S et al (2015) Australian wine consumer's acceptance of and attitudes toward the use of additives in wine and food production. Int J Wine Res 7:83. https://doi.org/10.2147/IJWR.S90802

Bernard JC, Bernard DJ (2009) What is it about organic milk? An experimental analysis. Am J Agric Econ 91(3):826–836. https://doi.org/10.1111/j.1467-8276.2009.01258.x

Berryman P (2014) Advances in Food and Beverage Labelling: Information and Regulations; Woodhead Publishing: London, UK. ISBN 9781782420934

Binker-Cosen MJ et al (2017) Palmitic acid increases invasiveness of pancreatic cancer cells AsPC-1 through TLR4/ROS/NF-κB/MMP-9 signaling pathway. Biochem Biophys Res Commun 484(1):152–158. https://doi.org/10.1016/j.bbrc.2017.01.051

Bornkessel S et al (2014) What determines ingredient awareness of consumers? A study on ten functional food ingredients. Food Qual Prefer 32(PC):330–339. https://doi.org/10.1016/j.foodqual.2013.09.007
Borrello M, Annunziata A, Vecchio R (2019) Sustainability of palm oil: drivers of consumers preferences. Sustainability 11(18):4818. https://doi.org/10.3390/su11184818
Boukid F, Castellari M (2021) Food and beverages containing algae and derived ingredients launched in the market from 2015 to 2019: a front-of-pack labeling perspective with a special focus on Spain. Foods 10(1):173. https://doi.org/10.3390/foods10010173
Brother International GmbH (2020) Clean label: Transparente Lebensmittel-Etiketten? | Brother. https://www.brother.de/blog/branchentrends/2020/clean-label. Accessed 16 May 2021
Brunner TA, Delley M, Denkel C (2018) Consumers' attitudes and change of attitude toward 3D-printed food. Food Qual Prefer 68:389–396. https://doi.org/10.1016/j.foodqual.2017.12.010
Bryant CJ et al (2019) Strategies for overcoming aversion to unnaturalness: the case of clean meat. Meat Sci 154:37–45. https://doi.org/10.1016/j.meatsci.2019.04.004
Busken DF (2013) Cleaning it up-what is a "clean label" ingredient? Cereal Foods World 60(2):112–113. https://doi.org/10.1094/CFW-60-2-0112
Caleja C, Barros L, Antonio AL, Carocho M, Oliveira MBPP, Ferreira ICFR (2016) Fortification of yogurts with different antioxidant preservatives: a comparative study between natural and synthetic additives. Food Chem 210:262–268. https://doi.org/10.1016/j.foodchem.2016.04.114
Câmara AKFI, Vidal VAS, Santos M, Bernardinelli OD, Sabadini E, Pollonio MAR (2020) Reducing phosphate in emulsified meat products by adding chia (*Salvia hispanica* L.) mucilage in powder or gel format: a clean label technological strategy. Meat Sci 163:108085. https://doi.org/10.1016/j.meatsci.2020.108085
Caparros Megido R et al (2014) Edible insects acceptance by belgian consumers: promising attitude for entomophagy development. J Sens Stud 29(1):14–20. https://doi.org/10.1111/joss.12077
Caparros Megido R et al (2016) Consumer acceptance of insect-based alternative meat products in Western countries. Food Qual Prefer 52:237–243. https://doi.org/10.1016/j.foodqual.2016.05.004
Carcelli A, Masuelli E, Diantom A, Vittadini E, Carini E (2020) Probing the functionality of physically modified corn flour as clean label thickening agent with a multiscale characterization. Foods 9:1105. https://doi.org/10.3390/foods9081105
Carpio CE, Isengildina-Massa O (2009) Consumer willingness to pay for locally grown products: the case of South Carolina. Agribusiness 25(3):412–426. https://doi.org/10.1002/agr.20210
Carrington D (2020) No-kill, lab-grown meat to go on sale for first time | Meat industry | The Guardian, 2 December. https://www.theguardian.com/environment/2020/dec/02/no-kill-lab-grown-meat-to-go-on-sale-for-first-time. Accessed 15 May 2021
Carvalho DC, Palhares RM, Drummond MG, Frigo TB (2015) DNA barcoding identification of commercialized seafood in South Brazil: a governmental regulatory forensic program. Food Control 50:784–788. https://doi.org/10.1016/j.foodcont.2014.10.025
Cassiday L (2017) Clean label: the next generation. Inform 28(8):6–10
Cavanna D, Righetti L, Elliott C, Suman M (2018) The scientific challenges in moving from targeted to non-targeted mass spectrometric methods for food fraud analysis: a proposed validation workflow to bring about a harmonized approach. Trends Food Sci Technol 80:223–241
Cheboi PK, Siddiqui SA, Onyando J, Kiptum CK, Heinz V (2021) Effect of ploughing techniques on water use and yield of rice in Maugo Small-Holder Irrigation Scheme, Kenya. AgriEngineering 3:110–117. https://doi.org/10.3390/agriengineering3010007
Cheung T, Junghans A, Dijksterhuis GB, Kroese F, Johansson P, Hall L, De Ridder D (2016) Consumers' choice-blindness to ingredient information. Appetite 106:2–12. https://doi.org/10.1016/j.appet.2015.09.022
Corciolani M, Gistri G, Pace S (2019) Legitimacy struggles in palm oil controversies: an institutional perspective. J Clean Prod 212:1117–1131. https://doi.org/10.1016/j.jclepro.2018.12.103

Cornish AR et al (2020) The price of good welfare: does informing consumers about what on-package labels mean for animal welfare influence their purchase intentions? Appetite 148:104577. https://doi.org/10.1016/j.appet.2019.104577

Coultate T, Blackburn RS (2018) Food colorants: their past, present and future. Color Technol 134(3):165–186. https://doi.org/10.1111/cote.12334

Curi R, Alvarez M, Bazotte RB et al (1986) Effect of Stevia rebaudiana on glucose tolerance in normal adult humans. Braz J Med Biol Res 19:771–774

Dang AK, Tran BX, Nguyen CT et al (2018) Consumer preference and attitude regarding online food products in Hanoi, Vietnam. Int J Environ Res Public Health 15(5):981. https://doi.org/10.3390/ijerph15050981

De Temmerman J et al (2021) The impact of the nutri-score nutrition label on perceived healthiness and purchase intentions. Appetite 157:104995. https://doi.org/10.1016/j.appet.2020.104995

Del-Mundo AM, Juliano BO (1981) Consumer preference and properties of raw and cooked milled rice. J Texture Stud 12(2):107–120. https://doi.org/10.1111/j.1745-4603.1981.tb01225.x

Delsignore M, Siddiqui SA (2022) Waste to Food. Chapter 8. From waste to food: legislative insights. Wageningen Academic Publishers, 197–208

Dickson-Spillmann M, Siegrist M, Keller C (2011) Attitudes toward chemicals are associated with preference for natural food. Food Qual Prefer 22(1):149–156. https://doi.org/10.1016/j.foodqual.2010.09.001

Ditlevsen K, Sandøe P, Lassen J (2019) Healthy food is nutritious, but organic food is healthy because it is pure: the negotiation of healthy food choices by Danish consumers of organic food. Food Qual Prefer 71:46–53. https://doi.org/10.1016/j.foodqual.2018.06.001

Domon B, Aebersold R (2006) Mass spectrometry and protein analysis. Science 312 212–217. https://doi.org/10.1126/science.1124619

Duasa J, Fatimah S, Mohamed Noor SF, Mohd M, Mohd Thas Thaker MA, Rahman MP (2021) The recombinant collagen-like protein as animal-based collagen substitution: a qualitative study. JCIS 6(2):17–50

Ekmekcioglu C et al (2018) Red meat, diseases, and healthy alternatives: a critical review. Crit Rev Food Sci Nutr 58(2):247–261. https://doi.org/10.1080/10408398.2016.1158148

El-Salhy M et al (2015) The relation between celiac disease, nonceliac gluten sensitivity and irritable bowel syndrome. Nutr J 14:92. https://doi.org/10.1186/s12937-015-0080-6

Endrizzi I et al (2019) Investigating the effect of artificial flavours and external information on consumer liking of apples. Molecules 24(23):4306. https://doi.org/10.3390/molecules24234306

Erika S, Ózsvári L (2020) Awareness of lactose-free products and pro-, pre- and synbiotics among consumers. J Food Investig special ed(1). https://www.researchgate.net/publication/345012485. Accessed 16 May 2021

Evans G, de Challemaison B, Cox DN (2010) Consumers' ratings of the natural and unnatural qualities of foods. Appetite 54(3):557–563. https://doi.org/10.1016/j.appet.2010.02.014

FAO/WHO (2007) Food labelling, 5th edn. Codex Alimentarius Commission, Rome

Fatima S et al (2019) Palmitic acid is an intracellular signaling molecule involved in disease development. Cell Mol Life Sci 76:2547–2557. https://doi.org/10.1007/s00018-019-03092-7

Francis FJ (1992) A new group of food colorants. Trends Food Sci Technol 3:27–30. https://doi.org/10.1016/0924-2244(92)90112-A

Galimberti A, De Mattia F, Losa A et al (2013) DNA barcoding as a new tool for food traceability. Food Res Int 50:55–63

Garvey EC, O'Sullivan MG, Kerry JP, Milner L, Gallagher E, Kilcawley KN (2021) Characterising the sensory quality and volatile aroma profile of clean-label sucrose reduced sponge cakes. Food Chem 342:128124. https://doi.org/10.1016/j.foodchem.2020.128124

Global Food Safety Initiative (2018) Tackling food fraud throughfood safety management systems. https://mygfsi.com/wp-content/uploads/2019/09/Food-Fraud-GFSI-Technical-Document.pdf

Godelmann R, Fang F, Humpfer E et al (2013) Targeted and nontargeted wine analysis by 1H NMR spectroscopy combined with multivariate statistical analysis. Differentiation of impor-

tant parameters: grape variety, geographical origin, year of vintage. J Agric Food Chem 61:5610–5619. https://doi.org/10.1021/jf400800d
Greer FR, Shannon M (2005) Infant methemoglobinemia: the role of dietary nitrate in food and water. Pediatrics 116(3):784–786. https://doi.org/10.1542/peds.2005-1497
Guyader S, Thomas F, Portaluri V et al (2018) Authentication of edible fats and oils by non-targeted 13C INEPT NMR spectroscopy. Food Control 91:216–224. https://doi.org/10.1016/j.foodcont.2018.03.046
Halagarda M, Suwała G (2018) Sensory optimisation in new food product development: a case study of polish apple juice. Ital J Food Sci 30(2):317–335. https://doi.org/10.14674/IJFS-960
Hamilton J et al (2000) Reduced fat products – consumer perceptions and preferences. Br Food J 102:494–506. https://doi.org/10.1108/00070700010336454
Hartmann C, Hieke S, Taper C, Siegrist M (2018) European consumer healthiness evaluation of 'free-from' labelled food products. Food Qual Prefer 68:377–388. https://doi.org/10.1016/j.foodqual.2017.12.009
Hashem KM, He FJ, Macgregor GA (2019) Labelling changes in response to a tax on sugar-sweetened beverages, United Kingdom of great Britain and Northern Ireland. Bull World Health Organ 97(12):818–827. https://doi.org/10.2471/BLT.19.234542
Hellberg RS, Hernandez BC, Hernandez EL (2017) Identification of meat and poultry species in food products using DNA barcoding. Food Control 80:23–28. https://doi.org/10.1016/j.foodcont.2017.04.025
Hemmerling S, Canavari M, Spiller A (2016) Preference for naturalness of European organic consumers: first evidence of an attitude-liking gap. Br Food J 118(9):2287–2307. https://doi.org/10.1108/BFJ-11-2015-0457
Hilbeck A et al (2020) GMO regulations and their interpretation: how EFSA's guidance on risk assessments of GMOs is bound to fail. Environ Sci Eur 32:54. https://doi.org/10.1186/s12302-020-00325-6
Hong E et al (2017) Modern analytical methods for the detection of food fraud and adulteration by food category. J Sci Food Agric 97(12):3877–3896. https://doi.org/10.1002/jsfa.8364
House J (2016) Consumer acceptance of insect-based foods in the Netherlands: academic and commercial implications. Appetite 107:47–58. https://doi.org/10.1016/j.appet.2016.07.023
House J (2018) Insects as food in the Netherlands: production networks and the geographies of edibility. Geoforum 94:82–93. https://doi.org/10.1016/j.geoforum.2018.05.011
Huang HW, Wu SJ, Lu JK, Shyu YT, Wang CY (2017) Current status and future trends of high-pressure processing in food industry. Food Control 72:1–8. https://doi.org/10.1016/j.foodcont.2016.07.019
Ibrahim SA, Ayivi RV, Zimmerman T, Siddiqui SA, Altemimi AB, Fidan H, Esatbeyoglu T, Bakhshayesh RV (2021) Lactic acid bacteria as antimicrobial agents: Food safety and microbial food spoilage prevention. Foods 10(12):3131. https://doi.org/10.3390/foods10123131
Isik A, Yasar MF (2015) Effects of brand on consumer preferences: a study in Turkmenistan. Euras J Bus Econ 8(16):139–150. https://doi.org/10.17015/ejbe.2015.016.08
Islam N, Ullah GM (2010) Factors affecting consumers' preferences on fast food items in Bangladesh. J Appl Bus Res 26(4):131–146
Islam MA et al (2019) Trans fatty acids and lipid profile: a serious risk factor to cardiovascular disease, cancer and diabetes. Diabetes Metab Syndr 13:1643–1647. https://doi.org/10.1016/j.dsx.2019.03.033
Jacquet JL, Pauly D (2008) Trade secrets: renaming and mislabeling of seafood. Mar Policy 32:309–318. https://doi.org/10.1016/j.marpol.2007.06.007
Jeppesen PB, Gregersen S, Poulsen CR, Hermansen K (2000) Stevioside acts directly on pancreatic β cells to secrete insulin: actions independent of cyclic adenosine monophosphate and adenosine triphosphate-sensitive K+-channel activity. Metabolism 49:208–214. https://doi.org/10.1016/S0026-0495(00)91325-8
Johnson R (2014) Food fraud and 'economically motivated adulteration' of food and food ingredients. www.crs.govR43358. Accessed 18 May 2021

Julia C, Hercberg S (2017) Nutri-score: evidence of the effective-ness of the French front-of-pack nutrition label. Ernahr Umsch 64(12):181–187. https://doi.org/10.4455/eu.2017.048

Kalogianni DP (2018) DNA-based analytical methods for milk authentication. Eur Food Res Technol 244:775–793

Kamel S, Al Otaibi H (2018) Trans-fats declaration, awareness and consumption in Saudi Arabia. Curr Res Nutr Food Sci 6(3):748–756. https://doi.org/10.12944/CRNFSJ.6.3.17

Katina K, Sauri M, Alakomi HL, Mattila-Sandholm T (2002) Potential of lactic acid bacteria to inhibit rope spoilage in wheat sourdough bread. Lebensm Wiss Technol 35(1):38–45. https://doi.org/10.1006/fstl.2001.0808

Kauppi, H. (2016) Insect economy and marketing: how much and in what way could insects be shown in packaging? SAARA-MARIA KAUPPI. https://aaltodoc.aalto.fi:443/handle/123456789/22808. Accessed 15 May 2021

Kauppi S-M, Pettersen IN, Boks C (2019) Consumer acceptance of edible insects and design interventions as adoption strategy. Int J Food Des 4(1):39–62. https://doi.org/10.1386/ijfd.4.1.39_1

Kawa LW, Rahmadiani SF, Kumar S (2013) Factors affecting consumer DecisionMaking: a survey of young-adults on imported cosmetics in Jabodetabek, Indonesia. SIJ Trans Ind Financ Bus Manage 1(5):175–180

Ketelsen M, Janssen M, Hamm U (2020) Consumers' response to environmentally-friendly food packaging - a systematic review. J Clean Prod 254:120123. https://doi.org/10.1016/j.jclepro.2020.120123

Khanum Mirza Safiullah Baig S, Sultan Kasim S (2018) Study of harmful effects of consuming food additives and public awareness. 2018 IJSRST | 4. www.ijsrst.com. Accessed 11 May 2021

Khetra Y, Kanawjia SK, Puri R (2016) Selection and optimization of salt replacer, flavour enhancer and bitter blocker for manufacturing low sodium Cheddar cheese using response surface methodology. LWT Food Sci Technol 72:99–106. https://doi.org/10.1016/j.lwt.2016.04.035

Köster EP (2009) Diversity in the determinants of food choice: a psychological perspective. Food Qual Prefer 20(2):70–82. https://doi.org/10.1016/j.foodqual.2007.11.002

Krga I, Milenkovic D (2019) Anthocyanins: from sources and bioavailability to cardiovascular-health benefits and molecular mechanisms of action. J Agric Food Chem 67:1771–1783. https://doi.org/10.1021/acs.jafc.8b06737

Krüger M et al (2014) Detection of glyphosate residues in animals and humans. J Environ Anal Toxicol 4(2):210. https://doi.org/10.4172/2161-0525.1000210

Kumar P et al (2021) Green information quality and green brand evaluation: the moderating effects of eco-label credibility and consumer knowledge. Eur J Mark. https://doi.org/10.1108/EJM-10-2019-0808

Le Lay C, Mounier J, Vasseur V, Weill A, Le Blay G, Barbier G, Coton E (2016) In vitro and in situ screening of lactic acid bacteria and propionibacteria antifungal activities against bakery product spoilage molds. Food Control 60:247–255. https://doi.org/10.1016/j.foodcont.2015.07.034

Lee JSH et al (2014) Environmental impacts of large-scale oil palm enterprises exceed that of smallholdings in Indonesia. Conserv Lett 7(1):25–33. https://doi.org/10.1111/conl.12039

Lensvelt EJS, Steenbekkers LPA (2014) Exploring consumer acceptance of entomophagy: a survey and experiment in Australia and the Netherlands. Ecol Food Nutr 53(5):543–561. https://doi.org/10.1080/03670244.2013.879865

Lent V (2019) Modeling value of natural food preservative using an example of innovative chitosan Orange Peel. Adv Econ Bus Manage Res 106:122–126. https://doi.org/10.2991/febm-19.2019.27

Leong HY, Show PL, Lim MH et al (2018) Natural red pigments from plants and their health benefits: a review. Food Rev Int 34:463–482

Li HS, Houston JE (2001) Factors affecting consumer preferences for major food Markets in Taiwan. J Food Distrib Res 32(1):97–109. https://doi.org/10.22004/ag.econ.26515

Looy H, Dunkel FV, Wood JR (2014) How then shall we eat? Insect-eating attitudes and sustainable foodways. Agric Hum Values 31(1):131–141. https://doi.org/10.1007/s10460-013-9450-x

Loureiro ML, McCluskey JJ, Mittelhammer RC (2001) Assessing consumer preferences for organic, eco-labeled, and regular apples. J Agric Resour Econ 26(2):404–416. https://doi.org/10.22004/ag.econ.31039

Magri TPR et al (2015) Interesterified fat or palm oil as substitutes for partially hydrogenated fat in maternal diet can predispose obesity in adult male offspring. Clin Nutr 34(5):904–910. https://doi.org/10.1016/j.clnu.2014.09.014

Malone T, Bailey Norwood F (2020) Gluten aversion is not limited to the political left. Agric Hum Values 37(1):1–15. https://doi.org/10.1007/s10460-019-09958-7

Manning L (2016) Food fraud: policy and food chain. Curr Opin Food Sci 10:16–21. https://doi.org/10.1016/j.cofs.2016.07.001

Markey O, Lovegrove JA, Methven L (2015) Sensory profiles and consumer acceptability of a range of sugar-reduced products on the UK market. Food Res Int 72:133–139. https://doi.org/10.1016/j.foodres.2015.03.012

Martínez-Zamora L, Peñalver R, Ros G, Nieto G (2021) Substitution of synthetic nitrates and antioxidants by spices, fruits and vegetables in clean label Spanish chorizo. Food Res Int 139:109835. https://doi.org/10.1016/j.foodres.2020.109835

Martins N et al (2016) Food colorants: challenges, opportunities and current desires of agro-industries to ensure consumer expectations and regulatory practices. Trends Food Sci Technol 52:1–15. https://doi.org/10.1016/j.tifs.2016.03.009

Mccann D et al (2007) Food additives and hyperactive behaviour in 3-year-old and 8/9-year-old children in the community: a randomised, double-blinded, placebo-controlled trial. Lancet 370:1560. https://doi.org/10.1016/S0140-6736(07)61306-3

McDonnell LM, Glass KA, Sindelar JJ (2013) Identifying ingredients that delay outgrowth of *Listeria monocytogenes* in natural, organic, and clean-label ready-to-eat meat and poultry products. J Food Prot 76(8):1366–1376. https://doi.org/10.4315/0362-028X.JFP-12-501

McGirr C, McEvoy CT, Woodside JV (2017) Vegetarian and vegan diets: weighing the claims. In: Nutrition guide for physicians and related healthcare professionals. Springer, New York, pp 203–212. https://doi.org/10.1007/978-3-319-49929-1_20

McGrath TF, Haughey SA, Patterson J et al (2018a) What are the scientific challenges in moving from targeted to non-targeted methods for food fraud testing and how can they be addressed? – spectroscopy case study. Trends Food Sci Technol 76:38–55

McGrath TF et al (2018b) What are the scientific challenges in moving from targeted to non-targeted methods for food fraud testing and how can they be addressed? – spectroscopy case study. Trends Food Sci Technol 76:38–55. https://doi.org/10.1016/j.tifs.2018.04.001

Mehdizadeh T, Langrood AM (2019) Chitosan coatings incorporated with propolis extract and Zataria multiflora Boiss oil for active packaging of chicken breast meat. Int J Biol Macromol 141:401–409. https://doi.org/10.1016/j.ijbiomac.2019.08.267

Merlo TC, Contreras-Castilloa CJ, Saldañaa E, Barancellia GV, Dargelioa MDB, Yoshidab CMP, Ribeiro EEJ, Massariolia A, Venturini AC (2019) Incorporation of pink pepper residue extract into chitosan film combined with a modified atmosphere packaging: effects on the shelf life of salmon fillets. Food Res Int 125:108633. https://doi.org/10.1016/j.foodres.2019.108633

Mertz O et al (2009) Swidden change in Southeast Asia: understanding causes and consequences. Hum Ecol 37:259–264. https://doi.org/10.1007/s10745-009-9245-2

Mezzomo N, Ferreira SRS (2016) Carotenoids functionality, sources, and processing by supercritical technology: a review. J Chem 2016:3164312

Michaelidou N, Hassan LM (2010) Modeling the factors affecting rural consumers' purchase of organic and free-range produce: a case study of consumers' from the island of arran in Scotland, UK. Food Policy 35(2):130–139. https://doi.org/10.1016/j.foodpol.2009.10.001

Mie A et al (2017) Human health implications of organic food and organic agriculture: a comprehensive review. Environ Health 16:1–22. https://doi.org/10.1186/s12940-017-0315-4

Milner L, Kerry JP, O'Sullivan MG, Gallagher E (2020) Physical, textural and sensory characteristics of reduced sucrose cakes, incorporated with clean-label sugar-replacing alternative ingredients. Innovative Food Sci Emerg Technol 59:102235. https://doi.org/10.1016/j.ifset.2019.102235

Monakhova YB, Kuballa T, Lachenmeier DW (2013) Chemometric methods in NMR spectroscopic analysis of food products. J Anal Chem 68:755–766

Monakhova YB, Ruge W, Kuballa T et al (2015) Rapid approach to identify the presence of Arabica and Robusta species in coffee using 1H NMR spectroscopy. Food Chem 182:178–184. https://doi.org/10.1016/j.foodchem.2015.02.132

Moons I, Barbarossa C, De Pelsmacker P (2018) The determinants of the adoption intention of eco-friendly functional food in different market segments. Ecol Econ 151:151–161. https://doi.org/10.1016/j.ecolecon.2018.05.012

Mooradian AD, Smith M, Tokuda M (2017) The role of artificial and natural sweeteners in reducing the consumption of table sugar: a narrative review. Clin Nutr 18:1–8

Moore LR (2014) "But we're not hypochondriacs": the changing shape of gluten-free dieting and the contested illness experience. Soc Sci Med 105:76–83. https://doi.org/10.1016/j.socscimed.2014.01.009

Mosier SL, Rimal A, Ruxton MM (2020) A song of policy incongruence: the missing choir of consumer preferences in GMO-labeling policy outcomes. Rev Policy Res 37(4):511–534. https://doi.org/10.1111/ropr.12391

Murley T, Chambers E (2019) The influence of colorants, flavorants and product identity on perceptions of naturalness. Foods 8(8):317. https://doi.org/10.3390/foods8080317

Musa M, Othman N, Fatah FA (2011) Determinants of consumers purchasing behavior for Rice in Malaysia. Am Int J Contemp Res 1(3):159–167

Nagdalian AA, Rzhepakovsky IV, Siddiqui SA, Piskov SI, Oboturova NP, Timchenko LD, Lodygin A, Blinov AV, Ibrahim SA (2021) Analysis of the content of mechanically separated poultry meat in sausage using computing microtomography. J Food Compos Anal 100:103918. https://doi.org/10.1016/J.JFCA.2021.103918

Nature (2021) Biosensors - latest research and news | Nature. https://www.nature.com/subjects/biosensors. Accessed 20 May 2021

O'Donovan P, McCarthy M (2002) Irish consumer preference for organic meat. Br Food J 104(3):353–370. https://doi.org/10.1108/00070700210425778

Octavianti F, Jaswir I, Noviendri D, Hasrini RF (2011) Carotenoids: sources, medicinal properties and their application in food and nutraceutical industry R-Phycoerythrin from Indonesian marine red algae and microalgae view project halal detection view project Dedi Noviendri Ministry of Marine Affairs and Fisheries Carotenoids: sources, medicinal properties and their application in food and nutraceutical industry. Artic J Med Plant Res 5:7119–7131. https://doi.org/10.5897/JMPRx11.011

Onwezen MC et al (2021) A systematic review on consumer acceptance of alternative proteins: pulses, algae, insects, plant-based meat alternatives, and cultured meat. Appetite 159:105058. https://doi.org/10.1016/j.appet.2020.105058

Oostenbach LH et al (2019) Systematic review of the impact of nutrition claims related to fat, sugar and energy content on food choices and energy intake. BMC Public Health 19(1):1–11. https://doi.org/10.1186/s12889-019-7622-3

Oselinsky K et al (2021) GMO food labels do not affect college student food selection, despite negative attitudes towards gmos. Int J Environ Res Public Health 18(4):1–19. https://doi.org/10.3390/ijerph18041761

Osorio CE, Mejías JH, Rustgi S (2019) Gluten detection methods and their critical role in assuring safe diets for celiac patients. Nutrients 11:2920. https://doi.org/10.3390/nu11122920

Paarlberg R (2010) GMO foods and crops: Africa's choice. New Biotechnol 27:609–613. https://doi.org/10.1016/j.nbt.2010.07.005

Pahmeyer MJ, Siddiqui SA, Pleissner D, Gołaszewski J, Heinz V, Smetana S (2022) An automated, modular system for organic waste utilization using heterotrophic alga Galdieria sulphuraria: Design considerations and sustainability, Bioresource Technology, 348:126800. https://doi.org/10.1016/j.biortech.2022.126800

Pardo G, Martínez Y (2019) Conservation agriculture in trouble? Estimating the economic impact of an eventual glyphosate prohibition in Spain. Planta Daninha 37:e019197994. https://doi.org/10.1590/s0100-83582019370100138

Park S, Kim YR (2021) Clean label starch: production, physicochemical characteristics, and industrial applications. Food Sci Biotechnol 30(1):1–17. https://doi.org/10.1007/s10068-020-00834-3

Patelakis E et al (2019) Prevalence of vegetarian diet among children and adolescents in Germany results from EsKiMo II. Ernhrungs Umschau 66:85–89. https://doi.org/10.4455/eu.2019.018

Pawar RS, Krynitsky AJ, Rader JI (2013) Sweeteners from plants-with emphasis on Stevia rebaudiana (Bertoni) and Siraitia grosvenorii (Swingle). Anal Bioanal Chem 405:4397–4407

Peano C et al (2019) Sustainability for food consumers: which perception? Sustainability 11(21):5955. https://doi.org/10.3390/su11215955

Petrakis EA, Cagliani LR, Polissiou MG, Consonni R (2015) Evaluation of saffron (Crocus sativus L.) adulteration with plant adulterants by1H NMR metabolite fingerprinting. Food Chem 173:890–896. https://doi.org/10.1016/j.foodchem.2014.10.107

Poole CF (2003) The essence of chromatography. https://books.google.de/books?hl=de&lr=&id=k4esJheTf5QC&oi=fnd&pg=PR9&dq=chromatography&ots=8MkdT6uxQ4&sig=YKccMLp1rq46Erc3JXU7fRgyj6Y#v=onepage&q=chromatography&f=false. Accessed 19 May 2021

Popek S, Halagarda M (2017) Genetically modified foods: consumer awareness, opinions and attitudes in selected EU countries. Int J Consum Stud 41(3):325–332. https://doi.org/10.1111/ijcs.12345

Post MJ et al (2020) Scientific, sustainability and regulatory challenges of cultured meat. Nat Food 1(7):403–415. https://doi.org/10.1038/s43016-020-0112-z

Primrose SB (2019) The role of DNA analysis in the determination of food authenticity. In: DNA techniques to Verify food authenticity: applications in food fraud. Royal Society of Chemistry, London, pp 1–9

Profeta A, Siddiqui SA, Smetana S et al. (2021) The impact of Corona pandemic on consumer's food consumption. Journal of Consumer Protection and Food Safety 16:305–314. https://doi.org/10.1007/s00003-021-01341-1

Radder L, Roux R (2005) Factors affecting food choice in relation to venison: a south African example. Meat Sci 71(3):583–589. https://doi.org/10.1016/j.meatsci.2005.05.003

Rahman MM, Simsek S (2020) Go clean label: replacement of commercial dough strengtheners with hard red spring wheat flour in bread formulations. J Food Sci Technol 57:3581–3590. https://doi.org/10.1007/s13197-020-04390-w

Rahnama H, Rajabpour S (2017) Factors for consumer choice of dairy products in Iran. Appetite 111:46–55. https://doi.org/10.1016/j.appet.2016.12.004

Reijnders L, Huijbregts MAJ (2008) Palm oil and the emission of carbon-based greenhouse gases. J Clean Prod 16(4):477–482. https://doi.org/10.1016/j.jclepro.2006.07.054

Rejman K et al (2019) Do Europeans consider sustainability when making food choices? A survey of Polish city-dwellers. Public Health Nutr 22(7):1330–1339. https://doi.org/10.1017/S1368980019000326

Rhein S, Schmid M (2020) 'Consumers' awareness of plastic packaging: more than just environmental concerns. Resour Conserv Recycl 162:105063. https://doi.org/10.1016/j.resconrec.2020.105063

Robson K, Dean M, Brooks S et al (2020) A 20-year analysis of reported food fraud in the global beef supply chain. Food Control 116:107310. https://doi.org/10.1016/j.foodcont.2020.107310

Rodriguez-Amaya DB (2019) Update on natural food pigments - a mini-review on carotenoids, anthocyanins, and betalains. Food Res Int 124:200–205. https://doi.org/10.1016/j.foodres.2018.05.028

Russo D (2015) Relationship between genetically modified foods and mass retail. Quality 16(146):93–96. https://eds.a.ebscohost.com/abstract?site=eds&scope=site&authtype=crawler&jrnl=15822559&AN=103102258&h=QIEAmrqsUB9Vhb1Hc3aOoc8sBQHhQB1YZyN%2

BiEfAy%2F1RjAc6ZGM3u1KdGQ%2FM8c0%2B2zvUyQyy%2FUodgduuT4EFw%3D%3D&crl=c&resultLocal=ErrCrlNoResults&resultNs=Ehost&. Accessed 17 May 2021

Ryan JM (2016) Food fraud. Elsevier, Academic Press, Boston

Ryan LA, Zannini E, Dal Bello F, Pawlowska A, Koehler P, Arendt EK (2011) Lactobacillus amylovorus DSM 19280 as a novel food-grade antifungal agent for bakery products. Int J Food Microbiol 146:276–283. https://doi.org/10.1016/j.ijfoodmicro.2011.02.036

Savarese M, Wismer W, Graffigna G (2021) Conceptualizing "free-from" food consumption determinants: a systematic integrative literature review focused on gluten and lactose. Food Qual Prefer 90:104170. https://doi.org/10.1016/j.foodqual.2020.104170

Scales ZM, Narbay E, Hellberg RS (2021) Use of DNA barcoding combined with PCR-SFLP to authenticate species in bison meat products. Foods 10:347. https://doi.org/10.3390/foods10020347

Scapin T et al (2021) Influence of sugar label formats on consumer understanding and amount of sugar in food choices: a systematic review and meta-analyses. Nutr Rev 9:788. https://doi.org/10.1093/nutrit/nuaa108

Schösler H, de Boer J, Boersema JJ (2012) Can we cut out the meat of the dish? Constructing consumer-oriented pathways towards meat substitution. Appetite 58(1):39–47. https://doi.org/10.1016/j.appet.2011.09.009

Searing A et al (2020) Alternative proteins: market research on consumer trends and emerging landscape. Meat Muscle Biol 4(2):16. https://doi.org/10.22175/mmb.11225

Shim SM et al (2011) Consumers' knowledge and safety perceptions of food additives: evaluation on the effectiveness of transmitting information on preservatives. Food Control 22(7):1054–1060. https://doi.org/10.1016/j.foodcont.2011.01.001

Siegrist M, Sütterlin B (2017) Importance of perceived naturalness for acceptance of food additives and cultured meat. Appetite 113:320–326. https://doi.org/10.1016/j.appet.2017.03.019

Siddiqui SA, Bahmid NA, Mahmud CMM, Boukid F, Lamri M, Gagaoua M (2022a) Consumer acceptability of plant-, seaweed-, and insect-based foods as alternatives to meat: a critical compilation of a decade of research. Crit Rev Food Sci Nutr 1–22. https://doi.org/10.1080/10408398.2022.2036096

Siddiqui SA, Ristow B, Rahayu T, Putra NS, Yuwono NW, Nisa' K, Mategeko B, Smetana S, Saki M, Nawaz A, Nagdalian A (2022b) Black soldier fly larvae (BSFL) and their affinity for organic waste processing, Waste Management 140:1–13. https://doi.org/10.1016/j.wasman.2021.12.044

Siddiqui SA, Snoeck ER, Tello A, Alles MC, Fernando I, Saraswati YR, Rahayu T, Grover R, Ullah MI, Ristow B, Nagdalian AA (2022c) Journal of Insects as Food and Feed, 1–20. https://doi.org/10.3920/JIFF2021.0162

Simon RA (2003) Adverse reactions to food additives. Curr Allergy Asthma Rep 3:62–66

Sinesio F et al (2019) Impact of sodium reduction strategies on volatile compounds, sensory properties and consumer perception in commercial wheat bread. Food Chem 301:125252. https://doi.org/10.1016/j.foodchem.2019.125252

Sivathanu B (2015) Factors affecting consumer preference towards the organic food purchases. Indian J Sci Technol 8(33):1–6. https://doi.org/10.17485/ijst/2015/v8i33/78261

Skreli E, Imami D, Chan C, Canavari M, Zhllima E, Pire E (2017) Assessing consumer preferences and willingness to pay for organic tomatoes in Albania: a conjoint choice experiment study. Span J Agric Res 15(3):0114. https://doi.org/10.5424/sjar/2017153-9889

Soares S, Grazina L, Costa J et al (2018) Botanical authentication of lavender (Lavandula spp.) honey by a novel DNA-barcoding approach coupled to high resolution melting analysis. Food Control 86:367–373. https://doi.org/10.1016/j.foodcont.2017.11.046

Spink JW (2019) Food fraud prevention: introduction, implementation, and management. Springer, New York

Spink J, Vincent Hegarty P, Fortin ND et al (2019) The application of public policy theory to the emerging food fraud risk: next steps. Trends Food Sci Technol 85:116–128. https://doi.org/10.1016/j.tifs.2019.01.002

Spraul M, Schütz B, Humpfer E et al (2009) Mixture analysis by NMR as applied to fruit juice quality control. Magn Reson Chem 47:S130–S137. https://doi.org/10.1002/mrc.2528
Tan HSG et al (2016) Tasty but nasty? Exploring the role of sensory-liking and food appropriateness in the willingness to eat unusual novel foods like insects. Food Qual Prefer 48:293–302. https://doi.org/10.1016/j.foodqual.2015.11.001
Te Morenga L, Montez JM (2017) Health effects of saturated and trans-fatty acid intake in children and adolescents: systematic review and meta-analysis. PLoS One 12:e0186672. https://doi.org/10.1371/journal.pone.0186672
Tobi RCA et al (2019) Sustainable diet dimensions. Comparing consumer preference for nutrition, environmental and social responsibility food labelling: a systematic review. Sustainability 11:6575. https://doi.org/10.3390/su11236575
Torretta V et al (2018) Critical review of the effects of glyphosate exposure to the environment and humans through the food supply chain. Sustainability 10:950. https://doi.org/10.3390/su10040950
Ua-Arak T, Jakob F, Vogel RF (2017) Influence of levanproducing acetic acid bacteria on buckwheat-sourdough breads. Food Microbiol 65:95–104. https://doi.org/10.1016/j.fm.2017.02.002
Ulberth F (2020) Tools to combat food fraud – a gap analysis. Food Chem 330:127044. https://doi.org/10.1016/j.foodchem.2020.127044
USDA (United States Department of Agriculture) (2018) Oilseeds: world markets and trade. https://apps.fas.usda.gov/psdonline/circulars/oilseeds.pdf. Accessed 5 May 2021
Vaclavik L et al (2012) Liquid chromatography-mass spectrometry-based metabolomics for authenticity assessment of fruit juices. Metabolomics 8(5):793–803. https://doi.org/10.1007/s11306-011-0371-7
Valentinov V et al (2019) The anti-GMO advocacy: an institutionalist and systems-theoretic assessment. Kybernetes 48(5):888–905. https://doi.org/10.1108/K-01-2018-0016
Van Gunst A, Roodenburg AJC (2019) Consumer distrust about E-numbers: a qualitative study among food experts. Foods 8(5):178. https://doi.org/10.3390/foods8050178
Van Huis A (2013) Potential of insects as food and feed in assuring food security. Annu Rev Entomol 58:563–583. https://doi.org/10.1146/annurev-ento-120811-153704
Vandenbroele J et al (2021) Mock meat in the butchery: nudging consumers toward meat substitutes. Organ Behav Hum Decis Process 163:105–116. https://doi.org/10.1016/j.obhdp.2019.09.004
Venkatraman V, Clithero JA, Fitzsimons GJ, Huettel SA (2012) New scanner data for brand marketers: how neuroscience can help better understand differences in brand preferences. J Consum Psychol 22(1):143–153. https://doi.org/10.1016/j.jcps.2011.11.008
Verbeke W (2015) Profiling consumers who are ready to adopt insects as a meat substitute in a Western society. Food Qual Prefer 39:147–155. https://doi.org/10.1016/j.foodqual.2014.07.008
Verbraucherzentrale Nordrhein Westfalen (2010) Bundesweite Markt-erhebung: 'Ohne Zusatzstoffe'-Clean Labeling: Werbeaussagen kritisch beleuchtet
Vergura DT, Zerbini C, Luceri B (2019) "Palm oil free" vs "sustainable palm oil": the impact of claims on consumer perception. Br Food J 121(9):2027–2035. https://doi.org/10.1108/BFJ-01-2019-0020
Villnow V, Rombach M, Bitsch V (2019) Examining German media coverage of the re-evaluation of glyphosate. Sustainability 11(7):1910. https://doi.org/10.3390/su11071910
Wang R, Ke C, Cui S (2021) Product price, quality, and service decisions under consumer choice models. Manuf Serv Oper Manag. https://doi.org/10.1287/msom.2020.0947
Webster JL et al (2010) Consumer awareness and self-reported behaviours related to salt consumption in Australia. Asia Pac J Clin Nutr. http://hdl.handle.net/10536/DRO/DU:30032089. Accessed 11 May 2021
van der Weele C et al (2019) Meat alternatives: an integrative comparison. Trends Food Sci Technol 88:505–512. https://doi.org/10.1016/j.tifs.2019.04.018
Winter CK, Davis SF (2006) Organic foods. J Food Sci 71:117. https://doi.org/10.1111/j.1750-3841.2006.00196.x

Wissgott U, Bortlik K (1996) Prospects for new natural food colorants. Trends Food Sci Technol 7:298–302

Wu Y, Li M, Yang Y et al (2018) Authentication of small berry fruit in fruit products by DNA barcoding method. J Food Sci 83:1494–1504. https://doi.org/10.1111/1750-3841.14177

Wunderlich SM, Gatto KA, Mangano M (2018) Labeling policy for genetically modified and organic food: impact on consumer choice. FASEB J 31:640.32. https://doi.org/10.1096/FASEBJ.31.1_SUPPLEMENT.640.32

Xie HK, Zhou DY, Liu ZY, Li DY, Tan ZF, Dong XF, Liu XY, Shahidi F, Zhu BW (2019) Effects of natural phenolics on shelf life and lipid stability of freeze-dried scallop adductor muscle. Food Chem 295:423–431. https://doi.org/10.1016/j.foodchem.2019.05.133

Xing B, Zhang Z, Sun R et al (2020a) Mini-DNA barcoding for the identification of commercial fish sold in the markets along the Taiwan Strait. Food Control 112 107143. https://doi.org/10.1016/j.foodcont.2020.107143

Xing RR, Hu RR, Han JX et al (2020b) DNA barcoding and mini-barcoding in authenticating processed animal-derived food: a case study involving the Chinese market. Food Chem 309:125653. https://doi.org/10.1016/j.foodchem.2019.125653

Xiong X, Yuan F, Huang M, Xiong X (2020) Exploring the possible reasons for fish fraud in China based on results from monitoring sardine products sold on Chinese markets using DNA barcoding and real time PCR. Food Addit Contam A Chem Anal Control Expo Risk Assess 37:193–204. https://doi.org/10.1080/19440049.2019.1694709

Xuan BB (2021) Consumer preference for eco-labelled aquaculture products in Vietnam. Aquaculture 532:736111. https://doi.org/10.1016/j.aquaculture.2020.736111

Yi J, Kebede B, Kristiani K, Buvé C, Loey AV, Grauwet T, Hendrickx M (2018) The potential of kiwifruit puree as a clean label ingredient to stabilize high pressure pasteurized cloudy apple juice during storage. Food Chem 255:197–208. https://doi.org/10.1016/j.foodchem.2018.02.052

Yuan X (2020) Three essays on the effects of information on consumer choice of genetically modified foods. Texas A&M University. https://hdl.handle.net/1969.1/191932. Accessed 17 May 2021

Zhang L et al (2019a) Optimal environmental quality and price with consumer environmental awareness and retailer's fairness concerns in supply chain. J Clean Prod 213:1063–1079. https://doi.org/10.1016/j.jclepro.2018.12.187

Zhang J, Celli GB, Brooks MS (2019b) Chapter 1: natural sources of anthocyanins. In: Food chemistry, function and analysis. Royal Society of Chemistry, London, pp 3–33

Zimmerman T, Siddiqui SA, Bischoff W, Ibrahim SA (2021) Tackling airborne virus threats in the food industry: a proactive approach. Int J Environ Res Public Health 18(8):4335. https://doi.org/10.3390/ijerph18084335

Index

C. M. Galanakis (ed.), *The Age of Clean Label Foods*,
https://doi.org/10.1007/978-3-030-96698-0

The manufacturer's authorised representative in the EU is Springer Nature Customer Service Centre GmbH, Europaplatz 3, 69115 Heidelberg, Germany. If you have any concerns regarding our products, please contact ProductSafety@springernature.com

Printed and bound by CPI Group (UK) Ltd, Croydon, CR0 4YY
30/09/2026
02225496-0005